Industrial Design Rights

Industrial Design Rights

An International Perspective

Second Edition

Edited by
Brian W. Gray
Rita Gao

Wolters Kluwer

Published by:
Kluwer Law International
PO Box 316
2400 AH Alphen aan den Rijn
The Netherlands
Website: www.wklawbusiness.com

Sold and distributed in North, Central and South America by:
Aspen Publishers, Inc.
7201 McKinney Circle
Frederick, MD 21704
United States of America
Email: customer.service@aspenpublishers.com

Sold and distributed in all other countries by:
Turpin Distribution Services Ltd
Stratton Business Park
Pegasus Drive, Biggleswade
Bedfordshire SG18 8TQ
United Kingdom
Email: kluwerlaw@turpin-distribution.com

Printed on acid-free paper.

ISBN 978-90-411-5950-2

Printed and Bound by CPI Group (UK) Ltd, Croydon, CR0 4YY.

Editors

Brian W. Gray, is a partner at Norton Rose Fulbright in Toronto, and has practised intellectual property law for over thirty-five years in Canada. He is a former co-chairman of the Intellectual Property and Entertainment Committee of the Intellectual Bar Association. He has published widely in the field, and has been selected by peer review publications in the United Kingdom, the United States and Canada as one of the leading and most frequently recommended intellectual property lawyers practising in Canada.

Rita Gao, is an associate at Norton Rose Fulbright in Toronto. As a lawyer in the Intellectual Property (IP) Group, she focuses on creating, enforcing and commercializing intellectual property assets around the world. She helps clients develop and manage national and international intellectual property portfolios and provides strategic advice for patents and industrial designs in various innovative areas. She deals with multinational corporations, mid-size companies, as well as start-ups. She provides freedom-to-operate and patentability opinions. Prior to her legal career, Rita worked at Deloitte Consulting and a leading Canadian telecommunications company. She has a Juris Doctor degree from Osgoode Hall Law School and a Bachelor of Applied Science degree in Electrical Engineering from University of Toronto. She is a member of Intellectual Property Institute of Canada (IPIC). She is an avid traveller and is fluent in Mandarin Chinese.

Contributors

Karen Abraham, jointly heads the Intellectual Property and Information Technology Department of the Malaysian firm Shearn Delamore & Co. She has more than twenty-five years of experience covering IP Litigation, Mediation, Enforcement and Brand Consultation and Management. Karen has crafted brand management programmes for many leading multinational companies and had designed anti-counterfeiting and anti-piracy programmes and strategies for global IP brands.

Karen strongly advocates the Mediation of IP disputes. She is on INTA's (International Trademark Association) Panel of Neutrals. She is an Accredited Mediator of the Malaysian Mediation Centre and an Association Member of the Chartered Institute of Arbitrators (UK). Karen is the first Malaysian woman appointed on the Board of Directors of INTA. She is also the first Malaysian to sit on the International Association for the Protection of Intellectual Property (AIPPI), where she holds the position of Assistant Secretary General. Karen also sits on the Malaysian Bar Council's IP Committee, ADR Committee and on the Disciplinary Committee Panel of the Advocate and Solicitors Disciplinary Board. Karen has authored several International publications and has presented papers at leading International IP congresses and conferences.

Karen has consistently been recognized as leading lawyer over the last decade in, amongst others, successive editions of the Asia Pacific Legal 500, Asialaw Profiles, Asialaw Leading Lawyers, Chambers Asia Pacific and Who's Who Legal.

Sudhir D. Ahuja, is a Registered Patent Attorney, and has more than thirty years of experience in Industry and Intellectual Property Law. He specializes in Patents and Designs matters.

Hiromichi Aoki, is a partner at Yuasa and Hara, Tokyo and has practiced Trademark Law and Design Law for over thirty years in Japan. He was a visiting professor of the school of law at Hokkaido University, co-chairman of the Design Committee of the Asian Patent Attorney Association and a chairman of the Design Committee of the Japan Patent Attorney Association. He authored various books on intellectual property, including the recently published *'Non-Traditional Trademarks and Trademark Infringement'* and *'Brand and Design as Intellectual Property Rights-Comparative Analysis in Japan, EU and U.S.A.'*.

Dr Pablo Käufer-Barbe, is a partner at the Argentina based law firm Estudio Käufer-Barbe & Vinent.

Arly Carlquist, is a partner at Bech-Bruun, Copenhagen, and has practiced intellectual property law for over twenty years in Denmark. He is highly experienced in all matters of IT law, such as system development, outsourcing, hosting, cloud computing, hardware, software etc., which includes thorough knowledge of various intellectual property rights related issues associated herewith. He also advises on the drafting of software and patent license agreements, as well as in connection with other technology related matters. Moreover, he has extensive experience in the enforcement of intellectual property rights particularly in relation to cases concerning the infringement of copyrights in software, music, games, movies etc.

Álvaro de Castro, is an associate, IP/IT Area at Gómez-Acebo & Pombo Abogados, S.L.P. He has a degree in Law and in Business Administration from Universidad Autónoma de Madrid (2008) and a Master's degree in Intellectual Property and Information Technology from ESADE, Barcelona (2010). Álvaro joined the IP department of Gómez-Acebo & Pombo in 2010, and since then has devoted himself to legal advice and litigation in relation to industrial designs, patents, trademarks and unfair competition. He is a member of professional bodies and associations such as: Madrid Bar Association (since 2008), LES Spain & Portugal. With six years of IP experience, he is (co)author of several publications in the IP field.

Mike Chen, After graduating from Beijing Mechanical Engineering University with a BSc in mechanical engineering, Mike J. Chen received his LLB from the University of Political Science and Law of China in 1995. In 2004, Mr Chen obtained his LLM from London Metropolitan University in the United Kingdom.

Mr Chen began his professional career in 1996, serving as a judge at the IP Tribunal of the Beijing First Intermediate Court for five years. After he left the court, Mr Chen was invited to be legal counsel for Beijing Founder Electronics Co Ltd, where he successfully litigated the famous Trap Evidence Collection case.

Mr Chen specializes in IP litigation and dispute resolution. He has accumulated abundant experience in the legal profession and has been distinguished as one of China's top IP practitioners. He advises clients on IP litigation across a wide range of industries, including mechanical engineering, electrical and electronic engineering, telecommunications, chemistry and biotechnology.

Rafael Cordoville Freire, is a Patent Engineer/IP Consultant at Daniel Advogados. He has an academic degree in Electric and Electronics Engineering. He is an Official Brazilian Intellectual Property Agent with more than fourteen years of professional experience in Intellectual Property and has full knowledge of the IP laws and rules in force. He specializes in patents and designs in the fields of mechanics, electrics, electronics, telecommunications and software. He has vast experience in drafting and filing patent applications, designs, conducting prior art and FTO investigations, analysis of patentability/infringement, analysis and compliance of official actions and nullity procedures. Rafael is fluent in Portuguese, English and Spanish and is an IP lecturer on courses administrated by the Brazilian Industrial Property Agents Association (ABAPI).

Juvelyn Cubilla, is an IP Specialist at mirandah asia. She is a licensed Chemical Engineer with extensive experience in patents, IP consultancy, IP management, Freedom to Operate Opinions and Enforcement Action.

Luis F. Elias, is the founder of Luis Elias & Asociados, Caracas, and has practised intellectual property law for over thirty years in Venezuela. He has lectured and written widely on the subject. He is a Guest Professor in post-graduate studies on intellectual property at major Venezuelan universities.

Sue Gilchrist, is the leader of the Herbert Smith Freehills Intellectual Property group in Australia. Sue specializes in patent, design, trade mark, copyright and other IP litigation. Sue has led significant cases for clients in all areas of IP, including patents, designs, copyright, trademarks, passing off and confidential information. Sue's wealth of experience is regularly recognized in IP legal directories where she is identified as a leading lawyer in IP. Sue's practice covers a range of industries including consumer electronics, telecommunications, pharmaceuticals, consumer goods, entertainment, advertising and financial services. Sue is Chair of the Intellectual Property Committee of the Law Council of Australia, and a member of the editorial panel and contributor to the Australian Intellectual Property Law Bulletin.

Melissa Goode, is a Senior Associate in the Herbert Smith Freehills Intellectual Property group in Australia. She specializes in intellectual property, trade practices, contract law and commercial litigation. Melissa has experience in advising on a wide range of intellectual property issues, as well as advertising, consumer protection matters and contractual disputes. Melissa's experience includes advising clients in a range of industries, including information technology, health, media and entertainment, and consumer goods.

Christopher N. Hunter, is a partner at Norton Rose Fulbright in Toronto. During the early days of the dot com boom he was drawn into legal practice from a career in engineering. Combining the best of engineering and law, his practice focuses exclusively on all aspects of intellectual property commercialization and prosecution, and includes a specialization in industrial design enforcement and prosecution. Recently, he has been involved in some of the largest intellectual property transactions in Canadian history. He is regularly ranked in peer reviewed listings of intellectual property lawyers in Canada.

Stuart Irvine, is a Senior Associate with Freehills Patent Attorneys. He is a solicitor and a registered patent attorney, and focuses on design and patent matters. With respect to patents, Stuart's technical background is in computer science and he specializes in software, hardware, computer implemented inventions more generally, and consumer products. Stuart's clients are based in Australia and around the world, and range from small start-ups to large established companies. His work involves drafting, filing and prosecuting design and patent applications, and managing large intellectual property portfolios. He also advises on complex validity and infringement matters and is experienced in providing specialized support in patent and design litigation.

Mr Yu Jiunn Kao, is a partner of Saint Island International Patent & Law Offices. He joined the firm in 1998 and has over thirty years of working experience in IP matters. His practice areas are patent and design prosecution and procurement. Mr Kao received his BS in Civil Engineering from Chung-Yuan University in 1975. He received his MS in Civil Engineering from University of Nebraska, USA in 1982 and MIP from Franklin Pierce Law Center USA in 1991. Mr Kao is a member of Asian Patent Attorney Association (APAA) and Taiwan Patent Attorneys Association.

Tamara Lee, is a legal assistant in Shearn Delamore & Co's IP department. She graduated with an LLB (hons) from the University of the West of England, Bristol. She was called to the Bar of England & Wales in 2010, and admitted to the Malaysian Bar in 2012. Tamara advises on protection and maintenance strategies of various IP rights, consumer protection, copyright, domain names and franchising. In addition, she also advises clients on antitrust and competition issues such as exclusive dealing arrangements, licensing and distributorship, territorial restrictions, pricing issues, retail and resale pricing, dominance and potentially abusive conduct relating thereto, discounts, tying and bundling, rebates and intellectual property rights ('IPR').

Mr Frank F.J. Liu, is a senior partner of Saint Island International Patent & Law Offices. Since joining the firm in 1992, Mr Liu has over twenty years of working experience in IP matters. His practice areas are IP litigation, licensing and unfair competition. Mr Liu received his LLB from Fu Jen Catholic University in 1984. He is a member of Taipei Bar Association, Hsinchu Bar Association, Asian Patent Attorney Association (APAA) and Taiwan Patent Attorneys Association.

Ms Jenny Li, received her Bachelor of Science Degree and her Master of Science Degree in Mechanical Engineering from Beijing Technology and Business University in 2002 and 2005 respectively.

Before joining Kangxin Partners, P. C., Ms Li worked as R&D Supervisor (assistant engineer) in Beijing Genecare Water Treatment Tech., Ltd, where she was also responsible for project research and began being acquainted with patent application.

Ms Li joined Kangxin Partners, P.C. in 2007 and collected rich experience in the field of Intellectual Property. As a qualified patent attorney, Ms Li specializes in patent application filings, drafting, invalidation cases and other patent prosecutions with a focus in the fields of mechanics and mechanical engineering.

Roberta Arantes Lopes, is a lawyer and partner of Daniel Advogados since 2013 and has over fourteen years of experience in the field of intellectual property. Her practice is focused in IP litigation and counselling mostly on trademark, trade dress and design infringement cases. Roberta has a master's degree in Public Law and also teaches in several post-graduation programmes in Brazil. Roberta is a member of INTA Enforcement Committee.

Uwe Lüken, is a partner at the Düsseldorf office of Bird & Bird and a member of the Intellectual Property Group. He is a highly regarded trademark specialist. Clients seek his strategic advice to protect their intangible assets especially in international conflicts and on colour trademarks. He specializes in non-technical intellectual property and, in particular, trademarks, competition, design and copyrights. He provides long term strategic advice to many well-known German and international companies. Uwe focuses on protecting his clients' intangible assets through a combination of monitoring and filing strategies and by representing clients in conflicts in and out of court. He has been involved in the protection of colour trademarks in national registries and at CTM level as well as in colour trademark enforcement. He provides strategic advice and represents clients in proceedings – from searches and filing to infringement litigation. Uwe works with clients on both domestic and multi-jurisdictional issues. Uwe studied law at the Universities of Passau and Bochum. He was admitted to the Regional Court Düsseldorf Bar in 1998. He is a frequent speaker on a variety of topics relating to non-technical intellectual property and is the co-editor and author of the '*Stöckel/Lüken Handbook on Trademark and Design Law*'.

Klaus Ewald Madsen, is a partner at Bech-Bruun, Copenhagen, and has practiced intellectual property law for nearly thirty years in Denmark. He has broad experience within intellectual property law including patents, trademarks and copyrights and related areas such as advertising and marketing practices. He is admitted to appear before the Supreme Court and is highly recognized for his patent litigation and arbitration skills. He has been involved in a number of major patent litigations primarily within the pharmaceutical area over the past twenty years and has gained broad international experience. He is also known for his regulatory and licensing work.

Gladys Mirandah, is the Director of mirandah asia Singapore, Malaysia, Vietnam, Thailand, Indonesia and Philippines. She has been admitted to practice in Singapore, UK and Brunei and brings with her more than thirty-five years of IP experience in Asia.

David Charles Musker, is a UK and European Patent, Trademark and Design Attorney. He divides his time between R G C Jenkins & Co, where he has been a partner since 1992, and Queen Mary University of London where he is Professor of International Design Law. He is Chair of the Designs and Copyright Committee of the Chartered Institute of Patent Agents (CIPA), and past-President of the Designs Commission of the

Union of European Practitioners in Intellectual Property (UNION). He is the author of several books and chapters on design law, and a founder-editor and contributor to the CIPA/ITMA the Community Designs Handbook, as well as sitting on the editorial board of JIPLP and ECDR. He filed the first European Community design, on 1 January 2003, amongst many other UK, European and International designs.

Alexandra Neri, is a partner at the Paris office of Herbert Smith Freehills. Alexandra's practice focuses on e-commerce, intellectual property, IT and cybercrime. She is recognized within and outside France as an outstanding trade mark litigation lawyer, especially in internet-related matters, where her high-profile clients have included Google. She has won awards including Best in Internet/IT (Legal Media Group 2011) and IP/IT Woman of the Year (Option Droit & Affaires 2010). Among her numerous accolades, Alexandra has been described as 'our most valued French counsel ... dedicated, hands-on, knowledgeable and an excellent litigator' (World Trademark Review 1000, 2012) and 'dynamic, tough and highly client-oriented' (Chambers Europe 2011). A French native speaker, Alexandra is also fluent in English, Greek and Italian.

Luigi Pavanello, is a partner at NCTM Studio Legale Associato, Verona office. He graduated in law from the University of Rome La Sapienza in 1985. He graduated from LL.M in general studies from New York University in February 1988. He was admitted in Vicenza, Italy Bar (1994), and New York Bar (1990). He was a Lecturer in Insurance Law University of Padua, Italy 2001–2011.

George D. Raynal, is a partner at Saidman Design Law Group, where his practice covers all aspects of design law, with a particular focus on U.S. and international design patent prosecution and enforcement. George helps individual designers, start-ups and major domestic and foreign companies, such as electronics, consumer product and heavy equipment manufacturers, obtain robust design protection and enforce their rights against knock-offs. George manages large design patent portfolios and coordinates U.S. and global filing for some of the foremost design-driven companies. George regularly speaks on developments in design patent prosecution and litigation.

Perry J. Saidman, is a recognized pioneer in design law. His boutique firm specializes in obtaining and enforcing US design patents, copyrights in the visual arts, and trade dress for three-dimensional product designs. He has been involved in dozens of design infringement actions, has substantial experience in appellate practice involving design patents and has a worldwide network of expert design lawyers. Perry has taught Design Law at George Washington University Law School in Washington, DC, was the founder of both the Industrial Designs Committee of the American Intellectual Property Law Association (AIPLA) and the Design Protection Section of the Industrial Designers Society of America (IDSA). He regularly presents on design patents at conferences throughout the world, has testified before the US congress on design legislation, authored dozens of articles on design law and consulted with and spoken to many designer and lawyer organizations.

Mike von Seidel, is a patent attorney and a partner at Von Seidels and has extensive experience in filing and prosecuting local and foreign patent and design applications. He advises clients on their intellectual property rights and how to protect their interests. Mike has extensive experience of over forty years in patent searching, assessing patentability, drafting, filing and prosecuting local South African and foreign patent applications and registered design applications. He has been involved to an extent in patent litigation; however, his work has focused more towards to serious involvement in the licensing of intellectual property. Mike specializes in advising clients on the broad spectrum of intellectual property rights available to them and how to best protect their business interests.

Rutger Stoop, is an associate with the Amsterdam based law firm Brinkhof. Specializing in intellectual property, he wrote his master's thesis on the enforcement of patents by non-practising entities, for which he studied the situation in the US and the Netherlands. Rutger is active as an attorney-at-law in intellectual property, focusing on the areas of trademarks and design rights as well as unfair competition and patent law.

Kurt Stöpetie, After having headed the IP department of one of the biggest Dutch law firms, Kurt Stöpetie (1961) founded the Amsterdam based boutique Brinkhof. Brinkhof has an international practice in patents, IP, TMT and privacy law. Kurt is an acclaimed and seasoned trademark and design litigator. He has been involved in numerous trademark, design and unfair competition court cases, as well as many press and advertising law cases. He litigated various cases before the European Court of Justice, including on the direct effect of European law in arbitration proceedings and the direct effect of the TRIPs treaty in national law. Chambers characterized Kurt as 'impressive' and 'noted for having helped fight his clients out of some desperate situations'. Kurt is also a proficient litigator of commercial disputes and a bright negotiator in convoluted matters. He also advises on developing, securing and exploiting intellectual property positions; advising on license and contract structures, from production to distribution; and assessing possible intellectual property risks or conflicts and designing strategic pathways.

Juan Carlos Suarez, Prior to joining Goodrich, Riquelme y Asociados in Mexico, he was employed in the telecommunication industry as a mechanical engineer working in electric and electronic equipment installation, and diagnostics. He also worked as an engineering group head in the refrigeration industry developing new products. Mr Suarez has worked in IP since 1990. He is currently the head of the patent team of Goodrich, Riquelme y Asociados. He drafts and prosecutes patents in the field of mechanics and other areas related to mechanical and electrical engineering. He specializes in patent prosecution and is presently handling and prosecuting over 1,000 patents every year. He has also experience of patent-related activities such as searches, patentability evaluations and preparation of legal opinions, including the areas of patent infringement and freedom to practice. He is an active member of the Mexican Association for Intellectual Property Protection.

Janet Toh, is a partner in the Intellectual Property Department of Shearn Delamore & Co. She graduated with an LLB (Hons) from the Bristol University and LLM from the National University of Singapore. She focuses on intellectual property and technology issues in a wide variety of transactions and has worked on a wide range of agreements including distributorship, licensing, outsourcing, service and consultancy agreements. She has advised clients on intellectual property protection and ownership issues; advertising issues; consumer protection; copyright; domain names; e-commerce; franchise; gaming; regulatory approvals for food & drug and telecommunications issues. She provides advice on intellectual property protection for leading multinational companies in the pharmaceutical, tobacco and Internet-related services industries around the world. Her practice also covers intellectual property and information technology related corporate work and she has conducted due diligence for various acquisition projects. She has written numerous articles in international IP publications. She is the Past President of the Malaysian chapter of the Licensing Executives Society and is listed in the 2015 Who's Who Legal: Trademarks 2015.

Gonzalo Ulloa, is the President and Head of the IP/IT Area at Gómez-Acebo & Pombo Abogados, S.L.P. He has a degree in Law and Master of Law from Universidad Complutense de Madrid (1975), and a postgraduate degree in Intellectual Property from University of Strasbourg. Gonzalo joined Gómez-Acebo & Pombo in 1974 and became a partner in 1981. He was Managing Partner between 1994 and 1997, and has been Chairman since November 2011. He is the Legal Adviser to Her Majesty's Ambassador of the United Kingdom in Madrid since 2014. He is a member of multiple professional bodies and associations such as: Madrid Bar Association (since 1975), Member of the Board of The Queen Sofía Spanish Institute (New York), LIDC, Techlaw Inc., INTA, AIPPI, EuroFranchise Lawyers, IBA (Vice Chair Committee L2-Patents), IPO, PTMG, LES Spain & Portugal. With thirty-eight years of IP experience, he is author of numerous publications in the IP field and is consistently ranked in the most relevant lawyer directories.

Ms Ying Zhang, received her Master's Degree in Civil Law and IP Law from Dalian Maritime University in 2010. Ms Zhang passed the Chinese Bar Exam in 2009, and also passed the Chinese patent attorney Exam in 2012.

Ms Zhang joined Kangxin Partners, P.C. in 2010. Ms Zhang is highly experienced in patent litigation, patent invalidation, patent license and other IP matters. She also has rich experience in handling various kinds of Patent related negotiations and contract reviewing.

Throughout Ms Zhang's career, she has successfully handled a large number of cases for both foreign and domestic clients. She successfully acted for volumes of patent infringement litigation cases for well-known enterprises like Swarovski Co. Ltd, SATA Gmbh & Co., Taiwan Mingmen and others. She also successfully acted for Zhejiang Supor Co., Ltd. and SATA Gmbh & Co. in patent invalidation cases. Ms Zhang also has rich experience in patent license work, such as acting for Unilin Corporation.

Summary of Contents

Overview

Organization of Individual Country or Region Chapters

- Introduction
  - Highlights of New Developments in This Jurisdiction since Last Edition
- Conventions and Legislation
- Subject Matter
  - Definition/What Constitutes a Protectable Design?
    - Shape, Configuration
    - Ornamental or Functional
  - Originality/Novelty
    - Test for Originality Generally
    - Novelty
      - Prior Use
      - Prior Publication
      - Prior Application or Registration
  - Unobviousness (if appropriate)
- Duration of Protection
- Infringement
  - Scope of Infringement Action
  - Test for Infringement/Criteria to Constitute Infringement
  - Standing to Sue
  - Remedies for Infringement
- Defences to Infringement
  - Validity
    - Not Novel/Original
    - Functional/Not Ornamental
    - Other Possible Criteria, Such as Non-obviousness
    - Faults in Application
  - Marking
  - Non-infringement
  - Other Defences

Preface

This is the second edition of a book first published in 2001 under the sponsorship of the International Bar Association. The chapters from that edition have been completely rewritten, in most cases by the original authors, and many new countries have been added.

Since the last edition, much progress has been made in respect of the harmonization of industrial designs. The Community Design Directive was enacted in 1998, leading to increased harmonization among the European Community countries commencing from that time.

This development has been followed by the modifications made to the Hague Agreement in Geneva in 1999 which have resulted in the increasing acceptance of the Hague system.

At the time of the first edition only approximately twenty-nine countries had accepted the Hague Agreement, notwithstanding that it had first come into force in 1928. Absent as signatories were Japan, the United States, Canada, the United Kingdom and all the Scandinavian countries.

At the time of this writing in July 2015, Japan, the United States, Denmark, Norway, Finland and Korea have all acceded to the Hague Agreement, with Canada and others soon to follow. The European Union joined in 2008.

By establishing a system of registration of a single home country application which can be filed in each signatory country, it is hoped that the Hague Agreement will do for Industrial Designs what the Patent Cooperation Treaty has done for patents. Further, although each country is free to apply its own examination standards, the term of protection will be unified at fifteen years and at least some administrative standards relating to the form and content of the international application will be harmonized.

Differences still remain. The form and content of drawings (a fundamental part of any design application) have not yet been harmonized. The United States, unlike the rest of the world, continues to treat designs under its patent statute and apply patent principles to such designs. In addition, in many countries, there continues to be a variety of possible sui generis ways to protect designs. As David Musker points out in his chapter on United Kingdom Design Law, one of the striking features of UK Design Law is the plethora of available rights. Setting aside trade marks, unfair competition, and patents, the set of design rights to be considered in the UK consists of: UK national registered designs; international designs designating the UK (coming in the near future); Registered Community Designs ('RCDs'); international designs designating the EU; Unregistered Community Designs ('UCDs'); UK Unregistered Design Right ('UK UDR'); and UK copyright. In addition, of course, there is still considerable overlap between the protection granted by the trademark and unfair competition provisions which are in place in many countries.

In addition to increased harmonization, we are in a period of increased awareness of the economic importance of Industrial Design. The Apple and Samsung wars over smart phone shapes have sparked attention. As Raynal and Saidman point out in their United States chapter, the recent US Court of Appeals for the Federal Circuit decision in *Apple v. Samsung* has highlighted many important advantages that design rights have over patents in the US, not the least of which is the authorization of an award of total profit from the article of manufacture bearing the patented design without apportionment.

As with the previous edition, we have tried to organize each chapter around a series of identical headings or topics for each jurisdiction so that the scope and subject matter of protection can be more readily compared. In addition, we have entered the modern age, as a digital version of the book will be produced which will also allow easier comparison across jurisdictions.

The labours of the individual contributors deserve recognition. The contributors, of course, are what make the book what it is. For the Canadian chapter, I owe thanks to my co-contributors Christopher N. Hunter and Rita Gao.

Rita Gao has also acted tirelessly as my co-editor. Rita has pushed this book through to completion with organization and dedication. I wish also to thank Jacqueline Byers and Bobby Leung at Norton Rose Fulbright Canada LLP who have helped with the final reviewing and editing process, and to my assistant Cindy Fung who has helped with the manuscripts.

Finally I would like to thank Kluwer, and in particular Miriam Weemhoff, for their patience.

The chapters are current as of 1 July 2015.

Brian W. Gray
Toronto, 1 July 2015

CHAPTER 1

Industrial Design Rights: Argentina

Pablo Käufer Barbe

§1.01 INTRODUCTION

[A] Highlights of New Developments in This Jurisdiction since Last Edition

The most substantial developments since the last edition are:

(a) In 2007, the Locarno Convention on Models and Designs signed in 1968 as amended on September 28, 1979 was accepted by Argentina, whereby the International Classification for Models and Designs (8th Edition) is adopted as obligatory for Argentine Registrations and the old national list is no longer valid, by Resolution P 266/2007 executed by the President of the Argentine National Institute for Industrial Property (INPI) on October 30, 2007.

(b) The adoption by executive decree of registration fees and charges, subject to amends, for the services of INPI.

(c) The update of the financial penalties for infringers, though this presents a peculiar problem because this schedule for fines was established by Presidential Decree together with the approval of the National Budget for 2011. That is, the establishment of the financial penalties is not part of a statute enacted by Congress, but included as a part of the National Budget, which is valid for only one year at a time.

§1.02 CONVENTIONS AND LEGISLATION

Argentina is party to the Paris Convention, to the Agreement on Trade-Related Aspects of Intellectual Property Rights (the TRIPS Agreement), and to the Berne Convention and Locarno Agreement, as well as to other international treaties.

Industrial design is covered by the Industrial Designs and Models Decree/Law 6673/63, ratified by Law 16478 and regulated by Decree 5682/65, Copyright Law 11.723, Patent law 24481, amended by law 24.572, and reordered by decrees 560/95 and 260/96 (B.O. March 23, 1996), Trademark Law 22.632, and Law 24.286 establishing new fines for the Argentine Penal Code.

§1.03 SUBJECT MATTER

[A] Definition/What Constitutes a Protectable Design?

The defining characteristic of an industrial design is that it does not exist by itself, but is always the accessory to an object. Although an industrial design may increase the charm or value of an object, it does not improve its use or function.

For this reason, specific legislation on the industrial applications of artistic or merely ornamental creations was enacted in 1963, in the form of the Industrial Models and Designs Law, about thirty years after the Copyright Law 11.723.

Industrial models and designs are defined by section 3 of the Industrial Models and Designs Law 6673/63 of 1963, as "*the forms or aspects incorporated or applied to an industrial product that confer upon it an ornamental character.*"[1]

However, section 1 of the Copyright Law 11.723 enacted in 1933, states that "*...models and works of art applied to commerce and industry are covered by its provisions.*"[2]

It may be concluded by simply reading and comparing both legal texts, that the concepts are closely related, and that the same creation is covered by both statutes. In practice, however, Argentine courts have limited the scope of section 1 of the Copyright Law 11.723, and excluded from it artworks with industrial applications that have no artistic merit, although they may have a decorative or ornamental character.

[B] Originality/Novelty

The Civil Appeals Court, Chamber B, of December 29, 1959 ruled:[3]

> the work of art applied to commerce or industry; protected by law 11.723 [the Copyright Law] is that work of art that has *original artistic elements*, such that even with the dispensing of its destination or application, *it has the value of a pure art work....*

1. Decree/Law 6673/63, B.0, 9/8/1963 ADLA-XXIII-B-989, ratified by Law 16.478 and regulated by Decree 5682/65 of 20/7/1965 published by the Official Bulletin – Boletin Oficial ("B.0.") on 2817/65, ADLA 1965-B-1452.
2. Copyright Law 11.723, published by the Official Bulletin on 30/9/1933, ADLA 1920-1940, page 443.
3. CNCiv, Sala B, 30/12/1959, in re "Hechter, Julio y otro *cl* Rico Pefia e Hijo, Revista La Ley To. 99," p. 26.

The views of the Italian scholar, Piola Caselli[4] were influential in the position adopted in the majority of court decisions. According to Mr. M.A. Laquis, "... the work of art applied to industry must keep, in spite of its association with an industrial product, *its representative character.*" Mr. Piola Caselli says that, "...the industrial design may qualify for protection as an artistic work under copyright law if it is an intellectual production which... could be conceived *as pure artwork*, when it is detached from the material elements that give it the characteristic of an industrial product."

M.A. Laquis finds the difference between both types of design in the *accessory condition of the industrial design*. Citing the opinion of a French scholar, M. Philidon:[5]

> the *artistic design has its own existence, is independent, exists by itself.* The fabric or paper where it is traced has no importance, no value, no use by itself. If by a chemical procedure the features of a design are erased, the material that is left *cannot perform any of the services that the erased design rendered*; on the contrary, if the designs applied to dishes are erased, the function or use of the dish is not lost... .

Both the Industrial Models and Designs Law and the Copyright Law require *originality and novelty*, but the former does not require artistic value or merit to grant protection.

Several Court decisions denying the right to protection under the Copyright Law to industrial models and designs were decided based *on the lack of originality or novelty*, even if there was a creation of *significant aesthetic value.*[6]

This particular interpretation was reaffirmed in 1982, with the Industrial Models and Designs Law already enacted, when a Federal Court decided that:

> the forms or aspects built in or applied to an industrial product are considered a model or an industrial design, when *they confer upon it an ornamental character*, meaning not only what functions as an adornment in a strict sense, but also what constitutes a new aspect or form incorporated into the product in such a way as to render it more agreeable and to impress the sense of sight and so to appeal to the consumer. For the protection of industrial models, *an extraordinary quantum of creativity is not required*, nor that the industrial model or design be a work of art – though it might well be one – *it is enough that the creation have the purpose or aim of changing the esthetic qualities of the object* to which it is applied, *because what it is all about is not to industrialize art, but to beautify industry.*[7]

[C] Unobviousness (if Appropriate)

Not applicable

4. Piola Caselli, "Trattato di! Diritto di Aurore": Marghieri e Torino, Unione Tip.-De.Torinese, Torino, 1927, cited by Manuel Antonio Laquis, "Derechos reales," Buenos Aires, Editorial Depalma, 1975.
5. Philidon, as cited by I. Satanosky in "Derechos Intelectuales," Editorial TEA, Buenos Aires, 1954.
6. CCiv. 2a.Cap, 28/9/1938, Revista La Ley To. 16, p. 378.
7. CNCivComFed, Sala III, 20/8/1982, in re "Pasolde Ltd c/ Fadete SA," El Derecho, To. 102, p. 264.

§1.04 DURATION OF PROTECTION

Section 7 of the Industrial Models and Designs Law grants an initial five-year period of protection for a registered model or design. Upon the payment of a fee with each renewal, the initial period may be renewed two more times for period of five years each. That is, the duration of protection may be up to fifteen years. "Each renewal is to be requested no later than six months before lapsing date and failure to do so will attract the penalty under section II of the Industrial Models and Designs Law." After the term of protection has lapsed, the model or design is considered to be in the public domain and can be used by anyone.

§1.05 INFRINGEMENT

[A] Scope of Infringement Action

The penalty established by section 21 of the Industrial Models and Designs Law is a fine ranging from ARS 3,000 to ARS 100,000 for those infringers who manufacture the infringing goods presenting the protected characteristics of a model or design by themselves or through third party industrial products. Paragraph 2 applies the same penalty to those individuals having knowledge of the illicit nature of the infringing product, or who sell, offer for sale, exhibit, import, export or in any other way do business with the infringing product.

Paragraph 3 of section 21 of the Industrial Models and Designs Law punishes any individual who maliciously has in possession products covered or protected by their manufacturers.

Paragraph 4 of section 21 of the Industrial Models and Designs Law penalises any individual who, without having a model or design duly registered, invokes it maliciously. And finally, paragraph 5 of the same section 21 penalises an individual who sells as his or her own, blueprints of a model protected by a third party's registration.

[B] Test for Infringement/Criteria to Constitute Infringement

Servile imitation is of course considered an infringement. If the more relevant and innovative parts of the registered model or design are copied, upon a technician's report, the judge may find that the compared product significantly infringes the right as protected.

The limit to that is if there are many other similar designs or models registered, whereby the priority in registration is crucial. So the test of originality is considered foremost when assessing the right obtained by the design or model registration.

Finally, the test of good faith is administered when considering the conduct of the parties in the confrontation and during the trial.

[C] Standing to Sue

The penal proceedings must be instituted by the interested party, and are considered of a private nature, so there is no action that can be bought by the public prosecutor. The party itself must file a penal action before the Federal Court and prosecute it at his or her own risk and expense.

[D] Remedies for Infringement

Besides the provisions of the law as to the final destiny of the infringing goods or materials, whereby the owner of the registration may obtain the infringing goods by the cost of its materials, or have them destroyed by the court, if this situation entails unfair competition, the registered owner of the model or design may request compensation for loss of trade and or any other damage it may have suffered by the deeds of the infringer. If competition law enters the picture, additional penalties may be imposed upon the infringers. Prison is also available, especially if the infringer has a history of similar breaches, as well as financial charges, the payment of legal fees, etc.

§1.06 DEFENSES TO INFRINGEMENT

[A] Validity

Since registration is obtained simply through application, with no further control being exercised by third parties or the authorities, any infringement action may involve a test for the validity of a registration.

The most common grounds of defense are therefore, lack of novelty in the design, or the use of a design already in the public domain. This means that prosecutions are carried out on a case by case basis and result in case answers which present a real problem for the enforcement of model and design rights by owners.

There is also a pre-emptive action by users enabling them to request the annulment of the registration of an industrial model or design.

The courts have held that the legislator has not detailed a different treatment for the different causes of annulment for infringement established by the Industrial Models and Designs Law, that is to say annulment in situations where the industrial model registered has been made from a copy of another model (paragraph a or b of section 6) or annulment for reasons of public policy (paragraph e of section 6). In both cases, it is possible to obtain the registration's annulment through the cancellation action established by the Industrial Models and Designs Law, having a statute of limitations of five years.[8]

8. Sections 17 and 18, the Industrial Models and Designs Law.

[B] Marking

Not applicable

[C] Non-infringement

Not applicable

[D] Other Defenses

Models and designs have not been published regularly for the past several years. This practice has been corrected with the establishment of the government agency named "Instituto Nacional de la Propiedad Industrial."

Regardless, as a result of inconsistent publication, the usual defense to infringement is to claim that, although the Industrial Models and Design Register is public, the defendant did not know of the existence of the model or design in question so that there is no malice on the part of the defendant. Such is not the case nowadays. All Industrial Designs and Models registered can be consulted at this office either in person or by Internet.

The punishment is, however, established on an absolute liability basis. Therefore, the question of whether there was an intention to breach the law has no importance in determining the legal infraction or crime.

§1.07 PROCEDURES FOR FILING APPLICATION FOR REGISTRATION

The protection for an Industrial Model or Design is obtained through an application which must conform to the formal requirements established in the Industrial Models and Designs Law, and its Regulatory Decree 5682/65[9] and the grant of title and registration by the Instituto Nacional de la Propiedad Industrial.

Industrial models are considered to be tridimensional objects, so it is required to furnish a written description of the object to be registered, together with drawings showing it from different angles. Up to sixty different alterations may be registered of the same basic design.

Registration is mandatory in order to obtain the protection of the law, and the first applicant is considered to be the author, except when proved otherwise.[10]

This registration consists simply of filing the application with the Instituto Nacional de la Propiedad Industrial, while fulfilling all the legal and administrative requirements.

Unlike patents or trademarks, the registration does not confer a right based on the registration control by the State. It has been said that it is a "weak" protection.

9. Regulatory Decree 5682/65, B.O. 28/7/1965, ADLA 1 965-B-1452.
10. Op. cit., endnote no. 1, ss 4 and 5 of Decree Law 6673/63.

§1.08 EXPUNGING/CANCELLING OR VARYING REGISTRATION

A registration of a design may be voided by a court upon the request of any interested party in a nullification action on several grounds, including for example, lack of novelty, malicious copy of an existing model, public policy reasons, and an immoral shape. The effects of such a decision apply from the date the registration was granted.

§1.09 OTHER

[A] Interface between Industrial Design and Trademarks

What is the scope of each type of legal protection? In comparing industrial models and designs with copyright, some legal scholars have proposed that the general coverage granted by the Copyright Law, inasmuch as it protects the intellectual work, "whatever its employment might be," also covers its application to industrial products.[11]

The legal doctrine more commonly accepted is that such creations may be covered by both legal systems, although their protection cannot be claimed simultaneously.[12]

Some authors think that the word "simultaneously" does not mean to say that these rights cannot be claimed successively.[13]

Of course, the issues and systems of protection are different and independent. They would not overlap with the definitions in the Industrial Model and Designs Law, if the model is not ideally separable from the industrial product. For example, if a cartoon character is applied without right or license on some product, like a watch or a pen, then the infringement may be prosecuted under the provisions of the Copyright Law. However, if the product's model or design is combined with the product in a form that cannot be divided or separated, like a car, a cup, furniture, etc., then the protection should be obtained under the Industrial Models and Designs Law. Similar discussions on this subject are found in French and German legal doctrine.

Not all industrial products fall outside the Copyright Law, which in a general way affords better protection through longer terms, stiffer penalties, public prosecution etc., than the Industrial Models and Designs Law. When the work is separable or detachable in concept, and does not conform by itself to the product, then the Copyright Law is also applicable, and the author can protect his or her creation for the duration of his or her life and the additional legal period. But if the author's creation is applied to an industrial product, and reproduced several times, then it becomes a model and is protected by the Industrial Models and Designs Law.

---

11. Author's right is a translation of Derecho de Autor which is the Spanish version of Copyright. It has been deleted and replaced by Copyright.
12. Op. cit., endnote no. 1, section 28 of the Industrial Models and Design Law.
13. E.D. Aracama Zorraquin, "El sistema Argentina de modelos y disefios industriales" article published in Revista del Colegio de Abogados de La Plata, Afio VIII, Julio-Diciembre 1965, p. 49, La Plata, 1965; and Miguel A. Emery, "La protecci6n de los modelos y obras de arte o ciencia aplicada al comercio o a la industria en la ley 11.723 y en el decreto 6673/63," article published in Revista La Ley, 1986-8-774.

There are additional interpretations on the differences between both creations. For example, according to one legal commentator:

> patent law protects the new product or the new industrial process obtained through the invention. The exclusive right on the patented invention is acquired by virtue of an administrative act, the patent. This right is granted whenever certain conditions are met, namely, that it is a new invention (in an objective sense) and that it renders a new result which has an industrial application.[14]

The Industrial Models and Designs Law protect ornamental forms. As in other industrial property rights (patents and trademarks), industrial models and designs are purely a creature of statute and created by registration with the administration authorities.

The different subjects included within intellectual rights define areas that are not always well separated. The elements common to all creative activity are interconnected, and sometimes overlap, as in the case of artistic works and industrial models and designs. Designs are a classic example. Notwithstanding that, on one hand, the Copyright Law protects designs as artworks without application and, on the other hand, the Industrial Models and Designs Law protects them as applied artworks, several laws allow for them to be protected by both systems (copyright and designs). There is however, generally a limitation that the two systems cannot be invoked simultaneously in a legal defense. Section 28 of the Industrial Models and Designs Law states that "when a registered model or industrial design, in accordance with this law, could have been the object of a deposit under law 11.723 [the Copyright Law], the author may not invoke them simultaneously to defend his or her rights." To date, the courts have not interpreted the new Utility Models Regulation[15] provided for in the new Patent Law.

Sometimes, if the characteristics of the model allow it, the model can be simultaneously registered as a trademark. The purpose in so doing is to obtain a right that can be indefinitely renewed every ten years, provided it is used and serves as a distinctive sign and not only as an ornament to render the product more appealing to consumers. This has been done with the shape of some bottles of soft drinks, alcoholic beverages and perfumes, some shoe soles and tire tread designs.

It should be stressed that if a trademark is not simultaneously applied for, registration may not be granted due to lack of novelty.

Section 1 of the Trademark Law 22.632[16] permits the registration of "...designs, stamped materials, images, containers, wrappers, engravings and carvings and reliefs with distinctive aptitude (capacity)" as trademarks. The Statement of Motives of the Argentine Trademark Law specifically mentions the use of reliefs in tires and shoe soles as trademarks.

---

14. D. Lipzcyc, Derechos de autor y derechos conexos, Editorial UNESCO-CERLALCI-ZAVALIA, Buenos Aires, 1993.
15. Op. cit., end note no. 8, Utility Models regulation means the provisions of Patent and Utility Models Law 24481/24572 and Decree 260/96 pertaining to Utility Models.
16. Trademark Law 22.362, B.O. 2/1/1981, ADLA XLI-A-58.

Even though the text of Trademark Law does not explicitly require novelty, it has been long standing judicial practice to limit the availability of registrations of bottles and containers to the novelty of their application to the products to which they are geared for, and to their own individuality which makes them distinguishable from the gender to which they belong.[17]

According to some Argentine scholars (J. Otamendi, P.C. Breuer Moreno and E. Aracama Zorraquin) and some foreign scholars (i.e., L. Van Bunnen, A. Braun, P. Mathely), the main feature of a bottle or container that confers characterization as a trademark is its "peculiar external appearance," bearing in mind that "...if that external appearance is purely vulgar or determined by the functionality of the object, the legal protection has no rational basis because no one may benefit from a legal monopoly, even temporarily, from what is of general usage...."[18]

Additionally, if it not sensible to request the registration of a bottle or container as a trademark that is not original in design, the Supreme Court has held that it is unacceptable to use another person's container or bottle to sell products of a different origin even when using its own stickers or labels.[19]

As a general rule, it can be said that a design that has no distinctive capacity between similar products cannot be imitated or plagiarised in the sense that it would enable its owner to request statutory protection, such as a drawing of a shoe on a can of shoe polish.[20]

Mr. J. Otamendi cites as an impediment to registering a model or design as a trademark, the prior common use of the bottle or container.[21]

Section 2(c) of the Trademark Law establishes that the form given to a product cannot be a trademark. In this sense it follows the principles of the former Trademark Law. There has been a long-standing controversy as to whether the concession of a certain form or shape grants a monopoly over a product, such monopoly being an object of the Industrial Models and Designs Law, or as the case may be, the Patent Law, and not the Trademark Law.

This reasoning can be addressed by stating that the monopoly is restricted only to the products covered by the trademark. The form of the product (soap, brush, etc.) would be limited to these products, and not extended to others, provided they are not the necessary shape of the particular product. One can imagine a handle of a common shape, in which case it would not be admitted as a trademark.

17. CNFed, Sala Contencioso-Administrativa, 10/10/1960, in re "Ponds Extract Co.," Revista Jurisprudencia Argentina, 1961-III-685.
18. CNCivComFed, Sala II, 13/12/1979, in re "Cannon S.A. *cl* Mendy S.A."
19. Fallos de la Suprema Corte de Justicia, 254-349 and several Federal Courts of Appeal decisions, like those cited at CNFedCapital, Rev. Jurisprudencia Argentina, To. 10, p. 216; CNFedLa Plata, Rev. De Jurisprudencia Argentina To. 57, page 568; also at Gaceta de! Foro To. 144, page l; CFedCapital, Jurisprudencia Argentina To. 13, p. 690; CFedCapital, Jurisprudencia Argentina To. 4, p. 97.
20. CFedCapital, Jurisprudencia Argentina, To. 26, p. 40.
21. Jorge Otamendi, "Derecho de Marcas," Ed. Abeledo Perrot, Buenos Aires, 1995. This was decided by the Argentine Federal Supreme Court, quoted as C.S., 7/4/1961 in re "S.R.L. Trebol *cl* SRLEADA," Fallos, 249-442 and in C.S., 8/6/1960 in re "Dubarry y Cia. Sucesores de M.E.Dubarry *cl* Peleteria Femina SRL," Fallos 247-271.

The courts have addressed this question and decided unequivocally that a design may be registered as a trademark.

In a decision in a case, the court held that a particular design on the pocket of a garment, deemed unnecessary, may be registered. The court noted:

> What the law does not allow is the registration of the necessary form or shape of a product as a trademark, where the meaning of a necessary form or shape entails its usual and common form, or that imposed by the nature of the product or service itself or its industrial function... .[22]

On the basis of this decision several other decisions were rendered, as in the case of a lid for a container,[23] the shape of certain parts of a shaver,[24] and the famous Parker clip for pens.[25] Finally, both scholars and the courts are admitting that the form which is distinctive and not.

[B] Protection under Copyright Law and Patent Law

Infringement of copyright is punishable with imprisonment from one month to six years,[26] as established by section 172 of the Argentine Penal Code except in those cases where sections 73 and 74 of the Copyright Law are applicable. In those cases copyright infringement is punishable with imprisonment for a period of one month to one year or by fine ranging from ARS 1,000 to ARS 30,000.[27]

The same punishment of imprisonment from one month to six years is applicable for those acts of infringement considered in section 72 of the Copyright Law.

The punishment also includes confiscation of the infringing materials, except when they belong to an innocent third party.

The infringing materials must be destroyed, except if the owner of the model registration decides to buy its components at cost. The materials left over may be used by provincial governments or the National War Department, without payment of any compensation to the defendant.

The parties may sue for damages, both material and moral. This action may be prosecuted through two venues, civil or criminal.

Under the well known legal principle "via non data, via non secuta," by choosing one avenue for obtaining damages, the other is automatically excluded and not available.

---

22. CNCivComFed, Sala II, causa 4335, 30/5/1986, in re "Levi Strauss *cl* Angelo Paolo."
23. CNCivComFed, Sala III, causa 3887 de] 9/12/1992 in re "Tecnopat S.R.L. *cl* Direccion Nacional de Propiedad Industrial's/ Denegatoria de Registro" and CNCivComFed, Sala II, causa 9134, 22/5/1992 in re "Tecnopat S.R.L. *cl* Direccion Nacional de Propiedad lndustrial's/ Denegatoria de Marca."
24. CNCivComFed, Sala II, causa 200, *30112192,* in re "N.V.Philips Gloelampenfabriek *cl* DNPI s/ Denegatoria de Registr"
25. CNCivComFed, Sala I, 29/6/1993, causa 1813/93, in re "The Parker Pen Company c/ DNPI s/ Denegatoria de Registro."
26. Op. cit., endnote no. 2, s. 71 of the Copyright Law 11.723.
27. Law 24.286, establishing new amounts for fines established in the Argentine Penal Code and other special punitive laws, B.0. 29/12/1993, ADLA LIV-A-1994-40.

The civil action for damages is grounded on the precepts of sections 1109 and/or 1077 of the Argentine Civil Code. It must be initiated before the statutory term of two years lapses, after the party is notified of the penal sentence, otherwise the statute of limitations, as regulated by section 4037 of the Argentine Civil Code applies.[28]

The criminal action for damages is grounded on the general rule of section 29, paragraph 1 of the Argentine Penal Code which establishes that the penal sentence may also include an order for the indemnity of the moral and material damage caused to the victim, his or her family or a third party.

The judge may establish a compensatory sum whenever evidence of the amount of damages is lacking, as well as the payment of judicial costs and expenses by the party losing the case.

Criminal judges are very conservative in establishing damages, so it is advisable to go through the civil courts when requesting compensation, although this is more expensive because the plaintiff has to pay a court tax of 3% of the amount claimed.[29] This court tax may be recovered from the defendant if the plaintiff is successful.

There have been different outcomes with respect to moral damages in favor of corporations or legal entities. In some cases, moral damages have been awarded considering that the "social damage" or loss of prestige is not included in the amount of reparation established by the judge. The answer as to whether moral damages are considered or not depend upon the jurisdiction where the claim is filed, which in criminal law depends on the place where the offense was committed. Criminal proceedings are instituted by the public prosecutor with the investigation being the responsibility of the investigating judge and/or public prosecutor.

The Patent Law[30] defines a patent as "... a right granted for any new invention of a product or a procedure, provided that it entails an inventive activity and is susceptible to an industrial use...," while an invention is defined as "... any human creation that permits the transformation of matter or energy for its use by man...".[31] Inventive activity and industrial application are defined in paragraph (d) and (e) of the same section respectively as "...inventive activity [exists] when the creative process or its results are not inferred from the state of the art by a person normally versed in the pertinent technical subject..." and "...industrial application [exists] when the object of the invention leads to the attainment of a result or an industrial product, where the word industry includes agriculture, the forest industry, cattle raising, fishing, mining, the transformative industries proper and services...."

The Patent Law[32] introduces Utility Model certificates to the Argentine legal system. Utility models are defined as the right granted for "... a new form or disposition obtained or introduced in tools, instruments, utensils, devices or objects that are

28. Section 4037, Argentine Civil Code.
29. Court Tax Law 23.898, s. 2, B.O. 29/10/1990, ADI.A L-D-3751.
30. Patent Law 24.481, B.O. 21/4/1995, ADLA LV-C-2948, amended by Law 24.572, B.O. 23/10/1995, ADLA LV-E-5892, reordered text by Decree 260/96, B.O. 22/3/1996, ADLA LVI-B-1764.
31. *Ibid.*
32. Op. cit., endnote no. 8 (Title III, ss 53–58).

already known and destined for a practical use, inasmuch as they entail a better use of the function that they perform...."

The same section establishes that this right is granted only for the new form or disposition such as it is defined, but it cannot infringe on a valid patent.

The main difference between patents on the one hand, and utility models on the other, is the "inventive stature" or "inventive importance" involved in each kind of invention, the patent being of course, more important.

The reason for distinguishing between utility models and patents on the one hand, and industrial models and designs on the other, is their different function and purpose. While utility models and patents have a functional industrial application or use, industrial models and designs have only specific ornamental purposes. If the industrial model or design has a use, as defined by the Patent Law, that exceeds its ornamental character, as for instance, the shape, then application should be made for a utility model, a patent or a trademark.

Section 28, second paragraph of the Industrial Models and Design Laws allows the conversion of an application for a patent (now under Law 24.481 -utility model) which is actually an industrial model or design and thus is made "in error" as a patent (or utility model) application into an application for an industrial model or design. In such cases, some experts think that the time limits for filing claims for priority under the Paris Convention are reduced to those available for Models.

[C] Protection under Trademark Law and Unfair Competition

Section 31 of the Trademark Law calls for imprisonment from three months to two years, while Law 24.286 imposes fines from USD 1,000 to USD 15,000 on a person who falsifies or dishonestly imitates a registered trademark or designation; uses a registered trademark fraudulently imitated, or belonging to a third party without its authorization; sells or offers to sell a product or service with a falsified registered trademark fraudulently imitated or belonging to a third party without its authorization; sells, offers to sell or in any way markets products or services with a falsified or fraudulently imitated trademark.

The penal proceedings are initiated ex officio by the Public Prosecutor (Fiscal) before the Federal Court. Given the nature of these infringements, it is advisable to inform the Public Prosecutor of the facts and to work closely with him and his office.

There is an ongoing debate on the changing of the Penal Procedure Code, and since this change would when established in about a year give a more active role to the Public Prosecutors (Fiscales) and diminish the active role of Judges, it has important effects on the several corruption cases to be tried, so the present government is keenly interested in having it functioning when it leaves power by the end of 2015. The other side of the coin is that it is actively nominating new Public Prosecutors with links to the political party now in power.

[D] Mercosur and the Legal Regime in Its Member States

Argentina, together with Brazil, Uruguay and Paraguay entered into a common market project, known as Mercado Comun del Sur ("MERCOSUR") which is in the process of being perfected.[33]

The national legislation will have to be coordinated, as established in the Treaty of Asuncion,[34] and more specifically in the Ouro Preto Protocol, sections 8, 9, 14 and 25.[35]

The industrial property legislation in the three other member countries is somewhat different from that of Argentina, so problems will undoubtedly result in modifying not only legal texts but long established patterns of operation. For instance, most disputes in Argentina are resolved before the judicial courts, while in the other countries; the administrative option is more commonly used. Also, Paraguay and Uruguay examine model applications while Argentina does not.

In the area of trademarks and models, difficulties will arise in cases of dual registrations in favor of different owners.

This process will be lengthy and will depend not only on economic but also on political conflicts of interests. The main legal systems to be considered are those of the largest members, namely Brazil and Argentina. In these countries further developments will have to be carefully monitored to ascertain the course of copyright and industrial property legislation for all the countries involved.

As of today, the only applicable provisions are of a very general nature and have not received legislative regulation in Argentina. However, through a new constitutional amendment enacted in 1994, common market and integration treaties are given an elevated place, over and above ordinary statutes.[36]

On January 5, 1995 Argentina ratified The International Agreement incorporating the Agreements Decisions Declarations and Ministerial Understandings of the Uruguay Round of the General Agreement on Trade and Tariffs, known as the Marrakech Agreement, which in the Intellectual Property field is known as the Trade Related Intellectual Property Agreement or "TRIP."[37] Part II, Article 9 of TRIP relates to the Berne Convention on Copyright, of which Argentina is also a member. In section 4 of TRIP (on industrial models and designs), Articles 25 and 26 are fully compatible with the provisions of the Argentine Copyright Law and with the provisions of the Argentine Industrial Models and Designs Law.

The only question that needs some consideration is the quantum of fines established by the Industrial Models and Designs Law. Due to the extraordinary

33. Law 23981, approving the Treaty for a Common Market, known as MERCOSUR signed by Argentina, Brazil, Uruguay and Paraguay in Asuncion, Paraguay, also known as the Treaty of Asuncion (B.O. 12/9/1991, ADI.A LI-C-2889), and Law 24.560 approving the Additional Protocol on the institutional structure of MERCOSUR also known as the Ouro Preto Protocol (B.0. 13/10/1995, ADI.A LV-E-5877).
34. *Ibid.*
35. Op. cit., endnote no. 33.
36. Section 75, paras 22 and 24 of the Argentine Constitution, as established on Aug. 1, 1994 by the National Constitutional Convention held at Santa Fe.
37. Law 24.425, B.O. 5/1/1995, ADLA, LV-A-29.

inflation suffered by Argentina some years ago, the amount of fines has become ridiculous. This matter is still a pending issue because when Congress updated fines for virtually all-punitive legislation, it did not include the Industrial Models and Designs Law. The question is addressed in the introduction because the fines amount was established by the executive decree which approved the National Budget, and some informal opinions doubt its legal force on several grounds, mainly the requirement of a law by the National Congress and the yearly character of the Budget.

BIBLIOGRAPHY (AVAILABLE IN SPANISH AND FRENCH)

Baylos Corroza, H. "*Tratado de Derecho Industrial,*" Edit. Civitas, Madrid, 1978.

Breuer Moreno, P.C. "*A Propósito de una crítica a los proyectos de leyes de la propiedad industrial,*" article published in Revista LA LEY, T. 101, page 1047.

Colombet, C, *"Propriété littéraire et artistique et droits voisins,"* Dalloz, Paris, 1986.

Chaloupka, P. "*La insoportable desigualdad de los iguales,*" paper published in *Derechos Intelectuales* 6, Editorial Astrea, Buenos Aires, 1994.

Di Guglielmo, P."*Observaciones y sugerencias relativas a los proyectos de leyes de propiedad industrial, sometidos a la aprobacion del Honorable Congreso de la Nacion,*" article published in Revista LA LEY, T. 100, page 932.

Desbois, Duchemin, Boutelet et Plaisant, "*La Protection des modelles d'art aplique a l'industrie,*" R.I.D.A., I, October 1933, page 91.

Emery, Miguel A., "La protecci6n de los modelos y obras de arte o ciencia aplicada al comercio o a la industria en la ley 11.723 y en el decreto 6673/63," article published in Revista La Ley, 1986-8-774.

Kemelmajer de Carlucci A. and Puerta de Chacon A., "*Registro de Modelos y Disefios Industriales,*" paper published in *Derechos Intelectuales 3,* Editorial Astrea, Bs. Aires.

Laquis, M.A. *Derechos Reales,* Editorial Depalma, Buenos Aires, 1975, 1988.

Ledesma, J.C. *Derecho Penal Intelectual,* Ed. Universidad, Buenos Aires, 1992.

Ledesma, J. "*Los dibujos y modelos ornamentales y los artículos aplicados a la industria,*" article published in Revista LA LEY, T. 85, page 747.

Lipzcyc, D. *Derecho de Autor y Derechos Conexos,* UNESCO-CERLALCI-ZAVALIA, Buenos Aires, 1993.

Mendilaharzu, E.F. "*La Protección Juridica de las Artes Figurativas Aplicadas al Comercio o a la Industria,*" article published in Revista LA LEY, T. 27, page 992 and independently published by the author, Buenos Aires, 1939.

Poli, I. *El Modelo De Utilidad,* Depalma, Buenos Aires, 1982.

Satanowsky, I. *Derechos Intelectuales,* Editorial TEA, Buenos Aires, 1954.

LIST OF CASES

Civil Appeals Court, Chamber B, of December 29, 1959 (LL, t. 99, p. 26).

(CNCiv, Sala B, 30/12/59 in re "Hechter, Julio y otro c/ Rico Pefia e Hijo, Raul," L.L. 99-26.

CCiv.la.Cap, 28/9/38.L.L. 16,378 in re "Suarez, Victor c/ Castiglia Hnos".

CNFed, CivCom, Sala III, August 20, 1982, in re "Pasolde Ltd. c/ Fadete S.A.," ED 102-264.

CNFed, Sala Contencioso Administrativo, October 10, 1960, in re "Ponds Extract Co.," J.A. 1961-III-685.

CNFed, Sala II Civil y Comercial, December 13, 1979, in re "Cannon S.A. c/ Mendy S.A.".

Corte Suprema de la Nacion CSN, Fallos, 254-349.

Corte Suprema de Justicia de la Nación Fallos; 249-442 in re "S.R.L. Trebol c/ SRLEADA".

Corte Suprema de Justicia de la Nación Fallos: 247-71 in re "Dubarry & Cia. sucesores de M.E. Dubarry c/Peleteria Femina S.R.L.," of June 8, 1960.

CFedCap. J.A., 10-216.

CNFed.La Plata, J.A. 57-658, G.F. 144-1.

CNFedCap., 13-690.

CFedCap., J.A.4-97.

CNFedCap., J.A. 26-40.

CNFed, Sala II, May 30, 1986, in re Levi Strauss *d* Angelo Paolo.

CNFed, Sala II, May 2, 1992, in re Tecnopat S.R.L. *d* DNPI. Unpublished.

CNFed,Sala II, December 20, 1992 inre N.V.Phillips Gloeillampenfabriken *d* DNPI, unpublished.

CNFed, Sala I, June 29, 1993 The Parker Pen Co. *d* DNPI, unpublished.

Note: The most commonly used collections of court decisions in Argentina are:

Revista Juridica Argentina LA LEY, usually cited as L.L.; Revista Jurídica El Derecho, cited as E.D.; Revista de Jurisprudencia Argentina, cited as J.A.; Gaceta del Foro, cited as G.F.; and the Publication of the former Direccion Nacional de la Propiedad Industrial "Patentes y Marcas" cited as P.M. The National Supreme Court of Justice publishes its decisions in "Fallos y Sentencias de la Corte Suprema de Justicia de la Nacion," cited as Fallos.

CHAPTER 2

Industrial Design Rights: Australia

Sue Gilchrist, Stuart Irvine & Melissa Goode

§2.01 INTRODUCTION

[A] Highlights of New Developments in This Jurisdiction since Last Edition

The Designs Act 2003 (the 2003 Act) commenced on 17 June 2004 and replaced the Designs Act 1906 (the 1906 Act). It represented the first major overhaul of Australia's designs legislation in almost 100 years. The reason for the change was summarized by the Government: 'The Australian Government wants to give designers a much better chance of protecting their creations from copying and changes to the Act will deliver that.'[1] In particular, the new infringement test in the 2003 Act was intended to broaden the scope of protection following concerns that the previous test for infringement had become so narrow that competing designs had to be virtually identical before the courts would find infringement.[2]

The 2003 Act brought about changes in three main areas: the registration process, design enforcement and new eligibility and infringement tests. Some examples of these are:

(1) The maximum term of protection is now *ten years* (under the 1906 Act the maximum term was sixteen years).

1. Comment by the Hon Warren Entsch MP, the then Parliamentary Secretary to the Minister for Industry, Tourism and Resources, on IP Australia website, http://www.ipaustralia.gov.au/ cited in Australian Government Advisory Council on Intellectual Property. *Review of the Designs System*, Issues Paper No. 1 (2013) ('*ACIP Issues Paper*') 30.
2. Australian Law Reform Commission, *Designs*, Report No. 74 (1995) ('*ALRC Report*') [1.2], [5.23], [6.6].

(2) A design must be *new and distinctive* (under the 1906 Act it had to be new or original).
(3) The registrability of a design is considered in light of the *similarities* between the design and the prior art (under the 1906 Act it was considered in light of the differences between the design and the prior art).
(4) The prior art base consists of documents *published anywhere in the world and acts done in Australia* that publicly disclose the design (under the 1906 Act the prior art base consisted of documents published in Australia and acts done in Australia publicly disclosing the design).
(5) To establish infringement, a product must be *identical or have a substantially similar overall impression* as a registered design (under the 1906 Act a design needed to be an obvious or fraudulent imitation of a registered design).
(6) Design applications *are registered on a formalities check only*. Substantive examination is optional, though must be requested and a design certified before an action for infringement can be brought (under the 1906 Act substantive examination occurred as a matter of course).

The change in legislation arose from the Australian Law Reform Commission's Report No. 74 *Designs* (Sydney, 1995) (the ALRC Report).[3]

A review of the Australian designs system was instigated in 2012 by Australia's Advisory Council on Intellectual Property (ACIP).[4] An issues paper was published in September 2013,[5] and an Options Paper was released in December 2014. In March 2015, ACIP released its Final Report,[6] and recommended a number of changes to the Australian designs system, many of which are focussed on increasing harmonization with international treaties and practice in design protection. The Final Report will be reviewed by the Australian Government and, if the recommendations are accepted, amendments to Australia's Designs Act and Regulations will be proposed.

§2.02 CONVENTIONS AND LEGISLATION

Australian designs are governed by the Designs Act 2003 and the Designs Regulations 2004 (the 2004 Regulations). These apply to all designs registered and design applications made under the 2003 Act or the 1906 Act.[7]

However, for designs and design applications originally made under the 1906 Act, certain provisions of the 1906 Act continue to apply, including those relating to validity, infringement, examination, and the term of the registration.[8]

3. Explanatory Memorandum, *Designs Bill 2002* (Cth) 1.
4. Australian Government Advisory Council on Intellectual Property, *Review of the Designs System* (22 Jul. 2014) <http://www.acip.gov.au/reviews/all-reviews/review-designs-system/>.
5. ACIP Issues Paper, *supra* n. 1; Australian Government Advisory Council on Intellectual Property, *Review of the Designs System*, Options Paper, December 2014 ('ACIP Options Paper').
6. Australian Government Advisory Council on Intellectual Property, *Review of the Designs System*, Final Report, March 2015 ('ACIP Final Report').
7. *Designs Act 2003* (Cth) s. 151.
8. *Designs Act 2003* (Cth) ss 151–156.

The discussion in this chapter relates to the 2003 Act. There are, however, still Australian designs to which the substantive provisions of the 1906 Act apply. A number of substantial differences exist between the 1906 Act and the 2003 Act, particularly with respect to the prior art base, the test for validity, and the test for infringement. Accordingly, if considering a design to which the 1906 Act may apply, it is critical that as a threshold issue it is established whether the 1906 Act applies and, if so, in what ways it applies.

Australia is a party to the Paris Convention for the Protection of Industrial Property (1883). See §2.07[A] below for discussion on filing applications in Australia which claim priority from foreign design applications.

Australia is also a party to the Agreement on Trade-Related Aspects of Intellectual Property Rights (the TRIPS Agreement) which contains specific obligations regarding the substantive protection of 'industrial designs'.

§2.03 SUBJECT MATTER

This section addresses designs registered under the 2003 Act only. For designs registered under the 1906 Act, see §2.02 of this chapter regarding the applicability of the 1906 Act.

[A] Definition/What Constitutes a Protectable Design?

The 2003 Act requires a design to be in relation to a product.[9] Specifically, a design in relation to a product is defined as the overall appearance of the product resulting from one or more visual features of the product.[10]

A 'product' is 'a thing that is manufactured or hand made'.[11] A product may be:

(1) a complex product comprising at least two replaceable component parts and permitting disassembly and reassembly;[12]
(2) a component part of a complex product, provided that component is made separately;[13] or
(3) a kit which, when assembled, is a particular product.[14]

A product may have one or more indefinite dimensions, though restrictions apply.[15]

9. *Designs Act 2003* (Cth) s. 8.
10. *Designs Act 2003* (Cth) s. 5.
11. *Designs Act 2003* (Cth) s. 6(1).
12. *Designs Act 2003* (Cth) s. 5.
13. *Designs Act 2003* (Cth) s. 6(2).
14. *Designs Act 2003* (Cth) s. 6(4).
15. *Designs Act 2003* (Cth) s. 6(3).

[1] Shape, Configuration

Visual features are defined by the 2003 Act to include shape, configuration, pattern and ornamentation of a product.[16]

The 2003 Act further specifies that the feel of a product and materials used in a product are not visual features.[17] This is intended to limit design protection to the appearance of a product only.[18]

Where a product has one or more indefinite dimensions, the indefinite dimension itself is not a visual feature. Further, if a product with indefinite dimensions has a repeating pattern, more than one repeat of that pattern is not a visual feature.[19]

[2] Ornamental or Functional

A visual feature of the design is to be taken into account whether or not the feature is functional or ornamental. The 2003 Act specifically provides that visual features may, but need not, serve a functional purpose.[20] The ALRC Report considered and recommended against an exclusion of designs dictated by function.[21] This recommendation was accepted by the Government.

In this regard, however, the 2003 Act requires that when assessing validity or infringement, the freedom of the creator of the design to innovate must be taken into account.[22] To this end, if a feature is purely functional and defined, for example, by standardization, mechanical or physical constraints, the Designs Office or a court may give less weight to that feature when considering validity and infringement.

See §2.05[B][3] below for further discussion on how freedom to innovate influences questions of validity and infringement.

[B] Originality/Novelty

[1] Test for Originality Generally

In order to be valid, a design must be new and distinctive over the prior art base as it existed before the priority date of the design.[23]

A design will be new unless it is identical to another design in the prior art base.[24]

---

16. *Designs Act 2003* (Cth) s. 7(1).
17. *Designs Act 2003* (Cth) s. 7(3)(a)–(b).
18. Explanatory Memorandum, *Designs Act 2003* (Cth) 26.
19. *Designs Act 2003* (Cth) s. 7(3)(c).
20. *Designs Act 2003* (Cth) s. 7(2).
21. ALRC Report, *supra* n. 2, [4.49], Recommendation 25.
22. *Designs Act 2003* (Cth) s. 19(2)(d).
23. *Designs Act 2003* (Cth) s. 15(1).
24. *Designs Act 2003* (Cth) s. 16(1).

A design will be distinctive unless it is 'substantially similar in overall impression' to another design in the prior art base.[25] The test of substantial similarity in overall impression is also used to assess infringement and is discussed in detail in §2.05[B] below.

At a general level, the substantial similarity in overall impression test is intended to account for the fact that in design small differences in visual features can have a large impact on the overall impression created. The test is intended to operate such that minor or insignificant changes to a design are of limited relevance if the overall impression remains one of substantial or significant similarity.[26]

As discussed further in §2.05[B] below, determining whether a design is 'substantially similar in overall impression' to another design is a complex enquiry involving a number of competing considerations.

The 2003 Act requires an assessment of the *overall* impression (as noted above) and that greater weight must be given to similarities between designs than to differences between designs.[27]

When assessing substantial similarity in overall impression, the 2003 Act further requires that:

(1) Regard must be had to the state of development of the prior art base.[28] The ALRC Report stated that:

> The extent to which a design is substantially different will depend on the state of the development of the prior art. A more developed prior art base will mean smaller differences will be sufficient to result in a finding that there is no substantial similarity.[29]

See §2.03[B][2] below for further discussion on which designs form part of the prior art base. See §2.05[B][2] below for further discussion on the consideration of the state of development of the prior art base in the assessment of substantial similarity in overall impression.

(2) Particular regard must be had to any features identified as new and distinctive in a Statement of Newness and Distinctiveness (if any such statement was filed), though still within the context of the design as a whole. If no statement was filed the appearance of the design as a whole is considered.[30]

(3) Where only part of a design is substantially similar to another design, regard must be had to the amount, quality and importance of that part in the context of the design as a whole.[31]

25. *Designs Act 2003* (Cth) s. 16(2).
26. ALRC Report, *supra* n. 2, [5.15].
27. *Designs Act 2003* (Cth) s. 19(1).
28. *Designs Act 2003* (Cth) s. 19(2)(a).
29. ALRC Report, *supra* n. 2, [5.23].
30. *Designs Act 2003* (Cth) s. 19(2)(b).
31. *Designs Act 2003* (Cth) s. 19(2)(c).

(4) Regard must be had to the freedom of the creator of the design to innovate.[32] See §2.05[B][3] below for further discussion.

In assessing substantial similarity in overall impression the standard of a person who is familiar with the product (or similar products) to which the design relates is applied – the standard of the 'informed user'.[33] Further discussion of this standard is in §2.05[B][1] below.

In assessing whether a registered design is new and distinctive, the comparison is against a singular design that forms part of the prior art base. It is impermissible to 'mosaic' or produce a hypothetical amalgam of a number of prior art designs in order to defeat the registration of the design in issue.[34] The Federal Court of Australia has expressly provided that 'a design that combines various features, each of which can be found in the prior art base when considered as a whole but not in any one particular piece of prior art, is capable of being new or distinctive'.[35]

Although the assessment of newness and distinctiveness is made on a design-by-design basis (i.e., without mosaicing), regard is given to the state of development of the prior art base as noted above. This is intended to provide context to the analysis of the designs: a more developed prior art base will support the argument that smaller differences are sufficient to lend a design distinctiveness, while in a less developed prior art base larger differences may be necessary to confer distinctiveness on a design.

[2] Novelty

The 2003 Act provides that a design is registrable if it is new and distinctive when compared to the prior art base for the design as it existed before the priority date of the design.[36]

[a] Prior Use

The prior art base includes designs publicly used in Australia before the priority date.[37]

Where a design is derived from the owner (or the owner's predecessor in title) and disclosed without authorization, that disclosure can be disregarded provided an Australian design application is filed within six months of the date of the first disclosure.[38]

32. *Designs Act 2003* (Cth) s. 19(2)(d).
33. *Designs Act 2003* (Cth) s. 19(4).
34. *Keller v. LED Technologies Pty Ltd* (2010) 185 FCR 449, 503 [244]–[245] (Besanko J) ('*Keller*'); *LED Technologies Pty Ltd v. Elecspess Pty Ltd* (2008) 80 IPR 85, 93 [12].
35. *LED Technologies Pty Ltd v. Elecspess Pty Ltd* (2008) 80 IPR 85,93 [12],102 [55]; *Review 2 Pty Ltd v. Redberry Enterprise Pty Ltd* (2008) 173 FCR 450, 464 [60] ('*Review 2*').
36. *Designs Act 2003* (Cth) s. 15(1).
37. *Designs Act 2003* (Cth) s. 15(2)(a).
38. *Designs Act 2003* (Cth) s. 17(1)(b); *Designs Regulations 2004* (Cth) Reg. 2.01(3).

Australian law does not provide a general grace period by which a design owner's own disclosure can be ignored. This has been flagged as an issue for consideration in the review of the Australian designs system instigated in 2012.[39]

Provided an Australian application is filed within six months, disclosure at an officially recognized international exhibition can be disregarded.[40]

Where a design corresponds to an artistic work, Australian law also provides that in certain circumstances previous use of the artistic work can be disregarded.[41] These provisions are complex, but notably, no time limit attaches to the ability to disregard previous use of the artistic work. A design corresponds to an artistic work where the design embodies visual features of shape and configuration in a product that result in a reproduction of that work.[42] See §2.09[B] below for further discussion on the relationship between design protection and copyright in artistic works.

[b] Prior Publication

The prior art base includes designs published in a document anywhere in the world before the priority date.[43]

The expansion of the prior art base to beyond Australia in the 2003 Act means that a design will be compared against more information, thereby raising the eligibility threshold on what constitutes a registrable design that is both new and distinctive.[44]

§2.03[B][2][a] above discusses unauthorized disclosure, disclosure at an officially recognized international exhibition, and previous use of an artistic work. These provisions apply to prior publication as well as prior use.

[c] Prior Application or Registration

The prior art base includes any Australian design application with an earlier priority date (provided the application is eventually published by the Australian Designs Office).[45]

Foreign design or patent applications published after the priority date of the registered design do not form part of the prior art base, even if the foreign application has an earlier priority date.[46]

---

39. ACIP Issues Paper, *supra* n. 1, [3.2].
40. *Designs Act 2003* (Cth) s. 17(1)(a); *Designs Regulations 2004* (Cth) Reg. 2.01(1).
41. *Designs Act 2003* (Cth) s. 18.
42. *Copyright Act 1968* (Cth) s. 74(1).
43. *Designs Act 2003* (Cth) s. 15(2).
44. Commonwealth, *Parliamentary Debates*, House of Representatives, 11 Dec. 2002, 10081.
45. *Designs Act 2003* (Cth) s. 15(2)(c).
46. Foreign applications would not fall within s. 15(2)(c) because they are not a 'design application' as defined in the 2003 Act.

[C] Unobviousness (if Appropriate)

Not applicable

§2.04 DURATION OF PROTECTION

The maximum term for an Australian design to which the 2003 Act applies is ten years from the Australian filing date.[47] A single renewal fee is payable at five years from filing to extend the registration by an additional five years.

For designs under the 1906 Act, the maximum term is sixteen years. Given the 2003 Act commenced in June 2004, designs to which the 1906 Act applies could potentially remain in force until around June 2020.

§2.05 INFRINGEMENT

This section addresses designs registered under the 2003 Act only. For designs registered under the 1906 Act, see §2.02 of this chapter regarding the applicability of the 1906 Act.

The 2003 Act sets out the exclusive rights which the registered owner of a design has during the period of registration. The exclusive rights include the right to make or offer to make a product in relation to which the design is registered.[48]

At first instance, infringement proceedings are heard by a single judge. In Australia, no juries are used in design matters. The Federal Court[49] has adopted an individual docket system in which proceedings are allocated to a particular judge from commencement. Typically, that judge is then involved in all stages of the proceeding including the final hearing.

Appeals from first instance decisions are conducted before the Full Federal Court, which is usually made up of three judges of the Federal Court (not including the judge who heard the matter at first instance). There is an automatic right to appeal single judge final (but not interlocutory) decisions of the Federal Court to the Full Federal Court, except in the case of a single judge hearing an appeal of the decision of the Registrar, in which case leave is required to appeal the single judge decision to the Full Court of Australia.[50]

Parties may also seek to appeal decisions of the Full Federal Court to the High Court of Australia (which is the highest court in Australia). There is no automatic right of appeal. A party must apply for 'special leave to appeal'. Special leave is difficult to obtain and is rarely granted (although the High Court has shown recent interest in some intellectual property cases) and typically requires that the appeal involve a

47. *Designs Act 2003* (Cth) s. 46.
48. *Designs Act 2003* (Cth) s. 10(1)(a); see s. 10(1) of the *Designs Act 2003* (Cth) for the other exclusive rights.
49. The Federal Court (s. 83 of the *Designs Act 2003* (Cth)) and the Federal Circuit Court (s. 83A of the *Designs Act 2003* (Cth)) have jurisdiction to hear designs matters.
50. *Federal Court of Australia Act 1976* (Cth) s. 24; *Designs Act 2003* (Cth) s. 87(2).

genuine question of public importance, a difference of opinion between courts, or is otherwise required in the interests of justice. The appeal must also have significant prospects of success.

Infringement proceedings must be commenced within six years from the day on which the alleged infringement occurred.[51]

[A] Scope of Infringement Action

Under the 2003 Act, a person infringes a registered design if during the term of the registration of the design, and without the design owner's authorization, the person:

(1) makes or offers to make a product, in relation to which the design is registered, which embodies a design that is identical to, or substantially similar in overall impression to, the registered design; or
(2) imports such a product into Australia for sale, or for use for the purposes of any trade or business; or
(3) sells, hires or otherwise disposes of, or offers to sell, hire or otherwise dispose of, such a product; or
(4) uses such a product in any way for the purposes of any trade or business; or
(5) keeps such a product for the purpose of doing any of the things mentioned in paragraph (3) or (4).[52]

[B] Test for Infringement/Criteria to Constitute Infringement

In order to determine whether a product infringes a registered design, the court assesses whether the allegedly infringing product embodies a design that is identical to, or substantially similar in overall impression to, the registered design.

This is the same test as the one used to determine whether a design is distinctive for the purposes of registration and validity, in line with the recommendations of the ALRC Report, which said that 'the infringement and distinctiveness tests should be the same so that an infringing design is not a distinctive design and vice versa'.[53] The symmetry between the test used for assessing the distinctiveness of a design and the test used to determine whether a design has been infringed, has been described as broadening the scope of protection.[54]

As noted in §2.03[B][1] above, the 2003 Act requires an assessment of the *overall* impression and that greater weight must be given to similarities between designs than to differences between designs.[55] This requirement arises from an accepted ALRC

51. *Designs Act 2003* (Cth) s. 71(4).
52. *Designs Act 2003* (Cth) s. 71(1).
53. ALRC Report, *supra* n. 2, [5.22].
54. *LED Technologies Pty Ltd v. Elecspess Pty Ltd* (2008) 80 IPR 85,107 [70].
55. *Designs Act 2003* (Cth) s. 19(1).

recommendation that common elements of designs be given more weight than differences.[56]

When assessing substantial similarity in overall impression, the 2003 Act further requires that:

(1) Regard must be had to the state of development of the prior art base (see §2.05[B][2] below for further discussion).[57]
(2) Particular regard must be had to any features identified as new and distinctive in a Statement of Newness and Distinctiveness (if any such statement was filed), though still within the context of the design as a whole. If no statement was filed the appearance of the design as a whole is considered.[58]
(3) Where only part of a design is substantially similar to another design, regard must be had to the amount, quality and importance of that part in the context of the design as a whole.[59]
(4) Regard must be had to the freedom of the creator of the design to innovate (see §2.05[B][3] below for further discussion).[60]

In assessing substantial similarity in overall impression, the standard of a person who is familiar with the product (or similar products) to which the design relates is applied – the standard of the 'informed user'.[61] See §2.05[B][1] below for further discussion.

In practice, the court makes a 'studied comparison' based upon these requirements.[62] This comparison is between the allegedly infringing product and the registered design[63] and this is done in 'the context of the whole appearance of the designs...the whole appearance is relevant as the context for the design even though, strictly, the design is only the distinctive visual features of the product. Greater weight is given to distinctive features than to other parts of the design, although still in the context of the whole of the appearance'.[64]

As noted by Yates J in *Multisteps* citing Jacobs J in *Procter & Gamble Co v. Reckitt Benckiser (UK) Ltd* [2008] FSR 8 at [3]:

> The most important things in a case about registered designs are:
> 1. the registered design;
> 2. the accused object; and
> 3. the prior art.

56. ALRC Report, *supra* n. 2, Recommendation 37: 'The new designs legislation should give guidance to the courts in assessing distinctiveness. It should be specified that common elements are to be given more weight than differences.'
57. *Designs Act 2003* (Cth) s. 19(2)(a).
58. *Designs Act 2003* (Cth) s. 19(2)(b).
59. *Designs Act 2003* (Cth) s. 19(2)(c).
60. *Designs Act 2003* (Cth) s. 19(2)(d).
61. *Designs Act 2003* (Cth) s. 19(4).
62. *Multisteps Pty Ltd v. Source and Sell Pty Ltd* [2013] 214 FCR 323, 334 [55].
63. As required by s. 71(1)(a) of the *Designs Act 2003* (Cth).
64. *LED Technologies Pty Ltd v. Elecspess Pty Ltd* (2008) 80 IPR 85, 108 [72].

> And the most important thing about each of these is what they look like. Of course parties and judges have to try to put into words why they say a design has 'individual character' or what the 'overall impression produced on an informed user' is. But 'it takes longer to say than to see' as I observed in *Philips Electronics NV v Remington Consumer Products Ltd (No 1)* [1998] RPC 283 at 318. And words themselves are often insufficiently precise on their own.[65]

[1] *Informed User*

The standard that is used is that of a person 'who is familiar with the product to which the design relates, or products similar to the product to which the design relates' and this is referred to as the standard of the informed user.[66] The 2003 Act introduced the notion of the informed user into Australian designs law.[67] This standard was recommended by the ALRC, and it appears that the ALRC was inspired by the developments in Europe to protect Community designs.[68]

The informed user standard had been described as follows:

> Although it would be dangerous to attempt some comprehensive statement of principles that might be applied to the concept, it is apparent that an *informed user*:
> 1. is reasonably informed; not an expert but more informed than an average consumer;[69]
> 2. is an objective standard. However, expert evidence may still be adduced in court to assist the court in applying the informed user concept; and
> 3. focuses on visual features and is not concerned with internal features or features that are not visible to the naked eye.[70]

In a case relating to a dress design, the Federal Court held in relation to the informed user:

> Who is the informed user? Plainly, the informed user must be a person who is familiar with the product to which the design in question relates. Moreover, the informed user must be a *user* of the class of product in question, in this case, ladies' garments, or perhaps, more narrowly, ladies' dresses. A designer or manufacturer of such garments is not an informed user merely because he or she designs or manufactures them. Further, this user is not simply an ordinary consumer: the user must be an *informed* user.[71]

65. Cited by Yates J in *Multisteps Pty Ltd v. Source and Sell Pty Ltd* (2013) 214 FCR 323, 332-333 [49].
66. *Designs Act 2003* (Cth) s. 19(2)(d).
67. Whereas, the standard of the informed user had been adopted in the *Registered Designs Act 1949* (UK). This is discussed further in *Review 2* (2008) 173 FCR 450, 455-457[19]-[27] (Kenny J); *Multisteps Pty Ltd v. Source and Sell Pty Ltd* [2013] 214 FCR 323, 335-358 [57]-[71] (Yates J).
68. ALRC Report, *supra* n. 2, [5.17], which described the informed user as 'a person who is reasonably familiar with the nature, appearance and use of products of the relevant kind.'
69. See also *Review 2* (2008) 173 FCR 450, 457 [26]; *LED Technologies Pty Ltd v. Roadvision Pty Ltd and anor* (2011) 90 IPR 532, 549 [89] (Finkelstein J). Note, the ALRC Report provided that '[d]istinctiveness should be determined by users of the design rather than design experts': ALRC Report, *supra* n. 2, [5.19].
70. *LED Technologies Pty Ltd v. Elecspess Pty Ltd* (2008) 80 IPR 85, 105 [59].
71. *Review 2* (2008) 173 FCR 450, 455 [19] (Kenny J).

Her Honour went on to conclude in that case:

> [W]hilst there will be some women who subscribe to fashion magazines (such as *Vogue* or *Collezioni* that illustrated the prior art in this case) and have particular knowledge of, and familiarity with, fashion trends, there will be many other women who lack such knowledge and familiarity. Precedent and the text of s 19(4) indicate that it is from the perspective of someone closer to the former group that the newness and distinctiveness of the Review Design and the similarity of the Redberry garments in terms of overall impression must be judged: compare *Architectural Lighting* at [50].[72]

In making its decision as to whether or not the allegedly infringing product is substantially similar in overall impression to the registered design, the Federal Court has held that the role of expert evidence is to be a guide only and 'the court does not simply adopt the [expert's] assessment as its own, but makes its own objective determination'.[73] As such evidence only assists the court, rather than being determinative. Justice Yates said in one case that the expert evidence:

> [H]as assisted my consideration of the comparison that the court is required to undertake under s 19 of the Designs Act. Having been so assisted, I have proceeded on the basis of my own evaluation of the visual significance of similarities and differences, not that of the experts .....[74]

This approach to expert evidence was echoed in the ALRC Report.[75]

[2] State of Development of the Prior Art Base

In the context of infringement, the state of development of the prior art base assists to determine the scope of a registered design. This principle, as explained below, was well-established under the 1906 Act and has been embodied in the 2003 Act:[76]

> Small differences between the registered design and the prior art will generally lead to a finding of no infringement, if there are equally small differences between the registered design and the alleged infringing article. On the other hand, the greater the advance in the registered design over the prior art, the more likely that the court will find common features between the design and the alleged infringing article to support a finding of infringement (see *Dart Industries Inc v Décor Corporation Pty Ltd* (1989) 15 IPR 403 at 409).[77]

---

72. *Review 2* (2008) 173 FCR 450, 457 [27] (Kenny J).
73. *Review Australia Pty Ltd v. New Cover Group Pty Ltd and Ors* [2008] 79 IPR 236, 243–244 [32] (Kenny J).
74. *Multisteps Pty Ltd v. Source and Sell Pty Ltd* (2013) 214 FCR 323, 332 [47] (Yates J).
75. ALRC Report, *supra* n. 2, [5.21]: 'Whether or not a design is distinctive is intended to be an objective test. The court decides the issue by placing itself in the position of an informed user. It is not a subjective test whereby an "informed user" assesses distinctiveness by reference to his or her own knowledge and the court simply adopts the informed user's assessment. It is still a matter for determination by the court. The informed user describes the standard to be applied not who can give evidence.'
76. *Keller*, (2010) 185 FCR 449, 501 [237] (Besanko J).
77. *Keller*, (2010) 185 FCR 449, 462 [46] (Emmett J, Besanko and Jessup JJ agreeing).

The state of development of the prior art base can be determinative of whether or not a product infringes a design. For example, in one case in which it was found that there was no infringement of a design registration for automotive LED lights, the Federal Court held that 'in terms of infringement, one of the most significant matters is that the registered designs appeared in a crowded field of prior art' and that 'many of the features of similarity between the registered designs and the respondents' lights were features that were common in the prior art.[78]

In the context of assessing whether a design is distinctive so as to be valid, the Federal Court has emphasized that the prior art base should not be considered as a whole, but instead the registered design should be compared to each piece of prior art separately.[79] In the context of considering the scope of a registered design for assessing infringement, the Federal Court has also noted that the prior art base comprises designs and not design features or concepts.[80] The cases have referred to the following 'rule of thumb' for assessing infringement, which takes the approach of considering designs in the prior art base individually against the registered design and then using this to assess whether the alleged infringement falls within the scope of the registered design:

> [O]ne measure is whether the difference in distinctiveness between the Design and any previously known design is greater than the difference between the Design and the allegedly infringing design: s 19(2)(a).[81]

However, the cases also appear to use the prior art base to identify the features of the registered design that are 'common' in the prior art base and the features that are distinctive over the prior art base.[82] It is not entirely clear how this approach should be applied without considering the prior art base as a whole, and while still ensuring the features of individual prior art designs are not 'mosaiced' or combined to assess the distinctiveness of the registered design. An example of where such difficulty may arise is in determining the scope of a registered design which is distinctive in the way it combines various features, each of which can be found in the prior art base when considered as a whole but not in any one particular piece of prior art.[83]

78. *LED Technologies Pty Ltd v. Roadvision Pty Ltd and anor* (2012) 199 FCR 204, 223 [84] (Besanko J, with Mansfield and Flick JJ agreeing).
79. *LED Technologies Pty Ltd v. Elecspess Pty Ltd* (2008) 80 IPR 85, 104 [55] (Gordon J) in the context of discussing the distinctiveness of a registered design, but referred to in *Multisteps Pty Ltd v. Source and Sell Pty Ltd* [2013] 214 FCR 323, 342 [87] by Yates J in considering the prior art base to determine the scope of a registered design.
80. *Multisteps Pty Ltd v. Source and Sell Pty Ltd* [2013] 214 FCR 323, 342 [87] (Yates J), citing *LED Technologies Pty Ltd v. Elecspess Pty Ltd* (2008) 80 IPR 85, 104 [55] (Gordon J) and *Keller* (2010) 185 FCR 449, 464, [53], 503 [244]-[245].
81. *Keller*, (2010) 185 FCR 449, 506 [260] (Besanko J quoting the primary judge). This rule of thumb was also endorsed by the ALRC Report, *supra* n. 2, [6.30] and this paragraph in the ALRC report was also cited by Besanko J in *Keller*, (2010) 185 FCR 449, 501 [237].
82. For example, *LED Technologies Pty Ltd v. Roadvision Pty Ltd and anor* (2012) 199 FCR 204, 223 [84] (Besanko J, with Mansfield and Flick JJ agreeing).
83. Such designs are accepted as capable of being distinctive, as discussed *supra* in s. 2.03[B][1].

[3] Freedom to Innovate

The court is required to consider the freedom to innovate when assessing substantial similarity in overall impression. This means that where there is limited scope for variation, more weight can be given to smaller differences as constituting an independent development.[84] If a feature is purely functional and defined, for example, by standardization, mechanical or physical constraints, this would be a technical constraint on the freedom to innovate and the Designs Office or a court may give less weight to that feature when considering validity and infringement. For example, Gordon J in *LED Technologies Pty Ltd v. Elecspess Pty Ltd* took into account that the features of the registered design which were distinctive over the prior art base were the product of conscious design choice, rather than compelled by industry-wide standards or technological constraints.[85]

The ALRC also considered 'the market for the product and marketing constraints' as factors that can limit design freedom.[86] The ALRC Report and the Designs Examiners' Manual of Practice and Procedure both refer to the European Commission Green Paper which included 'the necessity of taking into account deep-rooted marketing requirements by the clients' and 'features imposed by fashion' as relevant constraints that limit the freedom of the designer.[87] When considering a registered design for a cross-over ladies' dress in *Review 2,* Kenny J took into account that there were significant constraints on design innovation arising from the nature of the product, in that designs for ladies' garments are invariably limited by what women customarily wear and that the 'dress' had well-recognized conventional limits. It is unclear how far the expansive concept of 'marketing constraints' can be taken and whether it can be argued that a designer is 'constrained' to include particular design features merely because there are particular trends in the market.

Ultimately though, consideration of the freedom to innovate occurs in the context of assessing whether the designs being compared are substantially similar in overall impression.[88] A particular feature cannot be disregarded merely because there are functional or technical requirements for its inclusion, as the designer may still have freedom to decide how to incorporate that feature and how to combine that feature with other features to create the overall impression of the design.

84. ALRC Report, *supra* n. 2, [5.24].
85. *LED Technologies Pty Ltd v. Elecspess Pty Ltd* (2008) 80 IPR 85, 106 [65]. Gordon J's decision was upheld on appeal in *Keller.* (2010) 185 FCR 449.
86. ALRC Report, *supra* n. 2, [5.24].
87. Commission for the European Communities, *Green Paper on The Legal Protection of Industrial Design* Brussels, (1991) III/F/5131/91-EN [5.5.8.3], referred to in ALRC, *supra* n. 2, [5.24] and IP Australia, Designs Examiners' Manual of Practice and Procedure (2003 Act), (2014) D09.4.5.1 <http://www.ipaustralia.gov.au/pdfs/designsmanual/WebHelp/designs_examiners_manual.pdf> (accessed on 3 Dec. 2014) ('*Designs Examiners' Manual*').
88. *Designs Act 2003* (Cth) s. 19(2)(d); *Dyson Ltd v. Vax Ltd* [2011] EWCA Civ 1206,1211 [21]–[22].

[C] Standing to Sue

The registered owner of a registered design may bring infringement proceedings in the Federal Court, the Federal Circuit Court or another prescribed court.[89] In order to sue for infringement, the owner must first have the design examined and a Certificate of Examination issued.[90]

The Act provides that the court may grant the Registrar leave to intervene in infringement proceedings.[91] The Registrar may be concerned to do so where the infringement proceedings raise important matters of designs law in which the Registrar has an interest, or matters of procedure.

[D] Remedies for Infringement

The 2003 Act expressly provides that the relief granted by the court may include:

(1) an injunction subject to such terms as the court thinks fit;
(2) at the option of the plaintiff – damages or an account of profits;[92] and
(3) additional damages as it considers appropriate, having regard to the flagrancy of the infringement and all other relevant matters.[93]

The court may refuse to award damages, or reduce the damages or refuse to make an order for an account of profits where, in the case of primary infringement (making or offering to make an identical or substantially similar product), the defendant was not aware of the design registration and had taken all reasonable steps to ascertain whether the design was registered. Or, in the case of all other infringing conduct (secondary infringement), if the defendant could not reasonably be expected to know the design was registered.[94]

The court may choose to award additional damages and these are governed by principles similar to those for aggravated and exemplary damages at common law.[95] Penalty and deterrence are relevant in making an award of additional damages.[96] The purpose is to 'strip the infringer of all pecuniary benefits received from the infringement' and the sufficiency of the compensatory damages is relevant.[97]

89. *Designs Act 2003* (Cth) ss 5 (definition of 'prescribed court'), 73, 83, 83A, 84.
90. *Designs Act 2003* (Cth) s. 73(3).
91. *Designs Act 2003* (Cth) s. 76.
92. *Designs Act 2003* (Cth) s. 75(1).
93. *Designs Act 2003* (Cth) s. 75(3).
94. *Designs Act 2003* (Cth) s. 75(2). Note the 2003 Act essentially defines primary infringement as 'makes' or 'offers' and secondary infringement as 'imports', 'sells', 'uses' or 'keeps for the purpose [of selling or using]': *Designs Act 2003* (Cth) s. 75(5).
95. *Review Australia Pty Ltd v. New Cover Group Pty Ltd and Ors* [2008] 79 IPR 236, 249 [53] (Kenny J) (and the authorities Her Honour refers to therein).
96. *Ibid.*
97. *Ibid.*, (Kenny J) citing *Aristocrat Technologies Australia Pty Ltd v. DAP Services (Kempsey) Pty Ltd (in liq)* (2007) 157 FCR 564, 570 [42].

§2.06 DEFENCES TO INFRINGEMENT

This section addresses designs registered under the 2003 Act only. For designs registered under the 1906 Act, see §2.02 above regarding the applicability of the 1906 Act.

In addition to arguing that a product does not infringe a registered design, the other main defence raised to an allegation of infringement is that the relevant design is not valid.[98] The 2003 Act provides that a person may apply to the Federal Court for an order revoking the registration of a design.[99]

In order to do so, the design must have been examined and a Certificate of Examination issued.[100] The grounds for revocation are:

(1) that the design is not a registrable design;[101] or
(2) that one or more of the original registered owners was not an entitled person in relation to the design when the design was first registered; or
(3) that each of the original registered owners was an entitled person in relation to the design when the design was first registered, but another person or persons were entitled persons in relation to the design at that time; or
(4) that the registration of the design was obtained by fraud, false suggestion or misrepresentation; or
(5) that the design is a corresponding design to an artistic work, and copyright in the artistic work has ceased.[102]

[A] Validity

[1] Not Novel/Original

One of the commonly argued grounds of invalidity is that the relevant design is not a registrable design because it is not new or distinctive when compared with the prior art base for the design as it existed before the priority date of the design.[103] See discussion at §2.03[B] above, regarding this test for validity.

[2] Functional/Not Ornamental

See discussion at §2.03[A][2] above.

[3] Other Possible Criteria, Such as Non-obviousness

Not applicable

98. This is expressly provided for in s. 74 of the *Designs Act 2003* (Cth).
99. *Designs Act 2003* (Cth) s. 93(1).
100. *Designs Act 2003* (Cth) s. 93(2).
101. Note, this requirement refers to s. 15 of the *Designs Act 2003* (Cth).
102. *Designs Act 2003* (Cth) s. 93(3).
103. *Designs Act 2003* (Cth) s. 15(1).

[4] Faults in Application

Under the Act, a person may seek revocation on the grounds of entitlement.[104] That is, that a person was entitled to be a registered owner of the design and has not been recorded as such, or one or more of the registered owners of the design were not entitled to be.[105]

If the Registrar is satisfied that a person was entitled to be a registered owner of the design and they were not listed, then the Registrar must revoke the registration. The Registrar may declare that a person whom the Registrar is satisfied was an entitled person at the time the design was first registered is an entitled person.[106] A person who has been declared to be entitled may then file an application in respect of the design with the same priority date as the original application.[107] An appeal lies to the Federal Court or the Federal Circuit Court against a decision of the Registrar.[108]

[B] Marking

Marking a product (or packaging) which has been protected by a design is generally advisable. Under the 2003 Act, it is prima facie evidence that the defendant was aware that the design was registered if the product embodying the registered design, or the packaging of the product, is marked so as to indicate registration of the design.[109]

Awareness of the design is relevant when the court assesses remedies for infringement.

[C] Non-infringement

See §2.05 above.

[D] Other Defences

The 2003 Act introduced a 'spare parts' defence against design infringement also known as a 'right of repair' exemption.[110] This provides that a person does not infringe where a component part of a product embodying a design that is identical, or substantially similar in overall impression to a registered design, is used to repair a product so as to restore its overall appearance. The purpose of this exemption is 'to

104. *Designs Act 2003* (Cth) s. 51. Where an 'entitled person' is defined as being, in relation to a design, a person entitled under s. 13 to be entered in the Register as the registered owner of the design: *Designs Act 2003* (Cth) s. 5.
105. *Designs Act 2003* (Cth) s. 52(2).
106. *Designs Act 2003* (Cth) ss 53(2)-(3).
107. *Designs Act 2003* (Cth) s. 55.
108. *Designs Act 2003* (Cth) s. 52(7).
109. *Designs Act 2003* (Cth) s. 75(4).
110. *Designs Act 2003* (Cth) s. 72.

ensure effective competition in the spare parts market leading to lower prices for consumers, for example, motorists in the case of spare parts for motor vehicles'.[111]

§2.07 PROCEDURES FOR FILING APPLICATION FOR REGISTRATION

This section addresses designs registered under the 2003 Act only. For designs registered under the 1906 Act, see §2.02 above regarding the applicability of the 1906 Act.

In order to obtain design protection in Australia, a design application must be filed and registration requested. Registration occurs on the strength of a formalities check only. Requesting substantive examination of a registered design is optional, however must be done (and the design certified) before infringement proceedings can be commenced.

Australian design applications are filed electronically though the IP Australia eServices portal.

A design application may include multiple designs, provided the designs relate to the same product or products that fall within the same Locarno Agreement Class.[112] This does not, however, introduce any saving in official fees as these are levied on a per-design basis rather than a per-application basis.[113]

In order to gain an Australian filing date, an application must simply indicate that what is filed is intended to be a design application, include one or more drawings, and include details allowing the applicant/entitled person to be identified and contacted.[114]

If these requirements are not met an Australian filing date is not awarded.

[A] Priority

An Australian design application may claim priority to a basic application filed in a convention country. Convention countries are defined by Design Regulations 2004,[115] the list matching that of the countries that are party to the Paris Convention. In order to claim priority to a basic application, the Australian design must be filed within six months of the basic application's filing date.

There is some uncertainty under Australian law as to how similar or otherwise the design of an Australian application must be to the design of the basic application in order for the priority claim to be valid. The Act and Regulations provide that the basic application must be an application made in a Convention country 'for protection of the design'.[116] While the issue has not been judicially considered, the approach of the

111. Commonwealth, *Parliamentary Debates*, House of Representatives, 11 Dec. 2002, 10081 (Warren Entsch).
112. *Designs Act 2003* (Cth) s. 22.
113. *Designs Regulations 2004* (Cth) Sch. 4.
114. *Designs Act 2003* (Cth) s. 21; *Designs Regulations 2004* (Cth) Reg. 3.01.
115. *Designs Regulations 2004* (Cth) Reg. 1.05, Sch. 1.
116. *Designs Act 2003* (Cth) s. 27(1)(b); *Designs Regulations 2004* (Cth) Reg. 3.06(1).

Designs Office at least is that strict identity between a priority application and the Australian design is not required:

> The requirement that the foreign application be for protection of 'the design' does not require strict identity in the representations of the design. The question is an objective one of whether the designs are the same, not whether the representations of the design are the same. In assessing whether the designs are the same, one approach is to consider whether the design in the foreign application, if publicly available before the priority date, would give rise to the ground of revocation of the design not being new. If so, the foreign application is for 'the design' in question.[117]

In order for a priority claim to be valid, the Australian design application must be made by the applicant of the basic application or the assignee of the applicant of the basic application.[118]

There is a textual difference between the provision defining who is entitled to be registered as the registered owner of a design[119] and the provision defining who must apply for a design in order to validly claim.[120] In light of this discrepancy, if priority is to be claimed from a basic application, the prudent approach is to ensure that the applicant is the same as the applicant of the basic application or an actual assignment of the basic application to the applicant has been executed.

A copy of the basic application does not need to be provided on filing. If requested by the Designs Office, however, a copy of the basic application must be provided within two months.[121]

[B] Drawings

In order to be valid, a registered design must be reasonably clear and succinct. That is, the design must appear with reasonable clarity, and without necessity for an unreasonably prolonged or complicated series of deductions from the registered representations/drawings.[122]

Formally, there are few substantive requirements with respect to design drawings:[123]

(1) Descriptive text must be limited to words necessary to understand the drawings.
(2) Drawings must be in proper proportion to each other (except where different proportions are required for clarity).
(3) Photographs are permissible, provided they are clear and the design is positioned against a neutral and contrasting background.

117. Designs Examiners' Manual, *supra* n. 87, D06.2.4.1.
118. *Designs Regulations 2004* (Cth) Reg. 3.06(2).
119. *Designs Act 2003* (Cth) s. 13.
120. *Designs Act 2003* (Cth) s. 27(1)(b); *Designs Regulations 2004* (Cth) Reg. 3.06(2).
121. *Designs Regulations 2004* (Cth) Reg. 3.06(3).
122. *Keller* (2010) 185 FCR 449, 460 [35] (Emmett J).
123. *Designs Regulations 2004* (Cth) Schs 2, 5, 6.

Neither the 2003 Act nor the 2004 Regulations mention or discuss drawing conventions such as line shading or stippling. Both forms of contour depiction are accepted by the Designs Office, though neither is required. The only specific requirement mentioned in this regard is that cross sections must be indicated by oblique hatching.[124]

The use and interpretation of broken or dotted outline in the drawings is not covered by the 2003 Act or the 2004 Regulations. General convention is that such outlining is used to illustrate features of lesser importance in the design, and a Statement of Newness and Distinctiveness is typically used to reinforce this intention. As discussed in §2.07[C] below, however, Australian law does not provide any mechanism to completely disregard illustrated features (whether presented in broken outline or not).

[C] Design Part of Another Article/Environment of Design

An Australian design must be in respect of something that is manufactured or hand made. A part of another product can be a design in its own right, but only if it is separately made.[125]

In assessing whether a design is substantially similar in overall impression to another design (relevant to questions of both validity and infringement), a statement of newness and distinctiveness may be used to draw attention to one or more particular features. Specifically, where a statement of newness and distinctiveness is filed, particular regard must be given to any visual features identified by the statement. Such regard, however, is to be given in the context of the design as a whole.[126]

Accordingly, where protection for an integral part of a product is desired, the entire product should be represented and a statement of newness and distinctiveness used to draw attention to the particular parts of the product that are of interest. A common approach aimed at achieving this is to illustrate the specific feature(s) of interest in solid lines, the remaining features in broken lines, and to file a statement of newness and distinctiveness drawing particular attention to the solid line features.

It should be noted that there is no mechanism under Australian design law that can be used to entirely disregard visual features depicted in the drawings. Coupled with the requirement for a design to be in respect of a distinct product, this has the result that it is challenging to seek design protection for a part of a larger product which is integrally manufactured.

[D] Description of Design (Where Required)

A Statement of Newness and Distinctiveness can optionally be filed with a design application. The use of such a statement, however, is limited to emphasizing the

124. *Designs Regulations 2004* (Cth) Schs 2, 5(10).
125. *Designs Act 2003* (Cth) ss 5, 6(2).
126. *Designs Act 2003* (Cth) s. 19(2)(b).

feature(s) of a design that are new and distinctive rather than explaining a design per se.

If additional description is required to explain a design this should be included in the drawings themselves. Written text in respect of drawings is permitted, but only to the extent necessary for understanding the drawings.[127]

[E] Applicant

While any person may file an Australian design application,[128] the applicant should be the person entitled to be the owner of the design on registration. Generally speaking, a person entitled to be the registered owner of the design includes the designer (i.e., the person who created the design) or a person who derives title to the design from the designer (e.g., by employment or assignment).[129]

Determining the correct applicant for an Australian design application is of critical importance. Design applications can be registered in a matter of months, and a ground of revocation exists where the registered owner/s recorded at the time a design is registered is/are incorrect (whether due to a non-entitled owner being recorded, or an entitled owner not being recorded).[130]

Further, in order to claim priority to a basic application the Australian application must be filed by the applicant of the basic application or the assignee thereof.[131] This requirement is discussed further in §2.07[A] above.

[F] Registration

If design protection is to be obtained, a request for registration of the design must be filed within six months of the priority date of the design.[132] Given the convention deadline is six months, this typically means that for convention applications registration must be requested on (or shortly after) filing the Australian application.

Once registration has been requested, the application is subjected to a check that certain formalities are met.[133] These formalities include, *inter alia*, that the application is in the form approved by the Designs Office (dictated by the online filing system), the application includes representations of the design, and that the product in respect of which the design is to be registered can be identified.

If any formalities requirements are not satisfied, a formalities report is issued. There is then a two-month deadline for satisfying the Registrar of Designs that the application is in order for registration.[134]

127. *Designs Regulations 2004* (Cth) Schs 2(5), (15).
128. *Designs Act 2003* (Cth) s. 21.
129. *Designs Act 2003* (Cth) s. 13.
130. *Designs Act 2003* (Cth) s. 93(3)(b)-(c).
131. *Designs Regulations 2004* (Cth) Reg. 3.06(2).
132. *Designs Act 2003* (Cth) s. 33; *Designs Regulations 2004* (Cth) Reg. 3.14.
133. *Designs Act 2003* (Cth) s. 39; *Designs Regulations 2004* (Cth) Regs 4.04, 11.18, Sch. 2.
134. *Designs Act 2003* (Cth) s. 41; *Designs Regulations 2004* (Cth) Reg. 3.14(2).

A formalities report will also issue if the application relates to multiple designs but this has not been identified in the application, and/or the relevant fees for all designs have not been paid. In this case, the additional designs may be excluded from the original application and pursued in further applications.[135] Such further designs must be filed before the original design application either lapses or is registered/published.[136]

If the formalities requirements are satisfied the design application is registered.[137] This can occur quite quickly, and publication of the design occurs at the time of registration.[138]

[G] Certification

Examination of a registered design is optional. Examination may be requested at any time during the life of the design and may be requested multiple times.[139] Although optional, a design must be examined and a Certificate of Examination must issue before infringement proceedings can be commenced.[140]

Examination may be requested by the design owner or a third party, or be initiated by the Registrar of Designs. Examination may also be ordered by a court. Examination may not proceed if court proceedings in respect of a design are pending and, unless the court orders the examination to proceed, must be discontinued if court proceedings are commenced during examination.[141] Although the 2003 Act gives primacy to the court proceedings, the Federal Court has also recognized that it may be appropriate in particular circumstances to stay court proceedings to allow examination by the Registrar to run its course.[142] Examination involves the Designs Office considering whether the design is new and distinctive, as discussed above at §2.03[B][1].[143]

If objections are raised, an examination report issues. Generally speaking, if examination is not concluded within six months of the date the first examination report issues, the Registrar makes a decision to revoke the design.[144] Conversely, if no objections are raised, or objections are successfully overcome within the six-month period, the Registrar makes a decision to issue a certificate of examination.[145] In either

135. *Designs Act 2003* (Cth) s. 23.
136. *Designs Regulations 2004* (Cth) Reg. 3.02.
137. *Designs Act 2003* (Cth) ss 39(2), 40(2).
138. *Designs Act 2003* (Cth) s. 45.
139. *Designs Act 2003* (Cth) s. 63(a).
140. *Designs Act 2003* (Cth) s. 73(3).
141. *Designs Act 2003* (Cth) ss 63(3)-(4).
142. For example, in *Enviro Pak Pty Ltd v. New Horticulture Pty Ltd (No 2)* [2013] FCA 624, Griffiths J considered it appropriate to stay the Court proceedings and order the Registrar to continue examination including because there was a plain overlap between the two proceedings, the Registrar's proceedings were well-advanced, the costs of the proceedings before the Registrar were significantly lower, and it was commensurate with the importance and complexity of the matters in dispute.
143. *Designs Act 2003* (Cth) ss 15, 65(2); *Designs Regulations 2004* (Cth) Reg. 5.02.
144. *Designs Act 2003* (Cth) ss 65(3), 86; *Designs Regulations 2004* (Cth) Reg. 5.04.
145. *Designs Act 2003* (Cth) s. 67.

case, the Registrar must give relevant parties an opportunity to be heard, and the Registrar's decision can be appealed to the Federal Court.

§2.08 EXPUNGING/CANCELLING OR VARYING REGISTRATION

This section addresses designs registered under the 2003 Act only. For designs registered under the 1906 Act, see §2.02 above regarding the applicability of the 1906 Act.

[A] Expunging/Cancelling

Once a design has been registered (which occurs on the strength of a formalities check only), any party can request for the design to be substantively examined by the Designs Office at any time, even if the design has previously been examined.[146] See §2.07[G] above for further discussion on examination.

If examination of a design leads to the Registrar of Designs being satisfied that a ground of revocation has been made out, the registration can be revoked. In this case, however, the Registrar must give the design owner a chance to be heard, and the Registrar's decision can be appealed to the Federal Court.

In addition, if a design has been examined and a Certificate of Examination has been issued any person may apply to the Federal Court for an order revoking the registration of the design as discussed at §2.06 above.[147]

A design application may be withdrawn and a design registration surrendered at any time.[148]

[B] Amendment

The ability to amend an Australian design depends on its status and is quite limited.

Prior to being registered, amendment is permissible provided it does not alter the scope of the design application by including matter that was not in substance disclosed in the original application, representations, or other documents.[149]

Once registered, and other than amendment to correct a clerical error or obvious mistake, a design can only be amended during examination. Amendment to a registered design during examination will not be allowed if it alters the scope of the design application by including matter that was not in substance disclosed in the original application, representations, or other documents.[150] In addition, an

146. *Designs Act 2003* (Cth) s. 63(1). The Registrar may also examine a design at its own initiative: s. 63(2).
147. *Designs Act 2003* (Cth) s. 93.
148. *Designs Act 2003* (Cth) ss 32, 49.
149. *Designs Act 2003* (Cth) s. 28.
150. *Designs Act 2003* (Cth) s. 66(6)(b).

amendment to a registered design during examination will not be allowed if it increases the scope of the registration.[151]

The 2003 Act requires that amendment during examination can only be made 'if the Registrar is satisfied ... that a ground for revocation of the registration of the design has been made out'.[152] In practice, however, the Designs Office permits an amendment to be made without having first raised an objection.[153]

Although the Designs Act provides that when making an amendment reference may be had to 'other documents' filed with a design application, the current electronic filing system implemented by the Australian Designs Office does not currently allow for 'other documents' to be submitted.

Importantly, the content of a Statement of Newness and Distinctiveness is *not* entered on the Register.[154] Consequently, while a Statement of Newness and Distinctiveness can be amended before registration it cannot be amended once a design has been registered. This can be particularly problematic where a piece of prior art is identified following registration and there is a desire to focus the design registration on the features that distinguish it from that piece of prior art. No mechanism to do this exists.[155]

Once a design has been registered, a request can be made to amend the Register to correct a clerical error or obvious mistake.[156] For an amendment to be made on the basis of a clerical error it must be shown that the error arises, for example, in the process of transcription and results in something other than what was intended being entered in the Register. For an amendment to be made on the basis of an obvious mistake, both the error and the correction for the error must be obvious on the face of the Register.

Finally, a third person aggrieved by a defect in the Register (e.g., an error or omission in the register) may apply to the Federal Court for an order to rectify the Register.[157]

§2.09 OTHER

This section addresses designs registered under the 2003 Act only. For designs registered under the 1906 Act, see §2.02 above regarding the applicability of the 1906 Act.

[A] Interface between Industrial Design and Trademarks

In Australia, the Trade Marks Act 1995 (the Trade Marks Act) regulates the trademarks regime. The Trade Marks Act defines a trademark as a 'sign' used, or intended to be

151. *Designs Act 2003* (Cth) s. 66(6)(a).
152. *Designs Act 2003* (Cth) s. 66(1).
153. Designs Examiners' Manual, *supra* n. 87, D10.3.
154. *Designs Act 2003* (Cth) s. 111; *Designs Regulations 2004* (Cth) Reg. 9.01.
155. *Reckitt Benckiser Inc* [2008] ADO 1.
156. *Designs Act 2003* (Cth) s. 149(2)(e); *Designs Regulations 2004* (Cth) Reg. 9.05.
157. *Designs Act 2003* (Cth) s. 120(1).

used, to distinguish goods or services dealt with or provided in the course of trade by a person from goods or services so dealt with or provided by any other person.[158] A 'sign' is defined to include, relevantly, any device, label, aspect of packaging, shape, or colour.[159]

Therefore, some subject matter that can be registered as a trademark can also be registered as a design. Some examples of where such overlap may occur are perfume and other bottles/containers, toys, tools and medical preparations.

While there is a restriction with respect to the overlap between the patents register and the trademarks register,[160] there is not the same with respect to the overlap between the designs register and the trademarks register.

However, it should be noted that the objectives of design protection are quite different from the objectives of trademark law. While for designs, the 2003 Act provides a registered owner with a general monopoly in relation to the particular appearance of an article, the Trade Marks Act provides a registered owner a more limited monopoly in relation to the way the trademark is used. Nonetheless, despite these different focuses, practitioners should bear in mind the potential overlap between these two regimes.

[B] Relationship between Industrial Design and Copyright/Design Copyright

A difficult issue associated with Australian designs law is the relationship between design protection and copyright protection in artistic works.[161] The difficulties arise because the rights of copyright owners in a two-dimensional artistic work (such as architectural plans) include the right to reproduce the work in a three-dimensional form.[162]

For many years, the government has sought to limit the protection afforded by copyright for artistic works embodied in mass-produced items. This is because it is considered preferable for the Designs Act to protect the shape and configuration of mass-produced items and people should register their designs accordingly, rather than argue that the product is an infringing reproduction of a copyright work, such as a drawing. The main reasons for this are, 'the low innovation threshold required to

158. *Trade Marks Act 1995* (Cth) s. 17.
159. *Trade Marks Act 1995* (Cth) s. 6.
160. Section 25 of the *Trade Marks Act 1995* (Cth) provides that the trademark is taken to have ceased from the end of the period of two years after the patent expires or ceases.
161. Section 10 of the *Copyright Act 1968* (Cth) defines an artistic work to include paintings, sculptures, drawings, engravings, photographs, buildings, models of buildings and works of artistic craftsmanship.
162. Section 31(1)(b) of the *Copyright Act 1968* (Cth) grants the copyright owner the right to reproduce the artistic work in a material form. Section 21(3) of the *Copyright Act 1968* (Cth) deems an artistic work to have been reproduced in the case of a two-dimensional work in three-dimensional form.

qualify for copyright protection, the lengthy duration of such protection and the absence of any public register to record such protection'.[163]

Sections 74–77 of the Copyright Act 1968 (the Copyright Act) attempt to address this interaction between copyright protection and design protection. These provisions set out a regime for the statutory protection of manufactured articles which involve the application of artistic works as a design.

The central definition is 'corresponding design' which in relation to an artistic work, means visual features of shape or configuration which, when embodied in a product, result in a reproduction of that work.[164] Copyright is not enforceable in an artistic work in relation to products made to a corresponding design of that artistic work that has been applied industrially and where those products have been sold, let for hire, or offered or exposed for sale or hire.[165] Manufacturing fifty or more products is taken to be industrial application, although manufacturing fewer products can also amount to industrial application.[166] This industrial application, and dealing in the resulting products, can take place anywhere and is expressly not limited to Australia.[167]

Use of an artistic work does not prevent registration of the corresponding design under the 2003 Act, provided that use did not amount to industrial application.[168]

Accordingly, it would be prudent for a trader, upon making a decision to exploit an artistic work as a corresponding design, to not promote or market the product until an application to register the corresponding design has been filed and the Designs Office has confirmed that the application has met the minimum filing requirements.[169]

The legislation distinguishes between two-dimensional and three-dimensional items and operates, in brief, as outlined below:

(1) An artistic work retains copyright protection when it is applied as a two-dimensional surface design to an article. If the design is also registered under the 2003 Act then dual protection is afforded. This arises from the definition of 'corresponding design' which is limited to features of 'shape and configuration' therefore allowing purely two-dimensional designs to retain the ability to have dual protection under copyright and designs law. For example, if the

163. Janice Luck 'Section 18 of the Designs Act 2003: The neglected copyright/design overlap provision' (2013) 23 *Australian Intellectual Property Journal* 68, 68.
164. *Copyright Act 1968* (Cth) s. 74.
165. *Copyright Act 1968* (Cth) s. 77.
166. Section 18(3) of the *Designs Act 2003* (Cth) provides that 'applied industrially' has the same meaning given by regulations made under s. 77 of the *Copyright Act 1968* (Cth). The relevant regulation is regulation 17 of *the Copyright Regulations 1969* (Cth) which provides, relevantly, that for the purposes of s. 77, a design is taken to be applied industrially if it is applied to more than fifty articles or to one or more articles (other than handmade articles) manufactured in lengths or pieces. However, it has been said in cases and commentary that this provision sets the maximum limit at which a design will be taken to have been applied industrially, and in some circumstances the court may consider that industrial application has occurred where a design has been applied to fewer than fifty articles. For example, see Sam Ricketson & Christopher Creswell, *Law of Intellectual Property: Copyright, Design and Confidential Information,* [21.60], (3rd ed., Thomas Legal, 2006).
167. *Copyright Act 1968* (Cth) ss 77(1)(b)-(c).
168. *Designs Act 2003* (Cth) s. 18.
169. Luck, *supra* n. 163, 91.

features of pattern or ornamentation in a painting are applied to the surface of a t-shirt, the painting will retain its copyright protection. If the design for the t-shirt is subsequently registered, the painting will still retain copyright protection.

(2) An artistic work which is a building, a model of a building or a work of artistic craftsmanship retains copyright protection when applied industrially as a design.[170] However, copyright protection is forfeited if the item is registered as a design.[171]

(3) Artistic works other than those referred to in paragraphs (1) and (2) which are applied as three-dimensional designs, lose copyright protection if the design is applied industrially, regardless of whether it is registered as a design.[172] For example, if a painting depicting the shape of a vase is reproduced in three dimensions by applying the shape of the vase in the painting to an actual vase, and fifty such vases (or, possibly less than fifty) are subsequently produced, the painting will not be protected by copyright laws in respect of any three-dimensional applications of the painting.[173]

[C] Relationship between Industrial Design and Unfair Competition

In Australia, designs protection depends upon registration and there is no unregistered designs right. However, there can be overlap between designs and the registered trademarks system (as discussed at §2.09[B], above) and the common law right of passing off, and statutory provisions concerned with unfair trading, and these may have the effect of affording some protection for unregistered designs in the right circumstances. For example, the reputation attached to a design or to the get-up or packaging of goods, may be able to be protected through an action in passing off.

Similarly, there may be some very limited protection afforded to unregistered designs indirectly through the Australian Consumer Law (which is Schedule 2 to the Competition and Consumer Act 2010) (ACL). Section 18 provides that a person must not in trade or commerce engage in conduct that is misleading and deceptive or likely to be misleading and deceptive.

This provision has broad application to a range of conduct that impacts upon consumers, and it could also be used by a person seeking to protect their unregistered

170. *Copyright Act 1968* (Cth) s. 77(1)(a). The term 'work of artistic craftsmanship' is not defined in the *Copyright Act.* However, s. 10(1) provides that works of artistic craftsmanship may also come within other categories of artistic works. While s. 10(1) provides that the other categories come within the definition of artistic work, regardless of whether the work is 'of artistic quality or not', the provision is silent in this regard with respect to works of artistic craftsmanship, suggesting some requirement of artistic quality. Generally, these works include works of applied art such as jewellery, metal work, pottery and glassware.

171. *Copyright Act 1968* (Cth) s. 75.

172. *Copyright Act 1968* (Cth) s. 77.

173. It should be noted that copyright protection referred to in paras (2) and (3) above is only lost in respect of three-dimensional applications of the artistic work to an article. The two-dimensional artistic work retains copyright protection per se.

design. That is, the design owner could allege that it is misleading and deceptive for a third party to seek to use the unregistered design as if it were their own or to hold themselves out to consumers as owning the design.

There has been considerable support for the introduction of protection for unregistered designs.[174]

[D] Unjustified Threats

The 2003 Act provides for relief from unjustified threats of infringement proceedings. A person threatened with design infringement proceedings may apply to the court for a declaration that the threats are unjustified, an injunction against the continuation of the threats, and the recovery of damages sustained as a result of the threats.[175] This type of action is also available under the intellectual property laws of Australia in the areas of copyright, patents and trademarks.[176]

The threat may not necessarily be made directly but may be through means of circulars or advertisements.[177] If a certificate of examination has not been issued in respect of a design, a threat to bring infringement proceedings, or other similar proceedings, in respect of the design will constitute an unjustified threat.[178] However, the mere notification of the existence of a registered design will not constitute a threat of infringement proceedings.[179]

An unjustified threats action can be a powerful step for an alleged infringer to take because it forces the design owner into court to justify its allegations of infringement. Usually, the person sued for unjustified threats will cross claim for infringement.[180] In turn, the alleged infringer will cross claim for revocation of the design.[181] The person suing for unjustified threats will be granted relief unless the respondent can satisfy the court that the design concerned is registered, has been examined and a Certificate of Examination has been issued, and the acts about which the threats were made infringe, or would infringe, the registered design.[182]

174. V Bell 'Opening "Pandora's box": ACIP's review of the Australian designs system', (2014) 27(2) *Australian Intellectual Property Law Bulletin* 174, 176. However, this has been qualified in the recent ACIP Options Paper, *supra* n. 5, [5.8], ACIP does not consider that an unregistered designs right should be introduced into Australia at this time.
175. *Designs Act 2003* (Cth) s. 77(1).
176. *Copyright Act 1968* (Cth) s. 202; *Patents Act 1990* (Cth) s. 128; *Trade Marks Act 1995* (Cth) s. 129.
177. *Designs Act 2003* (Cth) s. 77(2).
178. *Designs Act 2003* (Cth) s. 77(3).
179. *Designs Act 2003* (Cth) s. 80.
180. And this is permitted under the *Designs Act 2003* (Cth) s. 79(1).
181. And this is also permitted under the *Designs Act 2003* (Cth) s. 79(2).
182. *Designs Act 2003* (Cth) s. 78.

LIST OF CASES

Aristocrat Technologies Australia Pty Ltd v. DAP Services (Kempsey) Pty Ltd (in liq) (2007) 157 FCR 564.
Dyson Ltd v. Qualtex (UK) Ltd [2006] EWCA Civ 166.
Dyson Ltd v. Vax Ltd [2011] EWCA Civ 1206.
Enviro Pak Pty Ltd v. New Horticulture Pty Ltd (No 2) [2013] FCA 624.
Keller v. LED Technologies Pty Ltd (2010) 185 FCR 449.
LED Technologies Pty Ltd v. Elecspess Pty Ltd (2008) 80 IPR 85.
LED Technologies Pty Ltd v. Roadvision Pty Ltd and anor (2011) 90 IPR 532.
Multisteps Pty Ltd v. Source and Sell Pty Ltd [2013] 214 FCR 323.
Reckitt Benckiser Inc [2008] ADO 1.
Review Australia Pty Ltd v. New Cover Group Pty Ltd and Ors [2008] 79 IPR 236.
Review 2 Pty Ltd v. Redberry Enterprise Pty Ltd (2008) 173 FCR 450.

LIST OF LEGISLATION

Copyright Act 1968 (Cth).
Designs Act 2003 (Cth).
Designs Regulations 2004 (Cth).
Federal Court of Australia Act 1976 (Cth).
Patents Act 1990 (Cth).
Registered Designs Act 1949 (UK).
Trade Marks Act 1995 (Cth).

CHAPTER 3

Industrial Design Rights: Brazil

Roberta Arantes Lopes & Rafael Cordoville Freire

§3.01 INTRODUCTION

[A] Highlights of New Developments in This Jurisdiction since Last Edition

The legal provisions for Industrial Designs are included as a separate chapter in current Brazilian Industrial Property Law of May 14, 1996. The provisions for Industrial Designs have not changed since then. Moreover, Industrial Designs are granted under a registration system similar to the United Kingdom system and not via patents as in the past legislation.

§3.02 CONVENTIONS AND LEGISLATION

The current Brazilian Law 9279 of May 14, 1996 provides the provisions for protection of Industrial Designs and this Law is supported by a Normative Instruction elaborated by The Brazilian Patent and Trademark Office (BPTO). This Normative Act was recently updated in 2013 (as Normative Instruction 13/2013) and contains normative requirements for the purpose of supplementing the provisions established in the Law, and also serving as a guide for the applicants regarding specific procedures related to filing of industrial design applications.

§3.03 SUBJECT MATTER

[A] Definition/What Constitutes a Protectable Design?

[1] Shape, Configuration

Industrial Designs as defined in Article 95 of Law 9279 of May 14, 1996:

> is considered to be any ornamental plastic form of an object or any ornamental arrangement of lines and colors that may be applied to a product, that provides a new and original visual result in its external configuration, and that may serve as a type for industrial manufacture.

The Industrial Designs may be represented in a three dimensional (3D) form which is the plastic form of an article, or in the two dimensional (2D) form which is the arrangement of lines and colors that may be applied to product as a print.

[2] Ornamental or Functional

Both the 2D and the 3D designs must reveal visual aspects which are to be considered ornamental. Designs created solely based on functional purposes are not accepted as Industrial Designs. For instance, a spherical design created to be applied to a soccer ball is not eligible for registration as it is a compulsory shape of the article without which the article would not perform its function correctly. The ornamental aspects may be understood as visual features observed in the design which make it distinctive from other related designs, especially from other related articles which perform exactly the same function.

[B] Originality/Novelty

The Industrial Design is considered novel when such design cannot be found in prior art before its filing date.

The exception to this rule is the so-called grace period applied to Industrial Designs. Accordingly, a disclosure of the design that has taken place within 180 days prior to its filing date (or its priority date) will not be considered as part of the state of the art if such disclosure was made by: (a) the inventor; (b) by the Brazilian Patent and Trademark Office (BPTO) through official publication of the design application without the consent of the inventor and based on information obtained from him or as the result of acts performed by him; or (c) by third parties through information obtained directly or indirectly from the inventor or as a result of acts performed by him.

Requirement for originality is the second requisite necessary for registration and is defined when the design results in a distinctive visual configuration in relation to other prior objects. In addition, the original visual effect may be a result of combination of known elements.

[1] Test for Originality Generally

Besides being considered novel (i.e., not present in prior art), for the purpose of protection the industrial design must also be considered original. The concept of originality is set forth in section 97 of the Brazilian IP Law which states the following:

> Art. 97 – An industrial design shall be regarded as original when it results in a distinctive visual configuration in relation to other existing objects.
> Sole paragraph – the original visual result may be derived from the combination of known elements.

The general test for originality entails that the design should NOT be comprised in the state of the art; should be considered distinct from prior existing ones (designs that substantially reproduce prior ones could be subject to annulment actions) and also cannot be considered a purely artistic design. The criteria thus especially when it comes to the originality of the design can be rather subjective.

Recent case law involving the annulment of industrial designs indicate that the originality test should not only be based on the differences in dimension or comparison of alterations on the general configuration of the design vis-à-vis the ones already in the market, but actually on a more thorough investigation on the degree of ornamental distinctiveness of a certain design when compared to similar products and if it is in fact capable of resulting in an effective distinctive design.

[2] Novelty

[a] Prior Use

Prior use of a design before the filing of the application may result in finding of a lack of novelty and/or originality of the design if such use is proven by historical evidences. For example, selling an article in the market place may be detrimental to the subsequent registration of its respective shape since third parties may collect evidences of prior sale or public demonstration of the product and use them to request the nullity of the design.

[b] Prior Publication

Any prior publication of other similar designs which took place in any media means where the date of publication can be officially evidenced may be harmful to the novelty and/or originality of the Industrial Design and be used against the design in a potential nullity procedure on the grounds of lack of novelty.

[c] Prior Application or Registration

Any prior publication and/or disclosure of other similar designs which took place officially by any Patent Institute around the world may also be harmful to the novelty and/or originality of the Industrial Design and also be used in a potential nullity procedure.

[C] Unobviousness (if Appropriate)

Not applicable

§3.04 DURATION OF PROTECTION

The validity of the Industrial Design is ten years from its filing date. However, this term may be extended by three successive periods of five years through filing of a request for extension of the term. Thus, the total term of validity for an Industrial Design may reach twenty-five years.

§3.05 INFRINGEMENT

[A] Scope of Infringement Action

Infringement actions are filed by the owner of the industrial design or authorized exclusive licensee provided there is a specific provision in the license agreement vesting the licensee with powers to take action in Brazil. The lawsuit is filed before a State Court of Law with jurisdiction where the infringement is taking place (the venue may change depending on other factors involved in the lawsuit).

The basic scope of the infringement action is to prove the existence of infringement, combined with a request for past and present damages that the use of the infringing product may cause. The plaintiff may also request preventively, or in the course of the litigation, injunctions to have the infringing products seized, or commercialization stopped, among other measures to reduce the damages caused by the commercialization of the infringing products.

[B] Test for Infringement/ Criteria to Constitute Infringement

In order to show infringement, a first criterion is having a valid registration of a design in Brazil. Considering that granting is automatic, it is advisable to request the BPTO to perform a substantive examination of the design in order to have the BPTO's official decision on the merits of the design before initiating the lawsuit. Second, the design must meet originality and novelty requirements. No evidence of the prior art is required, though an expert analysis is advisable. Next, grounds for infringement may be established through comparative analysis; technical examination by an expert in the art can also be helpful.

[C] Standing to Sue

As previously mentioned, infringement actions may be brought by the owner of the industrial design or authorized exclusive licensee provided there is a specific provision in the license agreement empowering the licensee to take action in Brazil.

[D] Remedies for Infringement

Preliminary injunctions of all kinds (seeking for instance evidence to be provided before Court; search and seizure orders; halt commercialization, among others). Injunctions can also be granted in the course of the litigation. The plaintiff may also sue for past and present damages that the use of the infringing product may cause.

§3.06 DEFENSES TO INFRINGEMENT

[A] Validity

In an infringement action involving a granted Industrial Design, a third party has the opportunity to try to invalidate the design by bringing an Administrative Nullity Procedure and the term for bringing this Procedure is five years from the granting publication of the Design. The reason for such a considerable long term is that Industrial Designs in Brazil are granted automatically where no substantial examination is performed regarding novelty and originality requirements.

The Administrative Nullity Proceedings against a design may also be brought ex officio by the BPTO when the examiner understands that the design is not novel or original.

The Administrative Nullity Procedure suspends the effects of the grant of a Design Registration if started within sixty days from the granting publication of the design.

After the Administrative Nullity Procedure is started, the owner will be notified through publication in the Official Bulletin to present arguments in favor of the maintenance of the design within a sixty-day term. Subsequently, the BPTO will issue a technical report with the opinion regarding the nullity procedure, to which the owner and the petitioner must respond within sixty days. Even if no response is presented, the process will be ruled upon by the President of the BPTO, who will issue the final decision for the administrative process. The Nullity Procedure continues even after the Design Registration becomes extinct.

If the five-years term has already lapsed, the petitioner has also the option to file a Nullity Action against a granted design at Courts and the action may be instituted anytime during the lifetime of the design.

[1] Not Novel/Original

Lack of novelty and or originality are the main arguments for invalidating an Industrial Design. The evidence is usually based on similar design references published or granted prior to the date of filing (or priority date) of the Brazilian design. Such evidence may also be collected as the result of publications made in any other media such as journals, articles, magazines or even on the Internet where the date of publication shall be clearly proved.

The construction of the arguments to be presented in a Nullity Procedure based on lack of novelty and/or originality is basically a detailed comparison between the registered design and the prior designs where the similar portions are highlighted.

[2] Functional/Not Ornamental

Another argument which may be used to invalidate an Industrial Design is that the shape of the object is the result of aspects which are solely related to the function of the article. In other words, the entire configuration adopted is indispensable for the article performing its function and no other ornamental aspect can be observed.

[3] Other Possible Criteria, Such as Non-obviousness

Another possibility of argument to be used in a Nullity Procedure is to allege that the design lacks industrial applicability. This is usually applied to Industrial Designs protecting artistic works as artisanal objects, for example, where it is not possible to produce identical specimens.

[4] Faults in Application

It is also recommended to review the file wrapper of a design since flaws in the application may also serve as the base to invalidate the design. These flaws may include non-compliance with specific deadlines, or for instance, non-compliance with formalities requirements such as submission of Power of Attorney, or the non-payment of a quinquennium fee, and so on.

[B] Marking

In Brazil, marking is not mandatory and it is up to the owner or retailer the decision to place any reference to the design registration or application in the product. It should be noted though that the unfair use of marking, namely, the reference to a non-existent application or registration, or reference to a registration that is still pending may be liable to criminal penalties, as the misuse of marking would characterize crime of unfair competition.

[C] Non-infringement

Most usual defenses are the ones that attack novelty and originality requirements or procedural flaws that could result in the invalidation of the design registration. However, the defense should also be based on the differences in the shape of the allegedly infringing product when compared to the registered design where it must be demonstrated that the shape of the allegedly infringing product (as compared to the registered design) may not lead the consumer to make a mistake when purchasing the product (confusion or undue association).

[D] Other Defenses

Besides material defenses, the defendant may offer several different procedural defenses to try to invalidate or suspend the lawsuit or even file counter claims, depending on the subject matter. For instance, often there may be a previous relationship with the plaintiff that can be raised in support of a bad faith claim; the authorship of the design may be challenged; existence of similar designs in the market that would weaken the protection afforded by the design, among others.

§3.07 PROCEDURES FOR FILING APPLICATION FOR REGISTRATION

[A] Priority

In the event that the design application claims priority from an earlier filed foreign application, it is necessary to file a certified copy of such application at the date of filing of the Brazilian Application or within a ninety-day term from the filing date. It is also necessary to submit a simple Portuguese translation of the cover sheet with the identifier data of the priority application such as number, date, country, applicant(s), author(s), and title.

When the applicant of the priority application is not the same as the applicant for the design in Brazil, it also necessary to submit the assignment of priority rights duly executed by both parties. The assignment must be submitted at the filing date or within a 180-day term from the filing, no notarization or legalization is necessary. A simple Portuguese translation of the assignment must also be submitted along with the document.

[B] Drawings

The design application must refer to a single article and a plurality of variations are allowed, long as each variation is destined to the same proposal and keeps the same preponderant distinctive characteristic. Each application is limited to have until twenty variations.

If the Brazilian examiners conclude that the plurality of variations of an industrial design are not intended for the same purpose and they do not possess the same predominant distinctive characteristic, the BPTO's examiners will issue an Official Action requesting to divide the industrial design application.

Applicant may submit the drawings in black and white or with colors.

The designs related to 3D objects shall be illustrated with regular and continuous traces where no dashed lines representation is permitted. The designs related to 2D patterns must also be illustrated with regular and continuous traces, however when representing the 3D object where the 2D pattern is applied, such object must be represented in dashed lines and the protection is only applied to the 2D pattern.

[C] Design Part of Another Article/Environment of Design

Part of objects are registered as Industrial Design as long as the article may be manufactured separately from the rest of the object.

Considering complex products which are constructed by joining several portions, in order to obtain the broadest scope of protection as possible, it is recommended to file a design application to each portion separately and another design for the entire product.

[D] Description of Design (Where Required)

The description of the views of the design is always recommended to render a better understanding of the views filed in the design application. However, the description becomes compulsory whenever Industrial Design includes variations for the main article.

§3.08 EXPUNGING/CANCELLING OR VARYING REGISTRATION

During prosecution, failure to take action within a term prescribed by the law shall result in abandonment of the application. Such circumstances include for example non-compliance with an Office Action or non-payment of a quinquennium or renewal fee during the given term. In addition, lack of presentation of the Power of Attorney during the specified term (in the event where an applicant delegates representation to an Attorney) brings the application for Industrial Design to definitive shelving. Under these circumstances the restoration of the Industrial Design is not possible according to law.

The Brazilian Law is silent regarding the possibility to file amendments to an Industrial Design application before granting. However, in practice, it is observed that BPTO will usually accept amendment petitions if filed before granting of the Industrial Design, provided that the modifications are within the scope of the respective priority design. Otherwise, any amendment will be rejected as being additional matter.

§3.09 OTHER

[A] Interface between Industrial Design and Trademarks

The Brazilian IP Law provides protection for figurative, composite and tridimensional marks, provided the intended sign is not: (i) an object that is protected by registration as an industrial design in the name of a third party or (ii) the necessary, common or usual form of a product or packaging, or also a form that cannot be dissociated from a technical effect. As a result, protection for a certain view of the design as a mark may be sought provided that it does not fall in the legal prohibitions above (items XXI and XXII of section 124 of the Brazilian IP Law) and also the sign performs the function of a trademark (e.g., is a design that would be used to distinguish a certain product/service in the marked).

In addition, the Brazilian IP Law also forbids, in item VI of section 124 of the law, the registration of signs that are generic, necessary, common, usual or simply descriptive when they are related to the products or services it intends to identify. Therefore, in order to be protected as a mark the design must be distinctive and also different from the prior ones already on file to be allowed into registration.

[B] Relationship between Industrial Design and Copyright/Design Copyright

The standard for the protection as a copyright is slightly different and though protection from both categories may overlap, the scopes of protection for copyright and for industrial design are different. In a copyright the protection would cover works of art that are non-identical to prior existing ones. The originality standard degree is thus lower than the one required for industrial design protection. The work of art must also be creative and designed by a natural person (even if the registration is applied for a company, an assignment from a natural person is required in order to seek protection as a copyright).

It should be noted that in Brazil copyrights are protected regardless of registration, though it is usually advisable to protect the work of art with a registration, which is granted automatically, as no scrutiny on the novelty/originality aspects are done by the Registrar, which simply conducts a review of the documents presented based on formalities.

[C] Relationship between Industrial Design and Unfair Competition

Besides the general rules that are applicable to combat unfair competition practices and could be thus claimed in an design infringement case, the Brazilian IP Law expressly provides in its section 195, item XIII that false marking of a design may be considered a crime:

> 195: A crime of unfair competition is committed by whoever:
> XIII – sells, displays or offers for sale a product falsely declaring that the product is protected by a pending or granted patent, or by an industrial design registration, or mentions in an advertisement or on commercial stationery that it is pending, patented or registered when this is not the case.
> Penalty: Imprisonment for three (3) months to one (1) year, or a fine.

CHAPTER 4

Industrial Design Rights: Canada

*Brian W. Gray, Christopher N. Hunter & Rita Gao**

§4.01 INTRODUCTION

Industrial designs in Canada are governed by the *Industrial Design Act* (the '*Act*').[1] This *Act* had remained relatively unchanged for over 100 years prior to 1988. However, significant amendments were made to the *Act* in 1988, 1993, and 1994, and further amendments made in 2014 are expected to come into force in 2015. Nevertheless, industrial designs still remain a hybrid somewhere between copyright and patents with the concepts applying to authorship, enforcement and validity being not quite those applied to either copyright or patents.

The 1988 amendments attempted to better define the subject matter of industrial design. Further significant changes were made in 1993 with the passing of the *Intellectual Property Law Improvement Act* ('*IPLA*')[2] and to implement the intellectual property provisions of the *North American Free Trade Agreement* ('*NAFTA*')[3] in Canada. These amendments brought the *Act* more in line with the industrial design legislation of the two other signatory countries, the United States and Mexico. For the most part the *NAFTA* amendments were consistent with the *IPLA*. However, in some cases, the implementation of *NAFTA* meant broader amendments, which superseded

* This chapter is current to 1 Jul. 2015. It is based in part on the first edition of this chapter authored by Brian W. Gray, Kathleen Murphy and Effie Bouzalas. For this new edition, the authors would like to thank Chelsea Nimmo and Bobby Leung, associates at Norton Rose Fulbright Canada LLP, for conducting legal research and proofreading the manuscript. The authors would also like to thank Jeffrey Kang for reviewing the manuscript and providing feedback.

1. *Industrial Design Act*, R.S.C. 1985, c. 1-9, as am. by R.S.C. 1985, c. 10 (4th Supp.); S.C. 1992, c. 1; S.C. 1993, c. 15, c. 44; 1994, c. 47; 2001, c. 34; 2014, c. 39.
2. *Intellectual Property Law Improvement Act*, S.C. 1993, c. 15.
3. *North American Free Trade Agreement Between the Government of Canada, the Government of Mexico and the Government of the United States*, 17 Dec. 1992, Can. T.S. 1994 No. 2, 32 I.L.M. 289 (entered into force 1 Jan. 1994).

those in the *IPLA*, particularly in regard to the test for infringement. Further amendments were made in 1994 by the *World Trade Organization Agreement Implementation Act*.[4]

On 16 December 2014, Bill C-43 received Royal Assent introducing substantial changes to the *Industrial Design Act* as part of the *Economic Action Plan 2014 Act, No. 2* (hereinafter 'Bill C-43').[5] Bill C-43 introduces changes allowing Canada to adhere to the *1999 Geneva Act of the Hague Agreement Concerning the International Registration of Industrial Designs* (the 'Hague Agreement').[6] While Bill C-43 has received Royal Assent, it has not yet been proclaimed in force. As such, any matter arising on or after the coming-into-force date in respect of a design registered before that date or an application whose Canadian filing date is before the coming-into-force date, must be dealt with and disposed of in accordance with the prior provisions, with a few exceptions relating mostly to administrative details.[7]

Bill C-43 also includes important changes to the test for novelty and in respect of self-collision of related applications (to be discussed below). However, because of the transitional provisions,[8] the current *Act* will remain important and governing and relevant for some time.

Accompanying the *Act* are the *Industrial Design Rules* (the '*Rules*').[9] While the *Act* addresses the substantive issues relating to industrial design protection such as subject matter and infringement, the *Rules* deal with the procedural aspects of filing an application for registration of an industrial design in Canada. These *Rules* were significantly amended in 1999. They are expected to be significantly amended again in the near future (likely in 2015), to conform to the Hague Agreement. However, as of 1 January 2015, proposed new rules have not yet been disclosed. This chapter will first address the substantive issues and will then briefly discuss the procedures and formalities of applying for a registration as they exist in Canada now, and as they are likely to be changed to conform to the Hague Agreement.

Finally, because Canada is a common law jurisdiction, it should be noted that case law, including that of the United Kingdom, is often relied on in interpreting the *Act*.[10] It will probably be some time before a body of Canadian case law interpreting some of the new provisions is available.[11]

---

4. S.C. 1994, c. 47.
5. Bill C-43, *A second Act to implement certain provisions of the budget tabled in Parliament on 11 February 11, 2014 and other measures*, 2nd Sess., 41st Parl., 2014 (assented to 16 Dec. 2014), S.C. 2014, c. 39.
6. EC, *Council Decision 2006/954/EC of 18 December 2006 approving the accession of the European Community to the Geneva Act of the Hague Agreement concerning the international registration of industrial designs, adopted in Geneva on 2 July 1999*, [2006] O.J. L 386/28 at 30.
7. Clause 112 of Bill C-43, amending s. 30 of the *Act*, and adding s. 32 to the *Act*.
8. *Ibid.*, amending ss 29 and 30 of the *Act*, and adding ss 31–33 to the *Act*.
9. *Industrial Design Rules*, C.R.C. 1978, c. 964, as am. by SOR/91-450; SOR/93-297; SOR/93-584, as repealed by SOR/99-459, as enacted SOR/99-460, as am. by SOR/2003-210; SOR/2007-92; SOR/2008-268; SOR/2014-225.
10. The *Act* was originally modelled after the legislation in the United Kingdom on industrial design, today called *The Registered Designs Act*, 1949.
11. There have been very few court decisions relating to the present *Act* in the past twenty years. See *infra* n. 26 and related text.

§4.02 CONVENTIONS AND LEGISLATION

As discussed above, industrial designs are governed by *Industrial Design Act*.[12] There are also provisions potentially relevant to industrial design issues in the *Copyright Act*[13] relating to copyright in artistic works that are not mass produced and in graphic representations applied to the face of an article, and in the *Trade-marks Act*[14] relating to trademark rights in signs, including three-dimensional shapes or packaging, i.e., get-up or trade dress.

Canada is a party to the *Paris Convention for the Protection of Industrial Property* (1883),[15] which governs priority claims from foreign design applications.[16]

Canada is also a party to the *Agreement on Trade-Related Aspects of Intellectual Property Rights* (the 'TRIPS Agreement'),[17] which contains specific obligations regarding the substantive protection of industrial designs.[18]

It is expected that Canada will ratify the Hague Agreement sometime during 2015.

§4.03 SUBJECT MATTER

Unlike copyright and trademark protection in Canada, protection for industrial designs may only be acquired through registration.[19] Therefore, the first question is whether a design is the proper subject matter for registration.

[A] Definition: What Constitutes a Protectable Design?

The *Act* defines 'design' or 'industrial design' as 'features of shape, configuration, pattern or ornament and any combination of those features that, in a finished article, appeal to and are judged solely by the eye'.[20] This definition is unchanged by Bill C-43. This definition is derived from prior UK and Canadian case law (discussed below).

'Ornamentation' is not defined in the *Act* per se, nor does the definition of 'design' require that the design supply ornamentation in every case as long as the design distinguishes the appearance of an article. The design does not have to beautify

12. *Industrial Design Act*, *supra* n. 1.
13. *Copyright Act*, R.S.C. 1985, c. C-42, as am. by R.S.C. 1985, c. 10 (1st Supp.); c. 1 (3rd Supp.); c. 41 (3rd Supp.); c. 10 (4th Supp.); S.C. 1988, c. 65; 1990, c. 37; 1992, c. 1; 1993, c. 15, c. 23; c. 44; 1994, c. 47; 1995, c. 1; 1997, c. 24, c. 36; 1999, c. 2, c. 17, c. 31; 2001, c. 27, c. 34; 2002, c. 26; 2003, c. 22; 2004, c. 11; 2005, c. 38; 2012, c. 20.
14. *Trade-marks Act*, R.S.C. 1985, c. T-13, as am. by S.C. 1990, c. 14; c. 20; 1992, c. 1; 1993, c. 15; c. 44; 1994, c. 47; 1995, c. 1; 1996, c. 8; 2001, c. 27; 2002, c. 8; SOR/2004-85; 2005, c. 38; 2007, c. 25, c. 26; 2014, c. 20.
15. *Paris Convention for the Protection of Industrial Property*, 20 Mar. 1883, 828 U.N.T.S. 305, 21 U.S.T. 1583 (as revised at Stockholm on 14 Jul. 1967, last amended on 28 Sep. 1979).
16. Articles 4 and 4*bis* of *Paris Convention for the Protection of Industrial Property*.
17. *Agreement on Trade-Related Aspects of Intellectual Property Rights*, 15 Apr. 1994, 1869 U.N.T.S. 299, 33 I.L.M. 1197.
18. Articles 25 and 26 of the TRIPS Agreement.
19. However, see comments *infra* on the interface between industrial design, copyright and trademark protection.
20. Section 2 of the *Act*. 'Article' is defined as 'anything that is made by hand, tool or machine'.

the article.[21] Similarly, the artistic merit of the design is not a consideration in determining whether it qualifies as proper subject matter for registration.[22]

[1] *Design Includes Shape, Configuration, Pattern or Ornament and Any Combination Provided Not Functional*

Until 1988 there was no definition of 'design' in the *Act*. Because of this, much of the case law focused on the proper subject matter of industrial design protection. In particular, there had been considerable debate over whether 'design' meant only a pattern or representation *applied* to an article, or whether it could include the shape or the design of the article itself. The aforementioned definition of 'design', as well as the introduction of other provisions in 1988 codified the state of jurisprudence at that time, namely, that the shape of an article may be the proper subject matter of industrial design protection.

[2] *Functionality*

The *Act* expressly provides that design protection does not extend to features applied to a useful article that are dictated *solely* by a utilitarian function of the article or to any method or principle of manufacture or construction.[23] 'Utilitarian function' is defined as 'a function other than merely serving as a substrate or carrier for artistic or literary matter'.[24] It has been held that the *Act* does not protect a functional article where the significant design features are hidden when the article is in use and were never intended to be admired by or sold to the public at large.[25] This suggests that a court may consider the intention of the owner of the article in determining whether a design 'appeals to the eye'. However, a leading author on industrial design law notes that the object that a designer has in mind when adopting a design or the useful purpose that the design is intended to serve, or does serve, is not relevant in determining whether the design is protected.[26]

21. *Datafile Ltd. v. DRG Inc.* (1991), 35 C.P.R. (3d) 243 at 252 (F.C.A.).
22. *Walker, Hunter & Co. v. Falkirk Iron Co.* (1887), 4 R.P.C. 390 at 391 (Ct. Sess.). However, a certain degree of 'originality' is required: see discussion, *infra.*
23. Section 5.1 of the *Act;* See also *Zero Spill Systems Inc. v. Heide* 2015 FCA 115 [*Zero Spill*]; *Carr-Harris Products Ltd. v. Reliance Products Ltd.* (1969), 58 C.P.R. 62 (Can. Ex. Ct.) [*Carr-Harris*]; *Cimon Ltd. v. Bench Made Furniture Corp.* (1964), 48 C.P.R. 31 (Can. Ex. Ct.).
24. Section 2 of the *Act.*
25. *Mainetti S.p.A. v. E.R.A. Display Co. Ltd.* (1984), 80 C.P.R. (2d) 206 (F.C.T.D.) [*Mainetti S.p.A.*]. The Federal Court noted that the ornamentation and design on the ends of skirt hangers were hidden until the skirt was removed. The Court seems to interpret 'appeal to the eye' based on the intention of the owner of the article as well as public perception, rather than on whether the design was dictated by the function of the hanger or whether it was original. See discussion on 'Originality' *infra.*
26. John S. McKeown, *Fox on Canadian Law of Copyright and Industrial Design*, looseleaf, 4th ed. (Toronto: Carswell, 2003), Ch. 31 at 5 [*McKeown*], quoting from *Hecla Foundry Co. v. Walker, Hunter & Co.* (1889), L.R. 14 App. Cas. 550 (H.L.) [*Hecla Foundry Co*].

In 2012, the Federal Court of Canada issued its first industrial design infringement decision in about two decades: *Bodum USA Inc. v. Trudeau Corp. (1889) Inc.*[27] The Federal Court in *Bodum* noted that 'industrial designs protect the visual features of an article, not its functionality'[28] and decided that since the double-wall aspect of the plaintiff's registered design of drinking glasses had a utilitarian function, the utilitarian function of the double wall glasses (i.e., the space between the walls) was to be disregarded in the Federal Court's analysis.[29]

In May 2015, the Federal Court of Appeal further confirmed the broad scope of protection for industrial designs in *Zero Spill Systems Inc. v. Heide,*[30] stating that only features of an industrial designs whose form is dictated solely by function are excluded from protection by paragraph 5.1(a) of the *Act*. Features may be simultaneously useful and visually appealing. The Court of Appeal further notes that the very purpose of the *Act* is to provide residual protection for functional designs that would, but for section 64 of the *Copyright Act* be subject to copyright protection.[31] Further the Court made clear that the burden of providing that a registered design was dictated solely by function was on the person attacking its validity.

Bill C-43 now separates the requirements for registration from issues of infringement. The requirements for registration relating to functionality have been subtly changed so that a design cannot be registered if it consists '*only* of features that are dictated *solely* by a utilitarian function of the finished article.'[32] However, when considering infringement, design protection, as under the *Act*, does not extend to 'features applied to a useful article that are dictated *solely* by a utilitarian function of the article or to any method or principle of manufacture or construction'.[33] It is submitted that the addition of the word 'only' in respect of features 'solely' functional probably makes no difference in practice where the Industrial Design Office of the Canadian Intellectual Property Office (CIPO) will allow registration (assuming the design has eye-appeal) unless every feature is dictated solely by functional considerations.[34] This application of functional features to designs is also in line with UK case law.[35]

As previously stated, the *Act* and Bill C-43 continue to provide that the protection afforded by a design registration does not extend to any method or principle of manufacture or construction.[36]

---

27. 2012 FC 1128, [2013] 3 S.C.R. 372 [*Bodum*].
28. *Ibid.*, at paras 38, 46.
29. *Ibid.*, at para. 51.
30. *Zero Spill, supra* n. 23.
31. *Ibid.*, at paras 23–25.
32. Clause 105 of Bill C-43, adding subs. 7(d) to the *Act*.
33. Clause 107 of Bill C-43, adding s. 11.1 to the *Act*; s. 5.1 of the *Act*.
34. *Interlego A.G. v. Tyco Industries Inc.*, [1988] R.P.C. 343 (P.C. Hong Kong); *Re Industrial Design Application 1998-2446* (2003), 25 C.P.R. (4th) 256 (Canada Patent Appeal Board and Patents Commissioner).
35. *Amp Inc. v. Utilux Pty Ltd.*, [1972] R.P.C. 103 (H.L.); *Dyson Ltd. v. Vax Ltd.*, [2010] EWHC 1923 (Pat), citing *Lindner Recyclingtech GmBH v. Franssons Verkstäder AB,* [2010] E.C.D.R. 1 (Office for Harmonization in the Internal Market (Trade Marks and Designs) (Third Board of Appeal)), aff'd *Dyson Ltd. v. Vax Ltd.*, [2011] EWCA Civ 1206. See also *Samsung Electronics (UK) Ltd. v. Apple Inc.*, [2012] EWCA Civ 1339.
36. Subsection 5.1(b) of the *Act*; Clause 107 of Bill C-43, adding s. 11.1 to the *Act*.

[B] Originality/Novelty

The *Act* does not expressly require that the design be 'original' to be entitled to registration. However, it does provide that the certificate of registration, in absence of proof to the contrary, is sufficient evidence of the originality of the design.[37] For this reason, courts have concluded that originality is a prerequisite to registration.[38]

However, Bill C-43 has now removed reference to the copyright concept of originality, and when Bill C-43 comes into force, the requirement for originality will be replaced by the patent concept of novelty.[39]

[1] Test for Originality/Novelty Generally

Under the current *Act*, the standard is 'originality', not 'novelty'. The question of the degree of originality required in a design has received much consideration. Generally, a higher degree of originality is required for industrial designs than for copyright, as the latter generally requires that a work be the author's and not be copied from someone else. Some degree of intellectual activity or 'spark of inspiration' on the part of the designer, either in creating an entirely new design or in hitting upon a new use for an old one is required.[40] In other words, 'every design which is original is new, but every design which is new is not necessarily original'.[41] The Federal Court of Canada has held that the test used for determining the originality of a design is that of a purely visual assessment of the features of a finished article, and that where the article is functional, the visual features of the article protected by an industrial design registration should not be solely the result of a function of the article.[42] In addition, manufacturing process, materials used and colours should be ignored in the analysis of originality.[43]

In addition to originality, the *Act* does require that the design be 'new' in that the design should not be published anywhere more than one year before the filing date and not 'so closely resembling any other design already registered as to be confounded therewith'.[44]

Under Bill C-43, the copyright concept of originality has been jettisoned in place of the patent concept of novelty. A design is 'novel' if the same design or a design not differing substantially from it, is applied to a finished article that is the same as or analogous to the finished article in respect of which the design is to be registered, and

37. Subsection 7(3) of the *Act*.
38. See for example, *Clatworthy & Son Ltd. v. Dak Display Fixtures Ltd.*, [1929] S.C.R. 429 [*Clatworthy & Son*].
39. Clause 105 of Bill C-43, adding subs. 7(b) to the *Act*.
40. *Bata Industries Ltd. v. Warrington Inc.* (1985), 5 C.P.R. (3d) 339 (F.C.T.D.) [*Bata Industries*]. See also *Rothbury International Inc. v. Canada (Minister of Industry)*, 2004 FC 578, [2005] 1 F.C.R. 130 [*Rothbury*].
41. Lamont J. in *Clatworthy & Son Ltd. v. Dak Display Fixtures Ltd.*, [1929] S.C.R. 429 at 431 quoting from *Re Rollason's Registered Design* (1897), 14 R.P.C. 909 at 914. Query, however, whether this can be true if the designer, while the originator of the design, has already published it.
42. *Rothbury*, *supra* n. 40 at paras 30, 37.
43. *Bodum*, *supra* n. 27 at para. 55.
44. Subsections 6(3) and 6(1) of the *Act*.

the design has not been disclosed so as to have become 'available to the public' before the priority date by a third party or more than twelve months before the priority date by or through the person who filed the application.[45] Bill C-43 thus establishes a novelty bar and a grace period in essentially the same terms prescribed for Canadian patents, except that the period is calculated from the priority date and not from the actual Canadian filing date.[46] It seems likely that Bill C-43, by removing 'originality' as a concept, has removed the need for a 'spark of inspiration' as set out in *Bata Industries*[47] and replaced it with 'novelty' *simpliciter*.

Generally, under the *Act*, the rule is that a substantial difference over previous designs is a condition for originality. For instance, in an appeal from a decision of the Commissioner of Patents refusing to register a design for a toy spaceship, the Patent Appeal Board, in affirming the Commissioner's decision, held that the overall appearance of the design closely resembled that of the well-known spaceship design from the television series and movie 'Star Trek'. The changes made in the overall design by the applicant were not substantial enough to warrant registration.[48] However, in another case, the Patent Appeal Board found that small differences from a similar design were sufficient to warrant registration for a design of a paving stone, where there was very little scope for adding design features to the paving stone, as the stone was primarily a functional article and most features of the stone were dictated by use.[49]

In *Bodum*,[50] the Federal Court found both of the plaintiff's design registrations to be invalid due to lack of sufficient originality. The Federal Court compared the plaintiff's designs to the prior art, focusing on interior and exterior lines of the drinking glasses. In particular, the Federal Court cited *Bata Industries*[51] and *Clatworthy & Son*,[52] and further emphasized that for a design to be registrable, it must substantially differ from prior art.[53] The Federal Court also noted that the expert witness did not think there was any 'spark of inspiration' in the shape of the plaintiff's double-wall glasses.[54]

Under the *Act*, old designs can be combined into a new design, provided that an original design results in a design which is substantially different from any of the old designs or any known combination.[55] However, the combination of ordinary trade

45. Clause 105 of Bill C-43, adding subs. 8.2(1) to the *Act*.
46. *Ibid*.
47. *Supra* n. 40.
48. *Re Paramount Pictures Corporation Industrial Design Application* (1981), 73 C.P.R. (2d) 273 (Patent Appeal Board and Commissioner of Patents).
49. *Industrial Design Application 1997-2244, Re*, (2001) 14 C.P.R. (4th) 59; See also *Zero Spill Systems (Int'l) Inc. v. 614248 Alberta Ltd.*, 2013 FC 616 at para. 135, 111 C.P.R. (4th) 317, where R. L. Barnes J. comments on infringement analysis of designs of pipe trays: '[w]hen an industrial design incorporates fundamentally functional features even small differences in ornamentation may be sufficient to take the second design out of the ambit of an earlier design registration'.
50. *Supra* n. 27.
51. *Supra* n. 40.
52. *Supra* n. 38.
53. *Bodum*, *supra* n. 27 at para. 96.
54. *Bodum*, *supra* n. 27 at para. 34.
55. Please refer to *Carr-Harris*, *supra* n. 23, a case involving alleged infringement of a design for a tent peg. The Court found that the defendant had worked from the plaintiff's tent peg in developing its ultimate original design.

variants, such as features of a shoe, is insufficient to make the design original, even if the overall design is new.[56]

Under the *Act*, originality is a question of fact that will be assessed having regard to the nature and character of the subject matter and the whole of the design. Originality is determined as of the date of creation of the design, unless there is insufficient evidence as to the date of creation, in which case it will be assessed as of the date of registration.[57] It is judged by the 'eye of the court' and can be aided by expert evidence.[58]

Under Bill C-43, the concept is novelty and not originality. However, the requirement that a design shows an actual substantial difference from a design previously disclosed for the same or an analogous article is retained.[59] Further, although not expressly a statutory requirement for registration, Bill C-43 makes clear that the exclusive right granted by the registered design protects that design and a design not differing substantially therefrom.[60] Bill C-43 further provides that the court may consider the extent to which the registered design differs from any previously published design.[61] It is submitted that this test for infringement of the exclusive right should and will be applied as well when assessing novelty. Thus, 'novelty' is likely to be judged in the context of the exclusive right provided by Bill C-43 whereby not just the exact design but designs not substantially different will be considered when assessing novelty and the court and CIPO will consider the extent to which the proposed design differs from any previously published designs.

Under Bill C-43, novelty will be judged in respect of third party disclosures as of the priority date (not the filing date or registration date) and in respect of disclosures by or through the applicant (or the applicant's predecessor in title) made more than twelve months before the priority date.[62]

[2] Specific Novelty Bars

The *Act* also imposes specific bars to registration relating to originality or novelty.[63] They are discussed below.

56. Please refer to *Bata Industries, supra* n. 40, an infringement case. The Court found that while the identical design of the shoe was not in the existing trade, each of the design elements of the shoe were ordinary trade variants. The Court therefore rejected the plaintiff's argument that the design of the shoe as a whole was validly registered. The Court stated that if the degree of originality of industrial designs was the same as that for copyright, the plaintiff's design would have met the test.
57. *Bata Industries, supra* n. 40 at 342–343. The test for infringement is also the 'eye of the court'. This is dealt with in greater detail in the section on 'Infringement', *infra*.
58. *R. v. Premier Cutlery Ltd.* (1980), 55 C.P.R. (2d) 134 (Ont. Prov. Ct. (Crim. Div.)).
59. Clause 105 of Bill C-43, adding subs. 8.2(1) to the *Act*.
60. Subsection 11(1) of the *Act*.
61. Subsection 11(2) of the *Act*.
62. Clause 105 of Bill C-43, adding subs. 8.2(1) to the *Act*.
63. These terms are often, unfortunately, used interchangeably, but under Bill C-43, 'originality' as a requirement will disappear, replaced by 'novelty', see earlier discussion.

[a] Prior Use

Under the *Act*, the applicant for a registration must declare that the design was not, to his or her knowledge, 'in use' by any other person at the time he or she adopted the design.[64] This requirement does not limit the 'use' of the design strictly to use in Canada.[65] It is unclear what 'in use' means. One leading author has equated it to 'publicly known'.[66] Bill C-43 eliminates the requirement for such a declaration but requires that the design be created by the applicant or the applicant's predecessor in title.[67]

[b] Prior Publication

Under the *Act*, an application for registration of an industrial design will be refused if the application is filed in Canada more than one year after the publication of the design in Canada or elsewhere.[68] It is publication within one year of the actual Canadian filing date, not the convention priority date, assuming one exists, which is relevant. 'Publication' is not defined in the *Act*, but has been held to mean 'offering or making the design available to the public' where 'public' has been interpreted as those who are apt to be interested in the design. Disclosure to the general public is not a requirement.[69] Whether there has been publication is a question of fact. However, publication will generally include publication in a printed document or publication by a prior user,[70] sale of the design,[71] and depending on the circumstances, display of a prototype rather than a commercial embodiment of the design.[72]

Under Bill C-43, the test is no longer 'publication' but as with Canadian patents, whether the design is 'available to the public' and, as previously mentioned, whether there has been a disclosure by third parties before the priority date or by the or through the applicant himself/herself (or his/her predecessor) more than twelve months before the priority date.

---

64. Paragraph 4(1)(b) of the *Act*.
65. See McKeown, *supra* n. 26, Ch. 31 at 13, n. 85. McKeown states that this provision requires originality to be universal, stating that 'one cannot copy a design that is or has been publicly known abroad and register it on the basis that it is original in [Canada]'. Prior to *NAFTA*, prior publication within a year of the application was only a bar to registration if publication occurred in Canada (see discussion *infra*). This was inconsistent with the worldwide novelty test for 'use' contained in s. 4.
66. *Ibid.*, but there is no authority supporting this position.
67. Clause 105 of Bill C-43, adding subs. 7(c) to the *Act*.
68. Subsection 6(3) of the *Act*. Prior to *NAFTA*, the restriction on publication was limited to publication in Canada more than a year before filing the application.
69. *Algonquin Mercantile Corporation v. Dart Industries Canada Ltd.* (1984), 1 C.P.R. (3d) 75 (F.C.A.), leave to appeal to S.C.C. refused (1984), 57 N.R. 392 n [*Algonquin Mercantile*].
70. McKeown, *supra* n. 26, Ch. 31 at 20 citing for example, *Carpenter v. Smith* (1841), 11 L.J. Ex. 213.
71. *Cookie Florist Canada Ltd. v. 13281 Canada Inc. (c.o.b. Monsieur Felix & Mr. Norton Cookies Inc.)*, (1996) 75 C.P.R. (3d) 379 (F.C.T.D.), aff'd (1998), 80 C.P.R. (3d) 447 (F.C.A.), citing *Ribbons (Montreal) Ltd. v. Belding Corticelle Ltd.*, [1961] Ex. C.R. 388 at 402.
72. *Algonquin Mercantile, supra* n. 69.

Arguably, once Bill-C43 comes into force, a design may be made 'available to the public' by the act of publishing, selling or offering to sell, public demonstration and so on, similar to the patent law concept of 'disclosure'.

[c] Prior Existing Canadian Registration or Application

Under the *Act*, an application for registration will be refused if it is found to be identical to or so closely resemble a design already registered as to be confounded therewith.[73] Generally, the examiner will conduct a search of the register to satisfy himself or herself that the design meets this standard. Hence a third party registration could prevent registration of the applicant's design, even if the applicant's application had an earlier priority date. This has also provoked an insistence by some applicants that similar designs owned by the applicant be registered on the same day to avoid self-collision.

Bill C-43 has completely eliminated this requirement. However, it imposes a requirement, 'subject to the regulations', that the design has not been disclosed in an application filed in Canada whose priority date is before the priority date of the application for the design in question.[74] This continues the possibility of self-collision, but now based on the priority date and not the registration date. Applicants may be forced to file all similar applications on the same day to avoid this. It remains to be seen if the regulations once drafted and in force avoid this, at least for commonly owned applications.

§4.04 DURATION OF PROTECTION

[A] Registrations Issued before 1 January 1994

For designs registered before 1 January 1994, the exclusive right to an industrial design is valid for a term of five years but may be renewed, on or before the expiration of the term, for a further period of five years or less, on payment of the prescribed fee. However, the entire duration of an exclusive right cannot exceed ten years. The last of such designs would have expired as of 31 December 2003.

[B] Registrations Issued after 1 January 1994

As a result of *NAFTA*, the initial five-year term and five-year renewal were replaced in the *Act* with one ten-year term beginning on the date of registration of the design. In order to maintain the exclusive right, the registrant is presently required to pay one maintenance fee before the expiry of the fifth year.

Where the maintenance fees are not paid within the time provided for by the regulations, the registration fee may be paid with a penalty within six months.[75]

73. Subsection 6(1) of the *Act*.
74. Clause 105 of Bill C-43, adding para. 8.2(1)(c) to the *Act*.
75. Subsection 10(3) of the *Act*; subs. 18(2) of the *Industrial Design Regulations*.

[C] Registrations Issued from Applications Filed after the Coming into Force Date of Bill C-43

It is not known when Bill C-43 will come into force. Until then, applications will continue to be dealt with under the *Act* as set out above. Applications filed after the coming into force date of Bill C-43 will have a term of the later of ten years after the date of registration and fifteen years after the filing date.[76]

[1] Effect of Registration

The *Act* provides that an exclusive right in an industrial design may be acquired by registration of the design.[77]

During the existence of an exclusive right, no person shall make, import, sell, rent or offer for sale or rent any article to which the design or a design not differing substantially therefrom has been applied, without the licence of the proprietor[78] of the design.[79]

Under the *Act*, a certificate of registration, in the absence of proof to the contrary, is sufficient evidence of the design, of the originality of the design,[80] of proprietorship, of the commencement and term of registration and of compliance with the *Act*.[81] These provisions have not been carried over into Bill C-43. The registration of a design results in a presumption of the validity of a registration unless and until adjudged otherwise.[82]

§4.05 INFRINGEMENT

The *IPLA* did not amend the provisions in the *Act* relating to infringement. *NAFTA* introduced significant changes in respect of the law relating to infringement. First, the scope of an action for infringement has been expanded. Second, the test for infringement has become more relaxed. However, the new rules relating to infringement are subject to certain transitional provisions in the *IPLA*.

[A] Scope of Infringement Action

Prior to the *NAFTA* amendments, an action for damages for infringement could only be brought against the person who actually *applied* the design or a fraudulent imitation to

76. Clause 106 of Bill C-43, amending s. 10 of the *Act*; Clause 112 of Bill C-43, amending s. 30 of the *Act* and adding s. 31 to the *Act*.
77. Section 9 of the *Act*.
78. See §4.05[C] Standing to Sue, *infra*. The *Act* uniquely defines authorship in terms of proprietorship. See also s. 12 of the *Act*.
79. Paragraph 11(1)(a) of the *Act*.
80. See discussion on 'Originality', *supra*.
81. Subsection 7(3) of the *Act*.
82. *Alkot Industries Inc. v. Consumers Distributing Ltd.* (1985), 6 C.P.R. (3d) 168 at 171.

the article, and only if the application was for the purpose of sale.[83] This would usually be a manufacturer, who was also prohibited from selling, exposing for sale, publishing or using the article(s). The proprietor had no civil recourse against those who sold, imported or used the offending article(s). The proprietor had the option of bringing criminal proceedings against these individuals. However, the maximum penalty was the forfeiture of CAD 120 to the proprietor.[84] No doubt for this reason and others such as the burden of proof in criminal trials, such proceedings were rarely initiated.

The *Act* now prohibits, absent a licence from the proprietor of the registration, persons from making, importing for the purpose of trade or business, selling, renting, or offering or exposing for sale or rent,[85] any article in respect of which the design is registered and to which the design or a design not differing substantially therefrom has been applied, or doing any of these things in relation to a kit if doing so would constitute an infringement if committed in relation to an article assembled from the kit.[86] Bill C-43 does not alter these provisions.

[B] Test for Infringement

Prior to the *NAFTA* amendments, infringement occurred when the design 'or a fraudulent imitation thereof' had been applied to the ornamenting of an article. 'Fraudulent' presupposed a knowledge of the registered design and an imitation deliberately based upon the registered design, although it did not necessarily mean an intention to steal the design or to be fraudulent.[87] Therefore, the burden of demonstrating infringement was rather onerous.

The test under the *Act* is whether the design or a design *not differing substantially therefrom* has been applied to an article.[88] The *Act* also expressly provides that in making this determination, a court may consider the extent to which the registered design differs from prior art,[89] something which courts had already been doing.[90] Bill C-43 does not alter these provisions. The question of infringement was, and no doubt will continue to be determined by the 'eye of the court as properly instructed by expert witnesses'.[91]

83. Former s. 11 of the *Act* and see *Societe Anonyme Les Cuirs Meillon v. Brumer* et al. (1976), 25 C.P.R. (2d) 153 (Qc. C.A.); *Global Upholstery Co. Ltd. v. Galaxy Furniture Ltd.* (1976), 29 C.P.R. (2d) 145 (F.C.T.D.) [*Global Upholstery*].
84. *NAFTA* repealed the section of the *Act* dealing with the criminal prosecution of these acts.
85. While importing and renting have been added, the prohibitions of using and publishing have been deleted.
86. Section 11 of the *Act*. 'Kit' is defined as 'a complete or substantially complete number of parts that can be assembled to construct a finished article'.
87. *Algonquin Mercantile, supra* n. 69.
88. This test was rejected by the Federal Court of Appeal in *Algonquin Mercantile, ibid.*, which has been over- turned expressly by statute. This test is the same as that of the *Registered Designs Act*, 1949 (U.K.).
89. Subsection 11(2) of the *Act*.
90. See *Dunlop Rubber Co. Ltd. v. Golf Ball Development Ltd.* (1931), 48 R.P.C. 268 at 277–278. See also *Bodum, supra* n. 27.
91. See *Algonquin Mercantile, supra* n. 69 at 80–81. The 'substantially different' test is found in the analogous legislation of the United Kingdom. The leading decision on the test for infringement

As per the recent *Bodum* decision, where the registered design claims the design in its entirety as opposed to in part, an article has to be 'quasi identical' to the registered design in order to find infringement.[92] Infringement is found if the allegedly infringing products do not differ substantially from the registered designs.[93] The proper approach is to conduct a comparison of the article in question with similar design(s), from the point of view of how the informed consumer would see things.[94] Moreover, the Federal Court in *Bodum* relied on an expert witness' evidence in conducting the comparisons between the registered designs, the allegedly infringing designs, and the prior art.

Prior to *Bodum*, courts occasionally applied a three-fold test of infringement, namely: (i) would the designs be confused; (ii) would the infringer's design have any existence apart from the registered design; and (iii) is the infringer's design nearer to the registered design than to any other prior design.[95] The Federal Court of Appeal did not adopt this approach[96] holding that the proper test is the 'appeal to the eye' test. The court may look at one or more of, the description of the design, the drawings[97] or the articles themselves, as well as prior art.

[C] Standing to Sue

As mentioned, for infringements occurring prior to 9 June 1993, only the proprietor (as then defined) or the assignee of a registration can sue for infringement. The *Act* provides that an action for infringement may be brought by the proprietor of the design or by an exclusive licensee, subject to any agreement between the proprietor and the exclusive licensee. If the exclusive licensee brings an action, the proprietor must be made a party.[98] The author of a design is the first proprietor of the design, unless the author has executed the design for another person for good and valuable consideration, in which case the other person is the first proprietor.[99] Bill C-43 does not alter these provisions.

[D] Limitation

The *Act* states that no remedy may be awarded for an act of infringement committed more than three years before the commencement of the action for infringement.[100] Bill C-43 does not alter this provision.

---

in the United Kingdom states that the eye of the court must be the judge of whether a design is substantially different from the registered design. Please see *Hecla Foundry Co*, *supra* n. 26 at 555.

92. *Bodum*, *supra* n. 27 at para. 50.
93. *Ibid.*, at para. 71.
94. *Ibid.*, at para. 80, citing *Rothbury*, *supra* n. 40.
95. See e.g., *R. v. Premier Cutlery Ltd.*, *supra* n. 58 at 143–144 and *Mainetti S.p.A.*, *supra* n. 25 at 222.
96. *Algonquin Mercantile*, *supra* n. 69. Interestingly, the Court also rejected the 'substantially different' test.
97. See *infra* for discussion on description and drawings.
98. Subsection 15(2) of the *Act*.
99. Section 12 of the *Act*.
100. Section 18 of the *Act*.

[E] Remedies for Infringement

Prior to the *NAFTA* amendments, the *Act* only provided that the proprietor could obtain damages for infringement. *NAFTA* amendments codified the other forms of relief that courts had determined were also available, namely injunction, delivery up, the recovery of damages or profits,[101] punitive damages and disposal of any infringing article or kit.[102] The calculation of damages may depend on the circumstances of each case.[103]

In situations where a plaintiff fails to adequately prove that it has suffered damages, the court nonetheless may award exemplary damages.[104]

However, the Federal Court does not have jurisdiction to award a declaration of non-infringement of a registered industrial design where there is no action for infringement.[105]

As discussed in greater detail below, the *Act* provides that if an infringer can establish that he or she was not aware, and had no reasonable grounds to suspect that the design was registered, the court shall not award any remedy other than an injunction.[106] However, this section does not apply if there is proper notice by marking as discussed below. Bill C-43 does not alter provisions relating to remedies.

§4.06 DEFENCES TO INFRINGEMENT

[A] Validity

Besides the obvious defences of denial of infringement or the lapse of a limitation period, the defence generally will be an attack on the validity of the registration.[107]

Under the current *Act*, registered industrial designs enjoy a prima facie presumption of validity.[108] However, this assumption is rebuttable. To be registrable, an industrial design must be substantially different from prior art. A simple variant is not sufficient.[109] The discussion of validity turns largely on originality (or lack thereof) of the design in question, discussed above. Manufacturing process, materials used,

101. For example, *Findlay v. Ottawa Furnace & Foundry Co.* (1902), 7 Ex. Ct. 338 at 349; *Global Upholstery*, *supra* n. 83 at 168.
102. Section 15.1 of the *Act*.
103. *Industries LUMIO (Canada) Inc. c. Dusablon*, 2007 QCCS 1204 at para. 230, R.J.Q. 1216.
104. *Dutailier Inc. v. Maribro Inc.* (1988), 21 C.P.R. (3d) 543 at 547 (F.C.T.D).
105. *Peak Innovations Inc. v. Meadowland Flowers Ltd.*, 2009 FC 661, 75 CPR (4th) 359.
106. Subsection 17(1) of the *Act*.
107. As noted in McKeown, *supra* n. 26, Ch. 31 at 45, some of the reasons cited for a design registration's invalidity include: the design is not original; the declaration accompanying the application for registration was false; the design is not the proper subject matter for registration. See also *Clatworthy & Son*, *supra* n. 38 at 435. Unless a proceeding for expungement or counterclaim for expungement is initiated (see discussion *infra*) the finding of invalidity will only be as between the parties to the infringement action.
108. Subsection 7(3) of the *Act*; *Bodum*, *supra* n. 27 at paras 92–93.
109. *Bodum*, *supra* n. 27 at para. 96, citing *Clatworthy & Son*, *supra* n. 38.

colours, and features dictated solely by functional elements are to be ignored in the originality analysis.[110]

In *Zero Spill*,[111] the Federal Court of Appeal held that a registered industrial design enjoys a blanket presumption of compliance with the entire Act. The burden of rebutting the assumption is squarely on a defendant resisting an infringement claim. Where a defendant is challenging the presumption of validity of the registered design, it is required to plead that the asserted features of the design are unprotectable because they are dictated solely by function and is required to adduce evidence to prove it.[112]

However, as elaborated above under heading §4.03[B] Originality/Novelty, Bill C-43 has replaced the requirement of originality with the requirement of novelty. Nevertheless, the requirement that a design shows substantial difference from a design previously disclosed for the same or an analogous article is retained.[113] Thus, 'novelty' is likely to be judged in the context of the exclusive right provided in Bill C-43 whereby not just the exact design but designs not substantially different will be considered when assessing novelty.

There are other requirements before a design can be registrable under the current *Act*, such as specific novelty bars including prior use, prior publication, and prior existing Canadian design registration or application. As discussed above, Bill C-43 has eliminated the requirement of no prior use and replaced the requirement of originality or lack of confusion with existing design registrations with the requirement that the design be novel.

[B] Marking

Of particular historical importance was the defence that the articles to which the design had been applied were improperly marked. Prior to the passing of the *IPLA*, section 14 of the *Act*, now repealed, required that in order for a design to be protected, it had to be registered and the name of the proprietor had to appear upon the article embodying the design.

The case law interpreted section 14 of the *Act* to mean that registered rights were preserved if 'substantial compliance' with the requirements of section 14 of the *Act* were shown.[114]

With the repeal of section 14 of the *Act*, marking or the absence thereof can no longer affect the validity of a registration. Rather, pursuant to section 17(1) of the *Act*, a plaintiff has recourse to no remedy other than an injunction if a defendant establishes that he or she was not aware of and had no reasonable grounds to suspect the existence of a design registration. However, the section 17(1) defence may be overcome if a proprietor marks all or substantially all of the articles or the labels or packaging associated with those articles to which the registration pertains. Marking consists of a

110. *Bodum*, *supra* n. 27.
111. *Zero Spill*, *supra* n. 23.
112. *Ibid.*, at paras 18 and 20.
113. Clause 105 of Bill C-43, adding subs. 8.2(1) to the *Act*.
114. *L.M. Lipski Ltd. v. Dorel Industries Inc.* (1988), 20 C.P.R. (3d) 226 (F.C.T.D.).

capital letter 'D' in a circle and the name or usual abbreviation of the name of the proprietor.[115] Bill C-43 does not alter these provisions.

§4.07 PROCEDURES FOR FILING APPLICATION FOR REGISTRATION

Currently, section 4 of the *Act* states that the proprietor of a design, whether the first proprietor or subsequent proprietor, may apply to register the design by paying the prescribed fees and filing an application with the Minister in the prescribed form. This includes a drawing or photograph of the design and a description of the design along with a declaration that the design was not, to the proprietor's knowledge, in use by any person other than the first proprietor at the time the design was adopted by the first proprietor ('No Prior Use Declaration').[116]

In addition to the No Prior Use Declaration, requirements for obtaining a filing date further include: (a) name and address of the applicant and the agent (if applicable); (b) a title identifying the article or set; (c) a description of the features that constitute the design;[117] and (d) a drawing or photograph of the design.[118]

Bill C-43 has amended the *Act* to require a 'representation of the design'[119] instead of 'a drawing or photograph of the design'. It will be interesting to see if the representation of the design can include a video, especially in light of section 24.1 of Bill C-43, which stipulates that subject to the regulations, any document may be submitted in any electronic form and by any electronic means that is specified by the Minister or the Commissioner of Patents.[120]

The *Rules* provide that an application shall relate to one design applied to a single article or set, or to variants.[121] Where the Commissioner of Patents finds that subsection 10(1) of the *Rules* is not complied with, the applicant must limit the application to only one design and any other design disclosed may be made the subject of a separate divisional application.[122] Provided that it is filed before the registration of a design based on the original application, the separate application will have the same filing and priority date as the original.[123] Divisional applications are at risk of rejection if the original application registers first. This self-collision may be avoided by requesting that the original and divisional applications register on the same day.

115. Subsection 17(2) of the *Act*.
116. Subsection 4(1) of the *Act*.
117. The description is typically limited to a brief recitation of the features of the design that are claimed (e.g., ornament, shape). In the past, practitioners drafted detailed descriptions of the designs highlighting important features. This is no longer done for fear of inadvertently narrowing the scope of the registration.
118. Sections 9 and 11 of the *Rules*.
119. Clause 104 of Bill C-43, amending subs. 4(1) of the *Act*.
120. Clause 110 of Bill C-43, adding s. 24.1 to the *Act*.
121. Subsection 10(1) of the *Rules*. Different embodiments of the design may not be included. If different embodiments are included in an application, the examiner may require that each embodiment be filed as separate divisional applications.
122. Subsections 10(2) and 10(3) of the *Rules*.
123. Subsection 11(2) of the *Rules*.

The *Rules* further provide that an applicant may at any time correspond with the Commissioner by electronic or other means of transmission specified in the Canadian Patent Office Record.[124]

[A] Priority

Pursuant to section 29 of the *Act*, a six-month priority right for industrial design applications applies to nationals of all countries that are members of the World Trade Organization. A request for priority under this section of the *Act* is made by in writing indicating the date the application for registration of the design was first filed in or for the foreign country, the name of the country and the number assigned by that country to the application.[125]

Bill C-43 has modified these provisions slightly to use language more consistent with the language in the Paris and the Hague Conventions but has not changed the request for priority in substance.

§4.08 EXPUNGING OR VARYING REGISTRATION

According to section 22 of the *Act*, an application may be made to the Federal Court of Canada by the Attorney General or any person aggrieved to request that an entry in the Register of Industrial Designs made or omitted to be made without sufficient cause, be made, varied or expunged, as the Court thinks fit.

The case law has interpreted 'person aggrieved' so as to include a defendant in an action for infringement of a registered industrial design. As noted above, it is common practice in an infringement action for the defendant to counterclaim for expungement of the registered industrial design.[126]

The Federal Court of Canada has the exclusive jurisdiction to hear proceedings brought pursuant to section 22.[127] Such proceedings are properly initiated by filing a statement of claim to the Federal Court (Trial Division) and not by means of an application.[128]

Furthermore, under section 23 of the *Act*, the registered proprietor of any registered industrial design may apply to the Federal Court of Canada to add or alter the industrial design in any non-essential particular. Sections 22 and 23 of the *Act* are not altered by Bill C-43.

---

124. Subsection 3(6) of the *Rules*.
125. Subsection 20(1) of the *Rules*.
126. See *Clatworthy & Son, supra* n. 38 at 435.
127. Subsection 22(4) of the *Act*.
128. See for example, *Gandy v. Canada (Commissioner of Patents)* (1980), 47 C.P.R. (2d) 109 (F.C.T.D.) and *B & E Furniture Manufacturing Co. v. Goldcrest Furniture Ltd.*, [1980] 1 F.C. 457 (T.D.).

§4.09 OTHER

[A] Interface between Industrial Design and Trademarks

Under the Canadian *Trade-marks Act,*[129] a trademark includes a 'distinguishing guise'. A distinguishing guise is defined as 'a shaping of wares or their containers, or a mode of wrapping or packaging wares' to distinguish one's goods or services from another's goods or services.[130] Therefore, it is possible to obtain a registration under the *Act* as well as a registration under the *Trade-marks Act* for the same design.

One advantage of a trademark registration is the duration of protection. Trademark registrations can be renewed indefinitely in Canada provided that the registration is not expunged by the Registrar for non-use or struck from the Register as a result of an action by a third party.

Another advantage of trademark protection in Canada is that rights in a trademark can arise without registration if it is shown that through use, the trademark has become distinctive of its owner. Therefore, it is possible for an owner of a distinguishing guise who does not have a trademark registration but can prove a reputation associated with the use of the distinguishing guise to prevent others from infringing that trademark.

[B] Relationship between Industrial Design and Copyright

Copyright arises automatically in Canada in, *inter alia,* every original artistic work provided that certain conditions in respect of authorship are met. The definition of 'artistic work' in the *Copyright Act*[131] includes engravings, sculptures and works of artistic craftsmanship.[132] Therefore, an industrial design can theoretically be the proper subject matter of copyright as an 'artistic work'. As a result, the author of a design may be entitled to the advantages of copyright protection. Among these advantages are that registration is not required and that the term of protection is generally the life of the author plus fifty years.[133]

However, the *Copyright Act* and the *Act* have historically reflected the policy that copyright should protect artistic works which have not been industrially applied to useful articles and that therefore exclusive rights in copyright should not be granted in respect of mass produced articles. To do so would result in an unreasonable limit on industry. Mass produced articles that are at least 'appealing to the eye' are afforded a limited monopoly under the *Act,* not under copyright.

The *Copyright Act* and the *Act* were amended in 1988 in an attempt to clearly delineate the scope of copyright protection afforded to industrial articles. The revisions to the *Act* addressed the issue of the subject matter of industrial designs and the

129. *Trade-marks Act, supra* n. 14.
130. Section 2 of the *Trade-marks Act.*
131. *Copyright Act, supra* n. 13.
132. Section 2 of the *Copyright Act.*
133. Section 6 of the *Copyright Act.*

amendments to the *Copyright Act* addressed the issue of the interface between copyright and industrial design in the context of copyright infringement. The amendments apply to designs created after 8 June 1988, the date on which the amendments came into force.

The definition of 'article'[134] was added to the *Copyright Act* in the 1988 amendments.[135] This was done to set out the circumstances under which copyright in a design applied to a useful article or in an artistic work[136] from which the design is derived would not be infringed.[137] In other words, the *Copyright Act* allows for copyright to subsist in an industrial design if it otherwise meets the requirements of the *Copyright Act*. However, in keeping with the policy that copyright is not intended to protect industrial articles, the scope of protection afforded to such designs or artistic works is restricted.

It is beyond the scope of this chapter to describe in detail the conditions under which copyright in a design applied to a useful article or an artistic work from which such design is derived will not be infringed. Generally speaking, where the owner of the copyright in Canada or *elsewhere*[138] or someone under the authority of the owner: (i) reproduces more than fifty of the articles, or (ii) where the article is a plate, engraving or cast, the article is used for producing more than fifty useful articles, it is not infringement of copyright or moral rights[139] to reproduce the design of the article or a design not differing substantially therefrom by making the article, or making a drawing or other reproduction of the article, or doing anything with the article, drawing or reproduction that the owner of copyright has the sole right to do with the design or artistic work in which the copyright subsists.[140]

There are several exceptions to the non-infringement rule that apply only to artistic works insofar as they are used as or for certain purposes. Many of these works would otherwise be 'useful articles' within the meaning of the *Copyright Act*. The rationale for these exceptions is that these designs are not considered ones that are likely to impede legitimate industry.[141] These designs can be described generally as graphic or photographic representations applied to the face of an article, trademarks, patterned materials, architectural works that are buildings or models of buildings,

134. See s. 64 of the *Copyright Act*, where 'article' means 'anything that is made by hand, tool or machine'. This definition is the same as that found in s. 2 of the *Industrial Design Act*.
135. Section 64 of the *Copyright Act*.
136. An example would be drawings or plans of the design.
137. Section 64 of the *Copyright Act*.
138. Subsection 64(2) of the *Copyright Act*. One noted Canadian intellectual property lawyer who participated in drafting the 1988 amendments points out that if this were not the case, an owner of Canadian copyright could get around this section by assigning his or her copyright to someone outside Canada, and then proceed to mass produce the useful article in another country without losing full copyright protection in Canada. See W.L. Hayhurst, 'Intellectual Property Protection in Canada for Designs of Useful Articles: Sections 46 and 46.1 of the *Copyright Act*' (1989), 4 I.P.J. 381 [the 'Hayhurst Article']. Sections 46 and 46.1 are now ss 64 and 64.1, respectively.
139. 'Moral rights' are generally speaking the author's rights to the integrity of the work and the right to be associated with the work as its author. Section 14.1 of the *Copyright Act*.
140. Sections 64 and 64.1 of the *Copyright Act*.
141. See Hayhurst Article, *supra* n. 138.

representations of real or fictitious beings, events or places that are applied to an article as a feature of shape, configuration, pattern or ornament, articles that are sold as a set unless more than fifty sets are made, or such other works or articles as may be prescribed.[142]

Finally, like the amendments to the *Act*, the *Copyright Act* now affirms that copyright protection does not extend to features of useful articles dictated solely by a utilitarian function of the article, or use of a method or principle of manufacture or construction. In other words, copyright may subsist in a design applied to a useful article, but the utilitarian features of that article can be freely copied. This amendment has retroactive effect.

142. Subsection 64(3) of the *Copyright Act*.

CHAPTER 5

Industrial Design Rights: China

Mike Chen, Jenny Li & Ying Zhang

§5.01 INTRODUCTION

[A] Highlights of New Developments in Chinese Patent Law

Since the first amendment in 1993, the Chinese Patent Law has been amended twice, once in 2000, and then in 2008. Furthermore, there have been some new updates in *Guidelines for Patent Examination.*

According to the Chinese Patent Law and the current Guidelines for Patent Examination:

> No more than 10 similar designs for one same product may be filed in one application, which encourage the designers or applicants to create a lot of new designs;
> A two-dimensional (2D) design that mainly serves as an indicator or identification is unpatentable, which reflects a stricter quality requirement for a new design;
> A Graphic User Interface (GUI) related design displayed on an eligible product may be protected by a Chinese design patent, which meet the development of the modern industry and encourage IP protection of GUI related products;
> At filing, a brief explanation is necessary; the drawings or photographs filed shall be sufficient and clear, which reflects a stricter requirement for filed documents to clearly define the protection of a design;
> A Chinese design application that obviously lacks novelty shall not be granted for patent right, which reflects an improvement on the preliminary examination system for a Chinese design application; and
> A Chinese design patent shall possess novelty and inventiveness, which reflects a higher requirement for the quality of a granted design patent.

In summary, all the amendments and/or developments are intended to enhance the Chinese design patent protection system.

§5.02 CONVENTIONS AND LEGISLATION

China is a member of Paris Convention.

According to Article 5 of Paris Convention, industrial design shall be protected in all the countries of the Union.

Thus, the following articles of the Chinese Patent Law regulate some requirement for the Chinese design application and/or patent:

> Article 2 of the Chinese Patent Law regulates that, the term 'design' means a new design of a product's shape, pattern or the combination thereof, or the combination of its colour and shape and/or pattern, that is aesthetically pleasing and industrial applicable.
> Article 23 of the Chinese Patent Law regulates that, any design for which a patent right may be granted must not belong to an prior design; nor has any entity or individual previously filed before the date of filing with the Chinese Patent Office an application on an identical design which was published in patent documents published after said date of filing. The design for which a patent right may be granted must be substantially different from prior designs or a combination of the features of prior designs. Any design for which a patent right may be granted must not be in conflict with any prior legal rights of any other person.
> Article 25 of the Chinese Patent Law regulates that, 2D designs of images, colours or combinations of the two that mainly serve as indicators shall not be granted for patent right.
> Article 27 of the Chinese Patent Law regulates that, when filing a Chinese design patent application, documents including a request, drawings or photographs of the design as well as a brief explanation of the design and should be submitted. The drawings or photographs submitted should clearly indicate the design sought to be protected by the patent.
> Article 31 of the Chinese Patent Law regulates that, two or more similar designs used on the same product may be submitted together in one application.

In the new Guidelines for Patent Examination, it is regulated that, during the preliminary examination of a design application, the examiner may examine whether the design application meets the requirements of Article 9 based on obtained identical design application and further the examiner may examine whether the design patent application obviously does not meet the requirements of Article 23.1 on the basis of obtained information of relevant prior design(s) or conflicting application(s).

The same Guidelines for Patent Examination regulates that a GUI displayed by a display component of a product, which is related to both human-machine interaction and implementation of product function, is patentable. Particularly, the GUI itself is not a subject matter protected by design patent, only the combination of a product with GUI, i.e., a product including GUI is eligible as a protectable subject matter of GUI. The 'product' is defined as a unit that is sufficient for both human-machine interaction and implement of product function, which includes an end product like mobile phone, washing machine, and it can also be a component that is fully detachable from the end product, e.g., a dash board. However, these eligible products do not include a product that is part of a human-machine interaction and implement of product function, for example, a screen panel displaying GUI.

In addition, according to Article 4 of Paris Convention, any person who has duly filed an application for the registration of a utility model, or of an industrial design, or of a trademark, in one of the countries of the Union, or his successor in title, shall enjoy, for the purpose of filing in the other countries, a right of priority during the periods hereinafter fixed; any filing that is equivalent to a regular national filing under the domestic legislation of any country of the Union or under bilateral or multilateral treaties concluded between countries of the Union shall be recognized as giving rise to the right of priority.

Therefore, in China, Article 29 of the Chinese Patent Law regulates that, where, within twelve months from the date on which any applicant first filed in a foreign country an application for a patent for invention or utility model, or within six months from the date on which any applicant first filed in a foreign country an application for a patent for design, he or it files in China an application for a patent for the same subject matter, he or it may, in accordance with any agreement concluded between the said foreign country and China, or in accordance with any international treaty to which both countries are party, or on the basis of the principle of mutual recognition of the right of priority, enjoy a right of priority.

Furthermore, according to Article 11 of Paris Convention, the countries of the Union shall, in conformity with their domestic legislation, grant temporary protection to patentable inventions, utility models, industrial designs, and trademarks in respect of goods exhibited at official or officially recognized international exhibitions held in the territory of any of them.

Therefore, in China, Article 24 of the Chinese Patent Law regulates that, any invention-creation for which a patent is applied shall not lose its novelty if, within six months before the filing date of the application, one of the following events has occurred:

(1) it was exhibited for the first time at an international exhibition sponsored or recognized by the Chinese Government;
(2) it was made public for the first time at a prescribed academic or technical conference; or
(3) it was disclosed by any person without the consent of the applicant.

§5.03 SUBJECT MATTER

[A] Definition/What Constitutes a Protectable Design?

According to Rule 3, Article 2 of the Chinese Patent Law, the term 'design' means a new design of a product's shape, pattern or the combination thereof, or the combination of its colour and shape and/or pattern, that is aesthetically pleasing and industrial applicable. In other words, a Chinese design patent may protect a three-dimensional (3D) product's shape, pattern, colour, or protect a 2D product's pattern, colour.

In addition, a partial design is unpatentable. That is, the product shown in the drawings or the photographs shall be an integral unit that may be manufactured or sold independently.

Moreover, a 2D design of image serve as indicators is unpatentable. Where an application for a design patent meets the following three conditions, the application shall not be allowable:

(1) the product for which the design is applied is a plane printed matter;
(2) the design is made of patterns, colours, or their combination; and
(3) the design serves mainly as indicator.

'Serves mainly as indicators' means that the main purpose of the design is to help the public identify the related product, the origin of the service, and etc.

Under the Chinese Patent Law, for a GUI-related design, the patentable subject matters generally include:

(1) product with dedicated user interface;
(2) product with general operating system interface;
(3) product with application user interface;
(4) product with web application; and
(5) product with icon.

Meanwhile, the unpatentable subject matters generally include:

(1) product with graphic-text layout of website and webpage;
(2) product with electronic screen wallpaper or screen saver;
(3) product with booting or shutdown animation; and
(4) product with game user interface.

[1] Novelty

Any design for which a patent right may be granted must not belong to a prior design. The prior design refers to design(s) known to the public before the date of filing in China or abroad (or the priority date, where priority is claimed). The prior design includes designs that are made known to the public by publishing, public use or other means in China or abroad before the date of filing.

[a] Prior Use

Disclosure by use means that by use a design is disclosed or placed in the state of being available to the public.

Means of disclosure by use include making, using, selling, importing, exchanging, presenting, demonstrating, exhibiting and the like that can make the design available to the public. So long as by the above means the relevant design's content is placed in such a state that the public can know it if they wish, disclosure by use can be established, and it is of no relevance whether the public had actually known it.

Moreover, disclosure by use also includes disclosure on an exhibition stand or in a shop window of informative materials that are readable by the public or directly visible materials, such as posters, drawings, photographs, specimens, and samples.

The date on which the product or process is available to the public shall be regarded as the date of disclosure by use.

[b] Prior Publication

Publications in the context of the Chinese Patent Law mean independently existing disseminating carriers of designing contents, which shall indicate or have other evidence to prove the date of public issue or publication.

The determination of whether a document is a publication shall not be affected by the place or language of issue, the manner of acquisition, or its age. The amount of distribution, whether it has been read, or whether the applicant is aware of it is of no relevance either.

[c] Prior Application or Registration

This is similar to '§5.03[A][1][b] Prior Publication' above. That is, when a prior application or registration is published, it constitutes part of Prior Publication.

§5.04 DURATION OF PROTECTION

According to Article 42 of the Chinese Patent Law, the duration of a Chinese design patent shall be ten years, counted from the actual filing date in China. Under the Chinese Patent Law, a design patent cannot be renewed.

§5.05 INFRINGEMENT

[A] Scope of Infringement Action

A design means any new design of shape, pattern, or their combination, or the combination of the colour with shape or pattern, of a product, which creates an aesthetic feeling and is fit for industrial application.

Under Article 11[1] of the Chinese Patent Law, infringement of a granted design patent includes, for production or business purposes, manufacturing, offering for sale, selling or importing the design patent products without the patentee's permission.

1. Article 11 of Chinese Patent Law: After the granted of the patent right for an invention or utility model, except where otherwise provided for in this Law, no entity or individual may, without the

According to Article 59.2[2] of the Chinese Patent Law, the extent of protection of the patent right of a design shall be determined by the product design as shown in the drawings or photographs. The brief explanation in the design may be used to interpret the design of the product as shown in the drawings or photographs.

Therefore, the scope of protection of a design patent is the product(s) as shown in the drawings or photographs, namely the issued drawings or photographs by State Intellectual Property Office of the PRC (hereinafter referred to as 'SIPO'), rather than the products which are made by the patentee after filing design patents. In other words, in judging whether infringement has occurred, the scope of protection for a registered design is limited to product(s) shown in the drawings or photographs issued by SIPO.

The brief explanation in a design patent is only used to comprehend or interpret the scope of the protection of the design patent. For example, if colour is to be protected in a design patent, it shall be indicated in the brief explanation. If there is no indication regarding the colour in the brief explanation, then the colour is not part of the scope of the protection of the design patent.

[B] Test for Infringement/Criteria to Constitute Infringement

As previously stated, under Clause 2, Article 11 of the Chinese Patent Law, after the grant of the patent for a design, for production or business purpose, any entity or individual is not allowed to exploit the design patent, which includes making, offering to sell, selling or importing a product incorporating the patented design, without the authorization of the patentee or legitimate licensee.

[1] Identical or Similar Type of the Design Product

When judging if infringement has occurred, it shall be determined that whether an allegedly infringing product and the product in the design patent belong to an identical or a similar type of products. If the allegedly infringing product is not the same as, nor a similar type of, the design patent product, infringement will not be found.

In general, determining whether a product is identical to or of a similar type of the design patent product is based on usage of the design product. The usage of a product may be determined with reference to the brief description of the design patent, the International Classification for Industry Design, the function of the product as well as the product's sales and actual usage, and so on.

---

authorization of the patentee, exploit the patent, that is, make, use, offer to sell, sell or import the patented product, or use the patented process, and use, offer to sell, sell or import the product directly obtained by the patented process, for production or business purposes.

After the granting of the patent for a design, no entity or individual may, without the authorization of the patentee, exploit the patent, that is, make, offer to sell, sell or import the product incorporating its or his patent design, for production or business purposes.

2. Article 59.2 of Chinese Patent Law: The extent of protection of the patent right for a design shall be determined by the design of the product as shown in the drawings or photographs. The brief explanation may be used to interpret the design of the product as shown in the drawings or photographs.

For example, a mechanical watch and an electronic watch have the same usage, while a toy and a knick-knack belong to a similar type of products.

[2] Through the Eyes of Ordinary Consumers

Infringement is judged through the eyes of ordinary consumers, based on the knowledge level and the cognitive ability of the ordinary consumers of the design patent product.

Different categories of design patent products may have different consumers. An ordinary consumer of a certain category of product incorporating a design shall have the following characteristics:

(1) Common knowledge of the designs incorporated in, and commonly used design methods related to the same or similar products as the design patent being examined before the filing date of the design patent. For example, as to a bicycle, the ordinary consumers shall know the television sold in the market and the information disclosed by the advertisement.

(2) Certain capability of distinguishing the differences in shape, pattern and colour between allegedly infringing products and the design patent being examined, but without notice to minor differences in shape, pattern or colour of products. In other words, the minor differences account for a little proportion, and are not observed by the ordinary consumers, which could not influence on the overall of the design. For example, when buying a bicycle, the ordinary consumers usually observe the overall shape, the components, the location and the shape of the components. While, it is difficult for the ordinary consumers to observe the tread pattern, because the change of the tread pattern is quite minor compared to the overall shape of the bicycle, unless the change of the tread pattern acts as a key point of the design.

[3] The Principle of Judgment

Comparison of the allegedly infringing product and the design patent shall be made through the approach of whole observation and comprehensive judgment. The approach of the whole observation and comprehensive judgment is a determination based on an observation of the allegedly infringing product and the design patent as a whole rather than on part or details of the products or designs.

Based on the above principle, further analysis of the allegedly infringing products is conducted to determine if the products are identical or similar to the design patent being examined. If so, the allegedly infringing products fall within the scope of the protection of the design patent, and thereby constitute infringement.

Note that design features mainly determined by technical function(s), and features that do not affect the overall visual effects such as product material, internal structure and so on, shall not be considered in the analysis of whether infringement has occurred.

The following factors usually have more impact on the overall visual effects of a design:

(1) Parts that are easily observed during ordinary use of a product versus other parts of the product; and
(2) Design features of the patented design distinguished from the prior designs versus other design features.

[C] Standing to Sue

In general, a patentee or an authorized licensee is entitled to commence litigation against an alleged infringer.

If the patentee learns about possible infringement, the patentee can commence litigation based on a valid design patent and relevant evidence. If the patentee does not take action, he can authorize the licensee to take action against the alleged infringer. Of course, the licensee has obtained the license right when the infringement occurs.

According to Article 119 of Civil Procedural Law, the following conditions must be met when a lawsuit is brought:

(1) the plaintiff must be a citizen, legal person or any other organization that has a direct interest in the case;
(2) there must be a definite defendant;
(3) there must be specific claims, facts and causes for the suit; and
(4) the suit must be within the scope of acceptance for civil actions by the people's court and under the jurisdiction of the people's court where the suit is entertained.

[D] Remedies for Infringement

The following measures can be taken against an infringer.

[1] Judicial Remedy

If the business scale of the infringer and the amounts of infringing products are large, the patentee can file a patent litigation, which is a strong measure to fight against the infringement.

Once an infringing act is found by the court, civil remedies including prohibitive injunction, compensation for damages, etc. are available. Furthermore, preliminary injunctions and evidence preservative measures may also be available before or during the litigation, which can help minimize the patentee's losses.

The judgment of the court has a stronger deterrent force, and via this procedure the patentee's rights and interests can be protected to the best extent. The main disadvantage of the procedure is that it may take more time.

[2] Administrative Remedy

In addition, upon discovery of possible infringement, filing a patent complaint before local IP bureau (in each of the provinces or major municipalities in China) is a relatively quick measure to stop infringement. Where an infringing act is established, the local IP bureau shall order the infringer to immediately stop the infringing act and make a decision on this basis. This administrative procedure is cheap and convenient, but no compensation for the loss of the patentee is awarded except that the IP bureau may mediate between the parties concerned. Another disadvantage of this procedure is that the administrative remedy cannot commence since there is no IP bureau in some small city or the officers in the small IP bureau are entitled to handle this infringement dispute.

[3] Other Remedy

[a] Applying for a Customs Recordal

If suspected infringing product(s) made by an infringer is exported or imported, applying for a Customs Recordal of the design patent is another choice of remedy. Once Customs finds any suspected infringing product(s) during routine examinations, it will notify the patentee, and detain and confiscate the suspected infringing product(s).

[b] Sending a Warning Letter

If the scale of the alleged infringement is quite small, sending a warning letter is another choice, the warning letter requesting the alleged infringer to stop the infringing act. However, if the alleged infringer ignores the warning letter, it may be an alert that the alleged infringer may continue to engage in infringement.

§5.06 DEFENCES TO INFRINGEMENT

[A] Validity

An alleged infringer can file request for invalidation before the Patent Re-examination Board to challenge the validity of the design patent, as a defence to allegations of infringement.

[1] Not Novel

According to the Article 23[3] of the Chinese Patent Law as amended in 2008, new examination standard for design patents is in force, including Article 23.1 (novelty) and Article 23.2 (non-obviousness). The novelty and non-obviousness standards for design patents are similar to the novelty and the inventiveness examination standards for invention patents and utility model patents.

The prior design referred to in this law means any design known to the public before the date of filing in China or abroad.

Novelty in a design patent is embodied in two aspects. On one hand, the patent concerned possesses the novelty means that the patent concerned shall not be a prior design, namely there is no design identical or substantially identical with the patent concerned in the prior design. On the other hand, it means that compared to the patent concerned, the conflicting application does not exist, that is, neither identical design application nor substantially identical design application was filed by any entity or individual with SIPO and published on or after the filing date of the patent concerned.

Where an identical or substantially identical design is found, the design patent lacks novelty.

(1) Identical design is found when the patent concerned and the comparative design are designs for the same category of products, and all the design elements of the patent concerned are identical with the corresponding design elements of the comparative design, wherein design elements refer to shape, pattern and colour of a design.
If the difference between the comparative design and the patent concerned lies in the dimension of the product, and the difference does not lead to any change in the design of the product, the comparative design and the patent concerned are still identical designs.
In other words, the patent concerned lacks the novelty compared to the comparative design.
(2) The judgment of substantially identical design is only limited to designs of the same or a similar category of products. Where the patent concerned and the comparative design are for products of neither the same nor a similar category, without making a comparison and judgment of the two, it can directly be determined that the patent concerned and the comparative design are not substantially identical.

3. Article 23 of Chinese Patent Law: any design for which patent right may be granted shall not be a prior design, nor has any entity or individual filed before the date of filing an application relating to the identical design disclosed in patent documents announced after the date of filing.
Any design for which patent right may be granted shall significantly differ from prior design or combination of prior design features.
Any design for which patent right may be granted must not be in conflict with the legitimate right obtained before the date of filing by any other person.
The prior design referred to in this law means any design known to the public before the date of filing in China or abroad.

If an ordinary consumer can conclude, based on an overall observation of patent concerned and the comparative design, that any difference(s) between the two simply falls within the following circumstances, then the patent concerned and the comparative design are substantially identical, and thus the design patent lacks the novelty compared to the comparative design:

(a) The difference lies in only slight changes in some fine details which cannot be noticed while paying normal attention, for example, the designs of venetian blind differ only in the number of slats.
(b) The difference exists in the parts which cannot be seen easily or cannot be seen at all when in use; however, an exception may be made in the circumstances where there is evidence showing that certain special design in parts that cannot be seen easily has nonetheless a notable visual effect for an ordinary consumer.
(c) The difference is a result of the substitution of one design element as a whole by said design element of a usual design commonly known for the category of product.
(d) The difference exists in that the patent concerned is simply a repeated and continuous arrangement or an increase/decrease in the continuous number of the comparative design as a design unit following the normal arrangement of the category of product.
(e) The difference exists in that the patent concerned and the comparative design are a mirror image. Namely the patent concerned is an image in a mirror of the comparative design.

[2] Functional/Not Ornamental

If the difference between the comparative design and the patent concerned is merely a result of commonly used materials substitution, or that the difference only lies in a function, internal structure, technical performance of the product, and that the difference does not lead to any change in the design of the product, the comparative design and the patent concerned are still identical designs.

A special shape exclusively determined by the function of a product generally does not notably influence the overall visual effect.

[3] Other Possible Criteria, Such as Non-obviousness

According to Article 23.2, any design for which patent right may be granted shall significantly differ from prior design or the combination of prior design features. If compared to the prior design or the combination of the prior design features, the patent concerned does not possesses significant differences, it also means the patent concerned is obviousness, and it shall not be granted. No significant differences refer to at least one of the following circumstances:

(1) The patent concerned is not significantly different from the prior design of the product of the same or a similar category. If an ordinary consumer finds that the difference between the patent concerned and the prior design does not notably influence the overall visual effect of the product through a comprehensive observation on them, then the patent concerned is not significantly different from the prior design.
    The following factors can be further considered comprehensively:
    (a) the design changes in the parts which can be seen easily when they are used have more notable influence on the overall visual effect than the design changes in the parts which cannot be seen easily or cannot be seen at all;
    (b) where designs in some parts of a product are proved to be usual designs commonly known for the category of product, the changes in other parts have more notable influence on the overall visual effect;
    (c) if the difference lies only in minor changes in some fine details, the difference is not sufficient to have notable influence on the overall visual effect, the two are not significantly different.
(2) The patent concerned is transformed from the prior design, the design features of the two are identical or only have very slight difference, and there's an inspiration for the specific transformation in the prior design of the product of the same or a similar category.
(3) The patent concerned is a combination of the prior design or prior design features, the prior design and the corresponding part of the patent concerned are identical or only have very slight difference, and there is an inspiration for the specific combination in the prior design of the product of the same or a similar category.

The following types of combination fall into the circumstances where there is an obvious inspiration for the combination, and designs arising therefrom are not significantly different from the combination of the prior design or its design features:

(a) the design which is produced by merely aggregating multiple prior designs or their slightly changed forms for the product of the same or similar category;
(b) the design which is produced by replacing the design feature of a design with another design feature or its slightly changed form of the product of the same or similar category;
(c) the design of product which is produced by merely aggregating the prior design of shape for the product with the prior pattern, colour or their combination; or by replacing the pattern, colour or their combination in one prior design with those in another prior design.

If the above-mentioned circumstance gives rise to unique visual effect (the patent concerned produced unexpected visual effect, comparing with the prior design), there could be an exception.

[4] *Faults in Application*

If a design patent does not comply with Article 2,[4] and Article 9,[5] or belongs to the situation of item (6) of Article 25[6] the design patent may be declared invalid.

According to Article 2, if a new design does not refer to the shape, the pattern, or their combination, or the combination of the colour with shape or pattern, of a product, which creates an aesthetic feeling and is fit for industrial application, the design shall not be granted design patent.

According to item (6) of Article 25, if a design refers to two-dimensional printing goods, made of the pattern, the colour or the combination of the two, which serve mainly as indicators, the design shall not be granted.

Since a design patent right does not go through the substantive examination, an identical invention-creation may occur. According to Article 9, identical invention-creation referred to in Article 9 means that for design, the claimed design for product shall be identical or substantially identical.

[B] Marking

According to the provision of the Patent Law, the patentee can mark patent product with the appropriate design patent number. This is the right of the patentee, rather than a liability. In other words, the patentee is entitled not to mark the product with any design patent number.

Therefore, an alleged infringer cannot use the defence that he did not know about the design patent due to the fact that the patentee did not mark the product with any design patent number. If the alleged infringer is found to have engaged in an infringing act, no matter whether the patentee marks the product with a design patent, the alleged infringer would bear the liability of the infringement.

[C] Non-infringement

Except for challenging the validity of the disputed patent, the following act shall not be deemed as infringement of the patent right:

---

4. Article 2 of Chinese Patent Law: design means any new design of the shape, the pattern, or their combination, or the combination of the colour with shape or pattern, of a product, which creates an aesthetic feeling and is fit for industrial application.
5. Article 9 of Chinese Patent Law: for any identical invention-creation, only one patent right shall be granted. Where an applicant files on the same day applications for both patent for utility model and patent for invention relating to the identical invention-creation, and the applicant declares to abandon the patent for utility model, which has been granted and does not terminate, the patent for invention may be granted.
6. Item (6) of Article 25 of Chinese Patent Law: for the designs of two dimensional printing goods, made of the pattern, the colour or the combination of the two, which server mainly as indicators, no patent right shall be granted.

(1) If the act of an alleged infringer meets one requirement provided in Article 69[7] of the Chinese Patent Law, this act shall not be deemed as infringement of patent right. These acts include as follows:
   (a) the defence of exhaustion of patent right, namely after the sale of a patented product or product obtained directly by a patentee or any entity authorized by patentee, any other person is entitled to offer to sell, sell or import the design product;
   (b) the defence of the prior use, namely any person has been making the design patent or been prepared for manufacture before filing date of the design patent;
   (c) the defence of temporary borders, which only refers to the foreign means of transport that temporarily passes through the territory, territory waters or territorial airspace of China uses the patent concerned, in accordance with any international treaty;
   (d) the defence of using the design patent solely for the purposes of scientific research and experimentation;
   (e) the defence of the principle of Bolar exemption, namely any person exploits the design patent for the purpose of providing information needed for the regulatory examination and approval.

   In practice, the common defence in the aspects listed above is the prior use provision. That means if an alleged infringer can provide satisfactory evidence demonstrating he had made the design in a product or been prepared for making the design in a product before the filing date of the design patent, the alleged infringer can defend against allegation of infringement based on this ground, and continue to use the design patent in the original scope only.

(2) The defence of prior design

   The act of the alleged infringer meets the requirement of Article 62[8] of the Chinese Patent Law, the alleged infringer is entitled to defend the infringement.

7. Article 69 of the Chinese Patent Law: none of the following shall be deemed as infringement of the patent right: (1) where, after the sale of a patented product or product obtained directly by a patented process by the patentee or any entity or individual authorized by the patentee, any other person uses, offer to sell, sell or imports that product; (2) where, before the date of filing of the application for patent, any person who has already made the identical product, used the identical process, or made necessary preparation for its making or using, continues to make or use it within the original scope only; (3) where any foreign means of transport which temporarily passes through the territory, territorial waters or territorial airspace of China uses the patent concerned, in accordance with any agreement concluded between the country to which the foreign means of transport belongs and China, or in accordance with any international treaty to which both countries are party, or on the basis of the principle of reciprocity, for its own needs, in its devices and installations; (4) where any person uses the patent concerned solely for the purposes of scientific research and experimentation; or (5) where for the purposes of providing information needed for the regulatory examination and approval, any person makes, uses or imports a patented medicine or a patented medical apparatus, and where any person makes, imports the patented medicine or the patented medical apparatus exclusively for such person.

8. Article 62 of the Chinese Patent Law: in a dispute of patent infringement, where the alleged infringer has evidence to prove that the technology or design exploited by him or her forms part of prior art or is prior design, such exploitation does not constitute infringement of patent right.

In general, if the alleged infringer provides evidence to prove that the accused infringing design product is identical or has no substantial difference with the prior design, the alleged infringer can utilize the prior design to defend against the allegation of infringement. Please note the prior design shall be one design patent, rather than the combination of the several design patents.

(3) The defence of none purpose of production and business
In addition, if the alleged infringer can provide evidence to prove that his action does not utilize the production and business purpose, he may also defend allegation of infringement based on this point.

[D] Other Defences

In practice, there is one approach to defend the infringement, namely the defence of legitimate source. In other words, if the act of the seller meets with the provision of Article 70[9] of the Chinese Patent Law, the seller is entitled to defend the infringement.

In fact, if a seller can provide evidence to prove a source of the alleged infringing products, and that he did not know the products have been sold without the authorization of the patentee, the seller does not bear the liability to compensate for damages even though he has engaged in the infringement.

§5.07 PROCEDURES FOR FILING APPLICATION FOR REGISTRATION

[A] Priority

Under the Chinese Patent Law, only foreign priority may be claimed for a Chinese design patent application, that is, where, within six months from the date on which any applicant first filed in a foreign country a design patent application, if he files in China a design patent application for the same subject matter, he may enjoy a right of priority.

Under the Chinese Patent Law, the subject matter shown in the Chinese application shall be contained in the priority application claimed. That is, generally it is not necessary that the drawings to be filed for the Chinese design patent application are completely identical to that used for the priority application, so long as the design (the subject matter) shown in the drawings to be filed for the Chinese design patent application and the design shown by the priority application is the same design.

Under the Chinese Patent Law, if a priority is claimed, the certified priority document should be submitted no later than three months from the filing date in China. Missing the three-month term will result in an invalid priority claim.

9. Article 70 of the Chinese Patent Law: any person, who, for production and business purpose, uses, offers to sell or sells a patent infringement product, without knowing that it was made and sold without the authorization of the patentee, shall not be liable to compensate for the damage of the patentee if he can prove that he obtains the product from a legitimate channel.

[B] Drawings

To file a Chinese design patent application, either line drawings or photographs are allowable. The relevant drawings or photographs submitted by applicant shall clearly indicate the design of the product for which patent protection is sought.

In so far as a product with a 3D design is concerned, the applicant shall submit at least one perspective view and the orthographic projection of six-side views.

In so far as a product with 2D design is concerned, the applicant may submit the orthographic projection view of the relevant side(s).

The applicant shall also submit, if necessary, the exploded view, cutaway view, sectional view, enlarged view and view of state of variation. In addition, the applicant may submit the reference views to indicate the purpose of use, method or places of use of the product incorporating the design.

If line drawings are to be filed, solid lines of even thickness shall be used to show the shape of the design. Shadow lines, indicative lines, dotted lines, central lines, size lines, and dot dash lines shall not be used. The line drawings shall show the design clearly.

If photographs are to be filed, the following requirements shall be satisfied:

(1) The photographs shall be clear, and avoid vagueness resulting from focusing problems, etc.
(2) The background of the photographs shall be plain, and avoid showing other contents except the product incorporating the design. The product and the background shall have appropriate distinction in brightness to show the design clearly.
(3) The photographs shall be taken in conformity with the rules of exact projection and avoid distortion caused by perspective.
(4) The photographs shall avoid strong light, blinking, shadow, reflection, etc.
(5) The product in the photographs should not be allowed to indicate generally accompanied with additional inside filling or liner, unless the additional inside filling or liner must be kept in order to show the design clearly.

[C] Design Part of Another Article/Environment of Design

Under the Chinese Patent Law, another article or the environment of design shall not be included in the drawings or photographs to be filed except for reference views.

[D] Description of Design (Where Required)

A brief explanation may be used to interpret the design of the product as shown in the drawings or photographs.

The brief explanation shall include the following contents:

(1) the title of the product incorporating the design;
(2) the use of the product incorporating the design, which is helpful to ascertain the classification;
(3) the essential features of the design, wherein the essential features mean that the shape, pattern or their combination, or the combination of the colour with shape or pattern, or the position, which is different from prior design; and
(4) which drawing or photograph is designated to best show the essential features of the design.

Furthermore, the following contents shall be indicated in the brief explanation:

(1) protection for colour is claimed or a view is omitted;
(2) where an application for a patent for design is filed relating to several similar designs of the same product, one of the designs shall be designated as the main design in the brief explanation;
(3) as regards 2D product such as printed cloth, wallpaper, shall describe the unit pattern of the plane product which is connected in two sides or four sides and has no fixed limits if necessary;
(4) as regards a long and thin product, shall indicate the length is omitted in the drawings or photographs if necessary;
(5) where the product incorporating the design is made of transparent material or new material which creates special visual effect, it shall be indicated in the brief explanation if necessary;
(6) where the design product is a product in set, the names of the items in the product set shall be indicated if necessary.

However, the brief explanation shall contain no language of commercial advertisement, and shall not be used to show the function of the product.

§5.08 EXPUNGING/CANCELLING OR VARYING REGISTRATION

According to Article 44[10] of the Chinese Patent Law, a design patent can cease before the expiration of its duration. There are two situations: one is that patentee voluntarily gives up the design patent through the means of submitting a written declaration before SIPO; the other is that patentee does not pay for the annual fee in a prescriptive period duly.

In addition, according to Article 47[11] of the Chinese Patent Law, once the design patent is declared invalidated by the Patent Re-examination Board, and the decision

10. Article 44 of the Chinese Patent Law: In any of the following cases, the patent right shall cease before the expiration of its duration: (1) where an annual fee is not paid as prescribed; (2) where the patentee abandons his or its patent right by written declaration.
Any cessation of the patent right shall be registered and announced by the Patent administration department under the State Council.
11. Article 47 of the Chinese Patent Law: any patent right which has been declared invalid shall be deemed to be non-existent from the beginning.

comes into effect, the design patent shall be deemed to be non-existent from the beginning.

In general, once the design patent is granted, registered, and published, the drawings or photos in the design patent cannot be varied. Unless the change of the legal state of the design patent occurs, the announcement shall be published by SIPO, such as any cessation of the patent right, the license and pledge and so on. Any person can check the varying information via applying for a copy of the patent register.

§5.09 OTHER

[A] Interface between Industrial Design and Trademarks

Following main differences exist between Industrial Design and Trademarks:

(1) A design is attached to the product, which belongs to the design of the product itself, and does not separate from the product; while a trademark belongs to a mark, which is relatively dispensable, and can be separated from the product.
(2) The design is used to decorate the product, increasing aesthetic appeal. The design is only applied to the product, rather than including the service; while the trademark is used to distinguish from different category of product, or to indicate an origin of the product. The trademark can be based on goodwill.
(3) The period of protection of the design patent is ten years. After ten years, the design will form part of the public domain; while the period of protection of the trademark is ten years, which can be renewed.

[B] Relationship between Industrial Design and Copyright/ Design Copyright

Below are some main differences between Industrial Design and Copyright:

(1) A design patent includes the right of the property; while copyright includes the right of the property and the personal right.

The decision declaring the patent right invalid shall have no retroactive effect on any judgment or mediation decision of patent infringement which has been pronounced and enforced by the people's court, on any decision concerning the handling of a disputed over patent infringement which has been complied with or compulsorily executed, or on any contract of patent license or of assignment of patent right which has been performed prior to the declaration of the patent right invalid; however, the damage caused to other persons in bad faith on the part of the patentee shall be compensated.

If, pursuant to the provisions of the preceding paragraph, the monetary damage for patent infringement, the fees for exploitation of the paten tor fees for the assignment of the patent right is not returned, but such non-return is obviously contrary to the principle of equity, all or part of the preceding payments shall be returned.

(2) The period of protection of the design patent is ten years. After ten years, the design will form part of the public domain; while the period of copyright protection is different. The term of protection of the rights of authorship, alteration, and integrity of an author shall be unlimited. The term of protection of the right of publication, the right of exploitation and the right to remuneration in respect of a work of a citizen shall be the lifetime of the author and fifty years after his death, expiring on 31 December of the fiftieth year after his death. In the case of a work of joint authorship, such term shall expire on 31 December of the fiftieth year after the death of the last surviving author. The term of protection of the right of publication, the right of exploitation and the right to remuneration in respect of a work where the copyright belongs to a legal person or entity without legal personality, or in respect of a work created in the course of employment where the legal person or entity without legal personality enjoys the copyright (except the right of authorship), shall be fifty years, expiring on 31 December of the fiftieth year after the first publication of such work, provided that any such work that has not been published within fifty years after the completion of its creation shall no longer be protected under this law.

(3) After creation, the designer shall file application before SIPO to obtain the design patent right; while copyright is attached upon creation.

(4) The design patent can be granted when it is in conformity with the Patent Law; while the copyright shall possess requisite originality.

[C] Relationship between Industrial Design and Unfair Competition

According to Article 5 of Unfair Competition of the People's Republic of China, operators shall not adopt any of the following unfair means to carry on transactions in the market and cause damage to competitors:

> (2) using, without authorization, the names, packaging or decoration peculiar to well-known goods or using names, packaging or decoration similar to those of well-known goods so that their goods are confused with the well-known goods of others, causing buyers to mistake them for the well-known goods of others.

There exist following differences between Industrial Design and Unfair Completion:

(1) The design patent can be granted when it is in conformity with the Patent Law; while the packaging and decoration do not require application and registration.

(2) The design patent requires the novelty and obvious difference from prior designs; while the packaging and decoration emphasizes the uniqueness and well-known.

(3) The design patent protects the design of the product itself, such as the shape or pattern of the product; while packaging and decoration protect the exterior packaging of the product, excluding the product itself.

CHAPTER 6

Industrial Design Rights: Europe

Kurt Stöpetie & Rutger Stoop

§6.01 INTRODUCTION

To properly understand intellectual property legislation in the European Union ('EU'), we should begin with a brief, general introduction to the EU. The start for deeper structural collaboration between the European countries was made in 1957, when the Treaty establishing the European Economic Community ('TEEC') was signed in Rome. Consisting of just Belgium, France, Italy, Luxembourg, the Netherlands and West Germany at the time, the TEEC mainly proposed customs measures and the creation of an internal market of goods, workers and services in its Member States. By way of the Maastricht Treaty in 1993, which deleted the word 'Economic' and also introduced provisions leading to a common currency for the Member States, the European Community was renamed to European Union[1] and its functioning was further detailed in the Treaty on the functioning of the European Union ('TFEU'),[2] which came into force in 2009 by the signing of the Treaty of Lisbon in 2007. As of 2013, the EU consists of twenty-eight Member States.

One of the major goals of the EU has been the harmonization of the laws of its Member States. The TFEU plays a considerable part in EU legislation on intellectual property, considering the competition implications of Articles 101 and 102 in particular. For instance, Article 101 – briefly put – declares all agreements void which have as their object or effect the restriction of competition within the internal market. Obviously, intellectual property can be a restriction on competition. However, to exclude (in

1. Despite this renaming, it is still referred to as the 'Community' where design and trademark law is concerned.
2. Treaty of Lisbon amending the Treaty on European Union and the Treaty establishing the European Community, signed at Lisbon, 13 Dec. 2007, as published in the Official Journal C 306, 17 Dec. 2007, pp. 1–271, entering into force on 1 Jan. 2009.

principle) agreements on intellectual property, a remedy is found in paragraph 3 of Article 101 for agreements which contribute to improving the production or distribution of goods or to promoting technical or economic progress. Article 102 is also relevant, because it prohibits exercising intellectual property as being incompatible with the internal market when any abuse of a dominant position by one or more companies within the internal market or a substantial part of that market affects trade between Member States.

[A] Highlights of New Developments in This Jurisdiction since Last Edition

The harmonization of intellectual property laws of the Member States of the EU has made quite a few steps in the last decade. National design laws have been harmonized by the Community Design Directive in 1998 ('CDD'),[3] although full harmonization was never a goal of the CDD – only those aspects which were considered relevant to the European internal market (i.e., harmonizing the conditions for obtaining a registered design right and to that end introducing a unitary definition of the notion of design and of the requirements as to novelty and individual character).

Specifically for European design law, Council regulation (EC) No. 6/2002 of 12 December 2001 on Community designs came into force on 6 March 2002 (Community Design Regulation, 'CDR'),[4] while Community Design Implementation Regulation (EC) No. 2245/2002 ('CDIR')[5] entered into force later in 2002. The CDR provides a single Community-wide design right which ensures protection in all the EU Member States. However, the existence of the CDR does not affect the possibility of obtaining design protection by national applications. Where the CDD will be discussed mainly in the national chapters of individual EU Member States, the CDR will star in this European chapter.

§6.02 CONVENTIONS AND LEGISLATION

As the Paris Convention for the Protection of Industrial Property ('Paris Convention')[6] is relevant to intellectual property in general, it is obviously relevant for industrial designs as well. The Paris Convention provides general rules for, *inter alia*, designs in

3. Directive 98/71/EC of the European Parliament and of the Council of 13 Oct. 1998 on the legal protection of designs, as published in the Official Journal L 289, 28 Oct. 1998 pp. 28–35, entering into force on 17 Oct. 1998.
4. Council Regulation (EC) No. 6/2002 of 12 Dec. 2001 on Community designs, as published in the Official Journal L 003, 5 Jan. 2002 pp. 1–24, entering into force on 6 Mar. 2002, lastly amended by Official Journal L 386, 29 Dec. 2006, pp. 14–16.
5. Commission Regulation (EC) No. 2245/2002 of 21 Oct. 2002 implementing Council Regulation (EC) No. 6/2002 on Community designs, as published in the Official Journal L 341, 17 Dec. 2002 pp 28–53, entering into force on 24 Dec. 2002.
6. Paris Convention for the Protection of Industrial Property of 20 Mar. 1883, as revised at Brussels on 14 Dec. 1900, at Washington on 2 Jun. 1911, at The Hague on 6 Nov. 1925, at London on 2 Jun. 1934, at Lisbon on 31 Oct. 1958, and at Stockholm on 14 Jul. 1967, and as amended on 28 Sep. 1979.

respect of priority rights (Article 5 Paris Convention) and regarding temporary protection at certain international exhibitions (Article 11 Paris Convention). Also, since the revision of Lisbon in 1958, parties to the Paris Convention are obliged to grant protection for industrial designs.

The Agreement on Trade-Related Aspects of Intellectual Property Rights ('TRIPs')[7] provides for a minimum protection threshold for industrial designs (Articles 25 and 26 TRIPs). For instance, it requires parties to TRIPs to protect independently created industrial designs that are new or original. Parties to TRIPs can require that designs are not new or original if they do not significantly differ from known designs or combinations of known design features, and can also require that such protection shall not extend to designs dictated essentially by technical or functional considerations.

For substantially more important legislation on design protection in the EU, one must look to the TFEU, as the application of Articles 101 and 102 may affect the enforceability of design rights (see the introductory remarks above).

The CDR aims at facilitating the free movement of goods and services, by breaking down existing intra-Community barriers of national legislation for design protection. However, and perhaps somewhat contradictory, those barriers have not been truly broken down. The Member States' national legislation for design protection continues to exist independently of the design protection offered by CDR, albeit in a harmonized way as a result of the CDD. This is to maintain an alternative for those having only a commercial interest in design protection in only a few Member States. Those seeking protection by way of national design rights should be aware of the fact that national protection may differ between individual Member States. For more information on this, reference is made to the chapters on the individual EU countries.

The CDR provides unified Community-wide design protection; both registered (registered Community design, 'RCD') and unregistered (unregistered Community design, 'UCD'). As a consequence, a single application for an RCD results in design protection in all of the countries within the Community. A special position is assigned to Belgium, the Netherlands and Luxembourg, as these three Member States are together regarded as a single Member State (the Benelux), with a single administrative body for intellectual property: the Benelux Office for Intellectual Property ('BOIP').

Article 2 CDR appoints the Office for Harmonization in the Internal Market (Trademarks and Designs) as the central EU administrative body carrying out the tasks specified in the CDR ('OHIM').[8] Situated in Alicante, Spain, OHIM was originally established pursuant to the Community Trademark Regulation ('CTR').[9] The OHIM is responsible for the register of Community designs (Article 72 CDR) and publishes in the *Official Journal* as well as in its own *Community Designs Bulletin* (Article 73 CDR). The

7. The TRIPs Agreement is Annex 1C of the Marrakesh Agreement Establishing the World Trade Organization, signed in Marrakesh, Morocco on 15 Apr. 1994, and entering into force on 1 Jan. 1995.
8. As per the proposed changes in EU trademark legislation, OHIM will be renamed the European Union Intellectual Property Office ('EU IPO'). For editorial purposes, we will continue to refer to OHIM in this edition.
9. Council Regulation (EC) No. 40/94 of 20 Dec. 1993 on the Community trademark, amended by Council Regulation (EC) No. 207/2009 of 26 Feb. 2009 on the Community trademark.

Official Journal mainly covers general EU implementing dealings, and thus also EU legislation relating to designs and trademarks, as well as case law of the General Court and the EU Court of Justice. The Bulletin deals specifically with information of, and OHIM decisions on, applications of Community designs and nullity requests against Community designs.

Each Member State has designated at least one Community design court in its territory, which can hear civil proceedings, such as those for infringement and invalidity (Article 80 CDR).

§6.03 SUBJECT MATTER

[A] Definition/What Constitutes a Protectable Design?

The CDR introduces a two-tier protection system, creating two distinct types of design protection: the UCD and the RCD (Article 1(2) CDR). The former arises without the need for specific formalities, by the simple fact that the design is disclosed to the public in the manner provided for in the CDR (see also §6.04). A registered design can only be obtained by registration in accordance with the registration procedure as set out in the CDR. National design rights coexist with a Community design. However, provisions are included to ensure that all the design rights are enforced consistently.

The two Community design rights complement each other. The first, a short-term unregistered design right, is mostly of use for industries producing large numbers of designs for products with a short lifecycle. These require shorter protection with fewer formalities and do not merit the expense of registration (e.g., fashion designs). Such designs are protected for a period of three years from the date on which the design was first made available to the public. A UCD therefore offers a relatively easy and attractive way of protection within the Community. The second form of protection, a longer-term registered design right, is intended for those sectors which need greater legal certainty and longer-term protection. Apart from the complementary nature of these two different protection regimes, they can also be used intelligently in the first twelve months after a design is disclosed to the public, by initially relying on the protection of a UCD and only registering this design as an RCD with OHIM if it has proven to be a commercial success that is being imitated by third parties.

For the purpose of the CDD and the CDR, a 'design' is defined as the appearance of the whole or a part of the product resulting from the features of, in particular, the lines, contours, colours, shape, texture and/or materials of the product itself and/or its ornamentation (Article 3(a) CDR). The list of features of appearance is not intended to be exhaustive. Further, Article 4(1) CDR provides that a design shall be protected to the extent that it is new and has individual character (see further §6.03[B]).

The term 'product' includes any industrial or handicraft item including parts intended to be assembled into a complex product, packaging, get-up graphic symbols and typographic typefaces, but excluding computer programs (Article 3(b) CDR). The definition makes it clear that colours can constitute part of a design. A design can cover both two- and three-dimensional features of appearance. In any case, according to

Article 9 of the CDR, a design contrary to public policy or to accepted principles of morality is excluded from protection.

[1] Shape, Configuration

The wording 'appearance' is somewhat unclear and seems to be restricted to the *tangible* features of a design, and it seems not to include aspects like taste, sound, smell and what is known as the 'look and feel' of a design. Recital 11 of the CDD makes clear that protection is granted for those design features of a product, in whole or in part, shown visibly in an application and disclosed to the public. In clarifying the overall impression of the filed design, the *PepsiCo*-judgment held that the actual traded product to which the design is applied can be taken into account in order to confirm earlier conclusions on the characteristics of the design as shown in the registration, because the person comparing the products is not the 'average consumer' but the 'informed user' (more on this in §6.03[B][1]).[10]

The definition of a design makes clear that a design can also consist of the appearance of a part of a product. Pursuant to Article 3(c) CDR, a complex product means a product which is composed of multiple components which can be replaced permitting disassembly and reassembly of the product. Protection is not extended to component parts which are not visible during normal use of a product or which themselves would not fulfil the requirements as to novelty and individual character (Article 4(2) CDR).[11] However, component parts do not have to be visible at all times during normal use by the end user ('normal use' does not cover maintenance, servicing or repair work - Article 4(3) CDR). Having said that, it should be noted that designs for the purpose of the repair of a complex product, so as to restore its original appearance, are excluded from protection (Article 110(1) CDR).

[2] Ornamental or Functional

Functional use is not a requirement. However, it remains unclear whether designs only suitable for ornamental use can be filed for a wide range of design categories, since protection is given to a product *in concreto*. However, the fact that a design has gained a new function does not simply mean that independent design protection can be obtained.[12] No specific guidance has yet been issued by the EU Court of Justice on this point of law.

Article 8(1) CDR provides that a design right shall not subsist of features of appearance of a product which are solely dictated by its technical function. Also excluded from Community design protection are features of the appearance of a product which must necessarily be reproduced in their exact form and dimensions in

10. EU Court of Justice 20 Oct. 2011, C-281/10 P, *ECLI:EU:C:2011:679* (*PepsiCo*), para. 73.
11. General Court EU 3 Oct. 2014, T-39/13, *ECLI:EU:T:2014:852 (Niewinski/BHIM)*.
12. Court of Appeal OHIM 25 Jan. 2008, R84/2007-3 (*Ferrari/Dansk Supermarked*); Supreme Court of Judicature Court of Appeal (Civil Division) 23 Apr. 2008, [2008] EWCA Civ 358 (Laundry balls) and Court of Appeal The Hague 17 May 2011, *IER* 59/442 (*I-Drain/ESS*), para. 10.

order to permit the product in which the design is incorporated or to which it is applied to be mechanically connected to or placed in, around or against another product so that either product may perform its function, but can subsist in a design serving the purpose of allowing the multiple assembly or connection of mutually interchangeable products within a modular system (Article 8(2-3) CDR).

[B] Originality/Novelty

[1] Test for Originality Generally

Protection is awarded to a design by the CDR to the extent that it is new and has an individual character (Article 4(1) CDR). These requirements apply equally to UCDs and RCDs. A design is considered to have individual character if the overall impression it produces on the informed user differs from the overall impression produced on such a user by any design which was disclosed to the public before the date of filing or the date of priority. In assessing individual character, the degree of freedom of the designer in developing the design is taken into consideration as well.[13]

Thus, the individual character is assessed by comparing the impression produced by the design on the informed user with the impression created by earlier designs. The informed user will make a direct comparison between the designs at issue because that type of comparison actually relates to the impression produced on that user by earlier individualized and defined designs, as opposed to an amalgam of specific features or parts of earlier designs.[14]

According to the *PepsiCo*-decision, the concept of an 'informed user' must be understood as lying somewhere between that of the average consumer, as defined in trademark matters, who need not have any specific knowledge and who normally makes no direct comparison between the trademarks in dispute, and the sector expert, who is an expert with detailed technical expertise and who is able to distinguish potentially minimal differences between existing designs. Thus, the concept of the informed user may be understood as referring not to a user of average attention, but to a particularly observant one, either because of his personal experience or his extensive knowledge of the sector in question.[15] As regards the informed user's level of attention: although the informed user is not the well-informed and reasonably observant and circumspect average consumer who normally perceives a design as a whole and does not proceed to analyse its various details, he is also not an expert or specialist capable of observing in detail the minimal differences that may exist between the designs in conflict. Thus, the qualifier 'informed' suggests that, without being a designer or a technical expert, the user knows the various designs which exist in the sector

---

13. General Court EU 7 Nov. 2013, T-666/11, *ECLI:EU:T:2014:852* (*Budziewska/OHIM*). Also see General Court EU 9 Sep. 2014, T-494/12, *ECLI:EU:T:2014:757 (Biscuit)*.
14. ECJ 19 Jun. 2014, C-345/13, *ECLI:EU:C:2014:2013 (Karen Millen Fashions Ltd v. Dunnes Stores and Dunnes Stores (Limerick) Ltd)*, paras 26 and 35, with reference to ECJ 20 Oct. 2011, C-281/10 P, *ECLI:EU:C:2011:679 (PepsiCo)*, para. 55, and ECJ 18 Oct. 2012, C-101/11 P and C-102/11 P, *EU:C:2012:641* (*Neuman and Others v. José Manuel Baena Grupo*), para. 54.
15. ECJ 20 Oct. 2011, C-281/10 P, *ECLI:EU:C:2011:679 (PepsiCo)*, para. 53.

concerned, possesses a certain degree of knowledge with regard to the features which those designs normally include, and, as a result of his interest in the products concerned, shows a relatively high degree of attention when he uses them.[16]

[2] Novelty

The novelty test is satisfied if no identical design has been disclosed to the public before the filing date or priority date within the Community. Designs are deemed to be identical if their features differ only in immaterial details (Article 5 CDR).

Both an RCD and a UCD require such novelty. In case of an RCD, such a design is considered new if no identical design has been disclosed to the public before the date of the filing of the design application. In case of a UCD, such a design is considered new if no identical design has been disclosed to the public before the design for which protection is sought has been made available to the public.

[a] Prior Use

Neither novelty nor individual character will be prejudiced by any disclosure during the twelve months leading up to the filing of an application (Article 7(2) CDR). The designer or his successor in title is given a twelve-month grace period in which to test the market with his design, but at any time during that period he can apply for registration without fear that he will have invalidated an application for registration by his own prior use. This is particularly useful for designers who wish to expose their efforts to the public to judge their reaction before deciding whether to incur the expense of registration. The same provision applies if the prior disclosure in the one year grace period was filed by a third party who obtained his design as a result of an abuse of confidence or fraud in relation to the designer or his beneficiaries.

During this twelve-month period, designs will be awarded protection without the need for registration, and for up to three years as an unregistered design. However, it should be noted that a UCD confers on its holder only the right to prevent the use of the design by any third party if that use results from copying the UCD (Article 19(2) CDR).[17]

[b] Prior Publication

For the purposes of Articles 5 and 6 CDR, a design is deemed to be disclosed to the public if published, exhibited, used in trade or otherwise disclosed in the normal course of business of the relevant sector within the Community, other than under obligations

16. ECJ 20 Oct. 2011, C-281/10 P, *ECLI:EU:C:2011:679 (PepsiCo)*, para. 59.
17. According to Dutch case law, if there is sufficient similarity between the designs, copying is presumed on the basis of the earlier design having been copied unintentionally (see Dutch Supreme Court 21 Feb. 1992, *NJ* 1993, 164 (*Barbie*)).

of confidentiality. This means that a design is not rendered invalid just because there was a similar design somewhere else in the world (Article 7 CDR).

The EU Court of Justice clarified that it is possible that a design may not reasonably have become known in the normal course of business to the circles specialized in the sector concerned, operating within the EU, even though it was disclosed to third parties without any explicit or implicit conditions of confidentiality, if it has been made available to only one company in that sector or has been presented only in the showrooms of a company outside the EU.[18]

[c] Prior Application or Registration

In accordance with the Paris Convention and TRIPs, the CDR states in Article 41 that a right of priority can be obtained of six months from the date of filing of the first application. Consequently, any subsequent filing in any of the other countries that signed the Paris Convention or TRIPs before the expiration of that six-month period shall not be invalidated by reason of any acts accomplished in this period. Rights acquired by third parties before the date on which the right of priority is based, are reserved in accordance with the domestic legislation of each country that signed the Paris Convention or TRIPs (Article 4B Paris Convention).

An RCD application or registration may be invalidated based on Article 25(1)(d) CDR if it gets in conflict with a prior design which has been disclosed after the date of filing of the application or, if a priority is claimed, the date of priority of the RCD, and which is protected from a date prior to the said date by an RCD or an application for such a design, or by a registered design right of a Member State or an application for such a right, or by a design right registered under the Geneva Act of the Hague Agreement concerning the international registration of industrial designs[19] or by an application for such a right.

§6.04 DURATION OF PROTECTION

For the duration of the protection of an RCD, the date of the filing of the application is determinative. The duration of the protection initially covers five years as of the filing date, but can be renewed for four additional and consecutive terms of five years, leading to a maximum protection period of twenty-five years (Article 12 CDR).

As the UCD sets a low formal threshold for obtaining such design, the maximum protection period is considerably shortened to a non-renewable three-year period. This period commences as from the date on which the design was first made available to the public within the Community (Article 11(1) CDR). Such a design is first made available to the public within the Community if it has been published, exhibited, used in trade or otherwise disclosed, except where these events could not reasonably have become

18. EU Court of Justice 13 Feb. 2014, C-479/12, *ECLI:EU:C:2014:75* (*Gautzsch/Duna*), para. 36.
19. Adopted in Geneva on 2 Jul. 1999.

known in the normal course of business to the circles specialized in the sector concerned, operating within the Community (Article 11(2) CDR).

An effect of this criterion is that the moment of first global introduction does not necessarily coincide with the commencement date of the protection period for an UCD. Indeed, the design has to be made available to the public *within the territory of the Community* in such a way that, in the normal course of business, these events could reasonably have become known to the specialists in the sector concerned (see Article 11(2) in conjunction with Article 110a(5) CDR).

§6.05 INFRINGEMENT

[A] Scope of the Infringement Action

Article 19(1) CDR provides that the holder of an RCD has the exclusive right to use his design. This use covers the making, offering, putting on the market, importing, exporting or making use of a product in which the design is incorporated or to which it is applied, or stocking such a product for those purposes. This right extends to any third party who uses a design that does not produce a different overall impression on informed users, including the third party holder of a Community design registered later.[20] In this context, it should be stressed that RCDs are not examined on material grounds, but only on compliance with formal requirements as set out in Article 36(1) (Article 45 CDR).

The same rights are conferred on the holder of a UCD, but only if the contested use results from copying the protected design. If the contested use is the result of an independent work of creation by a designer who may be reasonably thought to be unfamiliar with the design disclosed to the public by the holder, it will not be deemed to be the result from copying (Article 19(2) CDR). The burden of proof lies with the holder of the UCD,[21] but if the similarity between the designs is high, the Dutch courts will assume that the later design is the result of copying.[22]

[B] Test for Infringement/Criteria to Constitute Infringement

If a design, which includes a design registered later, does not produce a different overall impression on the informed user, this design falls within the protective scope of the Community design (Article 10(1) CDR). Article 8 CDR provides that the protection of the Community design does not apply to the parts of the design that are the result of the technical function. According to Article 10(2) CDR, the degree of freedom of the designer in developing his design shall be taken into consideration in assessing the scope of protection of the Community design.

20. EU Court of Justice 16 Feb. 2012, C-488/10, *ECLI:EU:C:2012:88* (*Celaya v. Proyectos*), para. 52.
21. EU Court of Justice 13 Feb. 2014, C-479/12, *ECLI:EU:C:2014:75* (*Gautzsch v. Joseph Duna*), para. 44.
22. Dutch Supreme Court 21 Feb. 1992, *NJ* 1993, 164 (*Barbie*).

In determining the overall impression of both designs, only the parts of the designs visible during normal use of the product to which the design is applied are considered. It is therefore recommended to bear the use of the product in mind when determining the protective scope of the design. Non-visible elements should not be used in assessing the originality of the later design.[23]

[C] Standing to Sue

Article 14(1) CDR provides that the designer or his beneficiary holds the right to the Community design. In case two designers have jointly developed a design, the right to the Community design shall vest in them jointly (Article 14(2) CDR). Article 14(3) CDR provides that, unless otherwise agreed or specified under national law, the design that is developed by an employee in the execution of his duties or following the instructions given by his employer, shall vest in the employer. The provision of Article 14(3) CDR does not apply to Community designs that have been produced as a result of a commission.[24]

The holder of a design right can bring infringement claims against any third party that uses his design without his consent (Article 19 CDR).

[D] Remedies for Infringement

A holder of a Community design has a wide range of remedies at his disposal in case of infringement or threatened infringement. Where a Community design court finds that the defendant has infringed or threatened to infringe a Community design, it will – barring special reasons for not doing so – order the following:

(i) an order prohibiting the infringer from continuing with the acts which have infringed or would infringe the Community design;
(ii) an order to seize the infringing products;
(iii) an order to seize materials and implements predominantly used in order to manufacture the infringing goods, if the owner knew the effect for which such use was intended or if such effect would have been obvious in the circumstances; and
(iv) any further order imposing appropriate sanctions by the law of the Member State in which the acts of infringement or threatened infringement are committed, including its private international law (Article 89(1)).

Equally important is that the Community design court will take measures in accordance with national law which are aimed at ensuring compliance with the orders made, such as the forfeiture of penalties in case of non-compliance.

Notably, under the CDR, Member States are allowed to provide even more remedies to holders of a Community design than provided for in the CDR. As a

23. General Court EU 9 Sep. 2014, T-494/12, *ECLI:EU:T:2014:757* (*Biscuit*), para. 31.
24. Court of Justice EU, 2 Jul. 2009, C-32/08, *ECLI:EU:C:2009:418* (*FEIA v. Cul de Sac*), para. 79.

consequence of such national legislation and combined with national rules of procedure, right holders can to some extent cherry-pick a Member State which best suits their needs for enforcing their design rights. For example, in the Netherlands the holder of a Community design not only has many more remedies at its disposal as compared to the CDR, but an infringer will in principle also be ordered to compensate the full legal costs of the right holder. In combination with easy court access for provisional relief, this makes the Netherlands an attractive and strategically important entry for enforcement of IP-rights in the EU.

§6.06 DEFENCES TO INFRINGEMENT

Defences are mainly focussed on the validity of the invoked Community design and on non-infringement. The defences regarding the validity of the Community design are mainly the same as the test which is done in case of the application of a Community design (see §6.03).

According to Article 24(1) CDR, an RCD can be declared invalid by a Community design court on the basis of a counterclaim in infringing proceedings or on application to the Office. Even if the Community design has lapsed or has been surrendered, Article 24(2) provides that the design can still be declared invalid. The case for UCDs is slightly different. UCDs can be declared invalid by an application to a Community design court or by a counterclaim in infringement proceedings (Article 24(3) CDR). According to Article 80 CDR, each Member State has a limited number of Community design courts.

Article 25 CDR sums up an exhaustive list of grounds for invalidity. The most important ones will be discussed below.

[A] Validity

The CDR contains an exhaustive list of validity defences. Article 26(1) CDR provides that if a Community design is invalidated, it is deemed to not have had, as from the outset, the effects specified in the CDR, to the extent that it has been declared invalid.

[1] Not Novel/Original

A possible defence to an infringement claim is to challenge the Community design's required novelty. The requirements for novelty are set out in Article 5 CDR. To determine the novelty of a design, it is compared with designs that have been disclosed earlier. Likewise, if the Community design does not meet the requirements for individual character (Article 6 CDR), it is not considered to be original, and can therefore not be a valid Community design. If the overall impression of both designs is the same, the contested design lacks originality.[25]

25. General Court EU 22 Jun. 2010, T-153/08, *ECLI:EU:T:2010:248* (*Shenzen Taiden/OHIM*), para. 75.

[2] Functional/Not Ornamental

Article 8(1) CDR provides that a Community design does not subsist in features of appearance of a product which are solely dictated by its technical function. Article 25(1)(b) CDR provides that such a design lacks validity and may be declared invalid.

[3] Other Possible Criteria

Furthermore, a Community design may be declared invalid if the design is, for instance, not the appearance of the whole or part of a product as defined in Article 3(a) CDR, or if it is contrary to public policy or to accepted principles of morality (Article 9 CDR), or if it is in conflict with prior proprietary rights (i.e., trade name or trademark rights, copyrights,[26] design rights, or any of the signs referred to in Article 6*ter* Paris Convention - see Article 25(1), d, e, f and g CDR). According to Article 25(3) CDR, these last grounds can be invoked solely by the applicant or holder of these prior rights.

[4] Faults in Application

As opposed to Benelux design law, for example, neither the CDR nor the CDD provides for consequences of faults in the application for a Community design right. However, it seems likely that the lack of clarity of an application can be invoked before a court so as to limit the scope of protection of a design right.

[B] Marking

By marking a design, the holder of that product can demonstrate to the public that its product is protected by a design right. Although the marking of a Community design is neither required nor implemented in Community law, it might be helpful to inform the public that a design is registered as such by using dedicated marking. The marking of a Community design can be helpful when claiming damages in an infringement action, because the marking makes it easier to prove that the infringing party should have been aware of the fact that a product was protected by a design right and that it therefore should have been aware of infringing the rights in that design. Note that in other countries it is more common to mark a design, or in some cases even obligated to use such marking (see for example, Canada, where awarding financial compensation in an infringement suit is dependent on the use of marking).

[C] Non-infringement

The exclusive right granted by a Community design is subject to limitation. Consequently, the right may not be exercised in respect of acts done privately, for

26. General Court EU 6 Jun. 2013, T-68/11, *ECLI:EU:T:2013:298* (*Kastenholtz/OHIM*), paras 79–81.

non-commercial purposes, acts done for experimental purposes or acts of reproduction for the purpose of making citation or of teaching (Article 20(1) CDR). Furthermore, Article 20(2) CDR puts a few other limitations on the exclusive right, for instance in certain situations with regard to equipment on ships and aircraft.

[D] Other Defences to Infringement

The following defences to infringement are not based on the validity requirements as laid down in the CDR but, depending on the circumstances, may be useful defences to infringement nonetheless.

Article 21 CDR provides that in case the product has been put on the market in the Community by the holder or with his consent, his exclusive right for that product is exhausted. This means that the holder of such a Community design cannot exercise its rights any longer against the marketing of such a product. The right to challenge products that have been put on the market outside the Community will stay intact, however.

Another defence to infringement exists for any third person who can establish that, before the date of filing of the application or the date of priority, he has in good faith commenced use within the Community, or has made serious and effective preparations to that end, of a design included within the scope of protection of an RCD which was not copied from that RCD (Article 22(1) CDR).

§6.07 PROCEDURES FOR FILING APPLICATION FOR REGISTRATION

As noted above, and unlike a UCD, an RCD can only be obtained through registration. Applications for such a registration can be filed with either the OHIM, the central industrial property office of a Member State, or, for the Benelux countries, the BOIP (Article 35 CDR). The application process is further codified in Title IV of the CDR, with requirements further specified in the CDIR.[27]

The OHIM has composed Examination Guidelines on Community Design ('Examination Guidelines'), in which it gives practical information on how it applies the application terms as set out in the CDR.[28] Although these guidelines are generally considered non-binding, there has been case law in the past in which the General Court has referred to these guidelines.

Article 36 CDR sets out the conditions to be satisfied by applications. The OHIM will examine the application as against these formal requirements (Article 45 CDR). In case of deficiencies which may be corrected, the OHIM will request the applicant to remedy within the prescribed period (Article 46 CDR).

Article 47 CDR sums up the grounds for non-registrability during examination. They concern whether the design meets the definition of a design in Article 3(a) CDR, and whether it is not contrary to public policy. Even if the OHIM considers one of these

27. Available at https://oami.europa.eu/.
28. Available at https://oami.europa.eu/.

two grounds to be fulfilled, it will allow the applicant to withdraw or amend its application, or submit observations (Article 47(2) CDR). If the application passes these examinations, the application is registered by the OHIM (Article 48 CDR) with the date of filing being the formal registration date (Article 38 CDR). Finally, the registration will be is published in the *Community Designs Bulletin*.

If an application contains deficiencies on the formal requirements, the examiner will notify the applicant and the applicant is given two months to remedy the deficiencies (Articles 45(2) and 46(3) CD, Article 10(3)(4) CDIR and paragraph 11.10 Examination Guidelines).

It follows that no material test is performed by the OHIM (no examination as to the novelty, individual character and functionality). Furthermore, an RCD is readily granted and the enforcement of a design can quickly be arranged within the Community. All the more so as an RCD is presumed to be valid (Article 85(1) CDR). While UCDs are also presumed to be valid, its holder has to prove that the contested use results from copying the protected design, and this may be a burden that the holder cannot bear.

[A] Priority

The applicant can also, in principle, claim a right of priority when applying for an RCD, pursuant to Articles 41–44 CDR. This is the case if the applicant duly applied for a design in or for any State that signed the Paris Convention, or to the Agreement establishing the World Trade Organization, for the purpose of applying for an RCD in respect of the same earlier design. This grants the applicant the right of priority for six months from the date of filing of the first application. The CDR also provides for an exhibition priority (Article 44 CDR, see also §6.03[B][2][c] above).

If the applicant for the RCD wants to take advantage of the priority of an earlier application, the applicant is required to add a copy of this earlier application to his application and may also be required to add a translation if the language of the earlier application is not one of the languages of the OHIM.

[B] Drawings

An application should contain a representation of the design suitable for reproduction (Article 36(1)(c) CDR). However, if the object of the application is a two-dimensional design and the application contains a request for deferment of publication in accordance with Article 50 CDR, the representation of the design may be replaced by a specimen (Article 36(1)(c) CDR).

The OHIM specifies this in paragraph 4.4 and 4.5 of its Examination Guidelines. It states that drawings, photographs (except slides), computer-made representations or any other graphical representation are accepted, provided they are suitable for reproduction. Further, the Examination Guidelines provide conditions for a representation to be considered suitable for reproduction for the purpose of allocating a filing date. Paragraph 4.5 provides cumulative conditions for supplying a specimen of the design. More formalities are specified in paragraph 11 of the Examination Guidelines, which are based on Article 4 CDIR, providing both practical requirements (e.g., filing on unstapled opaque white paper) and formal requirements (e.g., the representation may contain no more than seven different views per design). Article 2 CDIR provides a few other requirements for an application containing multiple designs, such as consecutively numbering the designs.

Article 37 CDR provides that several designs may be combined in one multiple application for RCDs (with a maximum of 100 designs). This possibility is subject to the condition that the products in which the designs are intended to be incorporated or to which they are intended to be applied are all designated for the same class of the International Classification for Industrial Designs of the Locarno Agreement, except in cases of ornamentation (class 32).

[C] Design Part of Another Article/Environment of Design

The representation of a design should be limited to the features for which protection is sought. However, the Examination Guidelines provide that the representations may comprise other elements that help to identify the features of a design for which protection is sought. Paragraph 11.4 allows the following identifiers:

(a) dotted lines to indicate the elements for which no protection is sought or to indicate portions of the design which are not visible in that particular view, i.e., non-visible lines;
(b) boundaries to surround features of the design for which protection is sought; and
(c) colouring on a black and white drawing to highlight the features of the design for which protection is sought only.

Furthermore, with regard to ornaments, a product to which the design is applied may be identified by dotted lines or the ornamentation may be surrounded by boundaries. The OHIM notes that it is the applicant's sole responsibility to use the dotted lines, boundaries and colouring in such a way as to make clear for which features protection is sought.

[D] Description of Design (Where Required)

The CDR provides the option of adding a description to the application, explaining the representation or specimen (Article 36(3)(a) CDR). This single description per design may not exceed 100 words, and must relate only to those features which appear in the reproductions of the design or the specimen. It may not contain statements as to the purported novelty or individual character of the design or its technical value (Article 1(2)(a) CDIR and paragraph 9 Examination Guidelines). Such a description will not, as such, affect the scope of protection of a Community design and it will not be accepted if it is submitted after the date of filing.

§6.08 EXPUNGING/CANCELLING OR VARYING REGISTRATION

Pursuant to Article 13 CDR, the holder of an RCD or any person expressly authorized by him is required to request the renewal of the registration. If the application for renewal is not submitted in good time, the OHIM will cancel the design from the register with effect from the day after the expiration of the existing registration (Article 22(6) CDIR). However, a grace term of six months is applicable (Article 13(3) CDR).

The holder of an RCD can surrender his registration by declaring this in writing to the OHIM (Article 51 CDR), which will only have effect when it has been entered in the register. If a registration which is subject to deferment of publication is surrendered, it will be deemed from the outset not to have the effects specified in the CDR (Article 51(2) CDR). However, considering the unitary character of the Community design, a right holder cannot surrender its RCD for part of the Community (Article 1(3) CDR).

An RCD may be partially surrendered provided that its amended form complies with the requirements for protection, and the identity of the design is retained (Article 51(3) CDR). In that case, the declaration of surrender must contain a representation of the amended design in accordance with Article 4 CDR (Article 27(1)(e) CDIR).

§6.09 OTHER

[A] Interface between Industrial Design and Trademarks

Two-dimensional designs and three-dimensional shapes can be protected as trademarks under the CTMR. As such, a trademark does not require novelty. This can be beneficial in relation to designs which, for instance, can no longer be regarded as new because of disclosure or because the design rights have expired.

According to Article 4 CTMR, a trademark may consist of any sign capable of being represented graphically, in particular words, including personal names, designs, letters, numerals, the shape of goods or of their packaging, provided that such signs are capable of distinguishing the goods or services of one company from those of other companies.

However, Article 7(1e) CTMR also provides that signs consisting exclusively of: (i) the shape which results from the nature of the goods themselves, (ii) the shape of goods which is necessary to obtain a technical result and (iii) the shape which gives substantial value to the goods, may not be registered or, if registered, may be declared invalid. The rulings of the EU Court of Justice show that these absolute grounds for refusal are not easily overcome.[29]

29. See EU Court of Justice 14 Sep. 2010, C-48/09, *ECLI:EU:C:2010:516* (*Lego/OHIM*); EU Court of Justice 18 Sep. 2014, C-205/13, *ECLI:EU:C:2014:2233* (*Hauck/Stokke*).

[B] Relationship between Industrial Design and Copyright/Design Copyright

All Member States are parties to the Berne Convention,[30] which gives considerable latitude as to the protection of 'works of applied art and industrial designs and models' (Article 2(7) Berne Convention). Such works may be protected under copyright law, under specific laws, or both. As stated in Recital 32 of the CDR's preamble, it is important to establish the principle of cumulation of protection under the CDR and under copyright law, in the absence of the complete harmonization of copyright law, whilst leaving Member States free to establish the extent of copyright protection and the conditions under which such protection is conferred. This cumulation is provided in Article 96(2) CDR, which states that Member States shall determine the extent to which, and the conditions under which, such a protection is conferred, including the level of originality required.

This 'cumulation principle' can also be found in Dutch national legislation, where both copyright protection and dedicated design protection is available for the same design. This protection under Dutch copyright law is available regardless of a registration.

[C] Relationship between Industrial Design and Unfair Competition

Protection against unfair competition is available in all Member States, but the scope of protection per Member State differs considerably. In general, protection requires not only objective acts of misbehaviour, but also the existence of a subjective element of fault or negligence by the offender and some degree of avoidable confusion amongst the relevant public. This prohibition is an exception to the principle that anything which is covered by a monopoly right should be fully available for reproduction once the exclusivity has expired.

In conclusion, the implementation of the CDD and the entry into force of the CDR brought about an easy and efficient method of design protection. Where protection is required only for a specific territory, national registration could still be the preferred choice for design protection. However, the RCD offers attractive protection for a very large number of countries with a single filing, and the UCD offers a similar protection regime, without the need for registration, in exchange for a shorter protection period.

30. Berne Convention for the Protection of Literary and Artistic Works, lastly amended on 28 Sep. 1979.

CHAPTER 7

Industrial Design Rights: France

Alexandra Neri

§7.01 INTRODUCTION

[A] Highlights of New Developments in This Jurisdiction since Last Edition

First, the French law on industrial design rights has been substantially amended by the Order No. 2001-670 of 25 July 2001, transposing into French law the Directive No. 98/71/EC of 13 October 1998 in relation to the legal protection of designs.

In accordance with the Directive, the main changes in French law concern the substantive requirements (novelty and individual character), the duration (period of five years, which may be extended by periods of five years up to a maximum of twenty-five years) and the scope of protection of designs.

However, the French legislature has not transposed the exact wording of the Directive even if the same basic concepts are used. As the European Court of Justice ('ECJ') requires that the provisions of national law be interpreted as far as possible in the light of the wording and the purpose of the Directive, it is sometimes necessary to refer to the concepts of '*informed user*', '*normal course of business*' or to features '*dictated solely by a technical function*' and their interpretation by the ECJ.

As an extension of this Directive, the Council Regulation No. 6/2002 of 12 December 2001 on Community designs establishes a cumulative protection at the French and European level.

As to the litigation of designs, the provisions of the French Intellectual Property Code ('IPC') have been amended many times, in particular by the Law of 1 August 2003 organizing procedures for detention under customs control or by the 'Perben II' Law of 9 March 2004 on the strengthening of infringement sanctions.

But it is primarily the Law of 29 October 2007 on the 'fight against counterfeiting', transposing the Directive No. 2004/83/EC of 29 April 2004 on the enforcement of intellectual property rights, which has profoundly changed the relevant rules in this matter.

This law amends the regime of seizure for counterfeiting and creates an interim measures procedure as well as a right to be informed on the origin and distribution channels of the infringing products. It also clarifies the definition of acts that can be qualified as infringement, sets additional civil penalties (recall of products, confiscation, destruction, publication of the judgment) and gives exclusive jurisdiction to the 'Tribunaux de grande instance' (Regional High Courts).

Finally, the last reform on designs comes from the 'Yung' Law of 11 March 2014 strengthening the fight against counterfeiting. It has strengthened, in particular, the rules on compensation for loss due to counterfeiting by requiring judges to take into account the shortfall as well as the loss suffered by the victim. It also includes in the assessment the profits made by the infringer (cost savings of intellectual, material or promotional investments resulting from infringement conduct). In addition, the period of limitation for civil infringement proceedings increases from three to five years.

Statistically speaking, the practical relevance of industrial designs and models[1] in France is confirmed with 66,277 designs registered in France in 2013 (52,756 by the standard procedure - 13,521 by the simplified procedure).

However, aesthetic creations continue in many cases not to be filed due to protection by copyright, which is facilitated by a relatively flexible originality concept used by French courts. Nevertheless, the filing of a design and model is useful in terms of legal certainty (identification of protected creations and holding of a title with a certain date).[2]

Moreover, since the 2001 reform, autonomy of industrial design rights compared to copyrights increased significantly due to the adoption of distinctly different criteria of protection. The developments that follow as well as the specific part dedicated to the interaction with the copyright (see below §7.09[B]) testify to this evolution.

§7.02 CONVENTIONS AND LEGISLATION

At an international level, France is a party to several agreements that were reached to harmonize national legislations and facilitate obtaining protection in several States of the world.

The Hague Agreement of 6 November 1925 for the protection of industrial designs is applicable to nationals and residents of the Member State and to companies established there. It organizes an international registration of industrial designs but does not create a unitary title: design remains subject to the provisions of each Member State. The design must be filed at the World Intellectual Property Office ('WIPO') in

1. The semantic distinction between designs and models has no practical impact (designs in 2D and models in 3D).
2. The creator may always seek the protection of the provisions of Book I, but must prove prior rights in his or her creation by establishing the date of its creation, for example, by virtue of stamped registers delivered by the French National Institute of Intellectual Property ('INPI') or envelopes especially designed for the purpose and known as '*envelopes SOLEAU*'.

Geneva and the International Bureau verifies that the filing complies with the formal requirements laid down but does not examine the requirements for protection. By contrast, each national office of the designated countries may undertake a substantive examination and has the right to refuse, on its territory, the protection of a design within a certain time limit (six to twelve months) based only on the substantive examination. In France, in the absence of such a review procedure, the French judge will assess the validity of the French part of the international deposit. Since 2008, it is also possible to obtain under this international procedure an unregistered Community design.

The Paris Convention for the Defence of Industrial Property of 20 March 1883, administered by WIPO since 1967, provides in particular for a six-month priority right from the filing in one EU Member State to ask protection in other Member States. International deposit also sets this priority period running. This priority right has a strong practical interest since the deposit in France creates a right based on registration: prior disclosures destroying in principle the novelty are neutralized within this six-month period (see below §7.03[B][2])

The Agreement on Trade Related Intellectual Property Rights ('TRIPS'), appended to the Marrakesh Agreement in 1994, states, in addition to the most favoured nation clause, that Member States of the World Trade Organization ('WTO') must protect designs independently created that are new or original for a minimum of ten years (twenty-five years in France). Member States may provide that the design is not new or original if it does not significantly differ from known designs and may exclude designs essentially dictated by technical or functional considerations. Finally, the Article 26 section 1 of the agreement provides that the monopoly conferred by a registered design allows only to prohibit actions '*undertaken for commercial purposes*', limitation included in Article L. 513-6 of the IPC.

Regarding the Berne Convention on copyright of 9 September 1886, its practical contribution was limited by the 'Tod's' judgment of the ECJ on 30 June 2005. Article 2 section 7 of the Convention provides that advantageous treatment, namely the benefit of dual protection by the law of designs as well as copyright, is denied to the authors of a work whose country of origin is another Member State which only protects this work by industrial design rights. The Court found that this rule constitutes an indirect discrimination on grounds of nationality prohibited by the Article 12 EC.

At European level, the Council Regulation No. 6/2002 of 12 December 2001 on Community designs establishes two regimes for the protection of industrial designs on the entire territory of the EU: registered Community designs and models, but also unregistered Community designs and models. Indeed, it is possible to obtain, without registration, protection for a design for a period of three years as from the date on which it was first made available to the public within the Community. For protection on a longer term (maximum of twenty-five years), it will be necessary to undertake a filing at the Office for Harmonization in the Internal Market ('OHIM') or the French National Institute of Intellectual Property ('INPI'). This filing involves the carrying-out of a formal examination as well as a limited substantive examination consisting in verifying that the design meets the definition of a design as set out in Article 3 (a) of the

Community Design Regulation ('CDR') and is not contrary to public policy or to accepted principles of morality.

At the national level, the requirements for protection by the law of design rights are laid down in Articles L. 511-1 to L. 522-2 of the IPC for the legislative part and in Articles R. 511-1 to R. 523-6 for the regulatory part (Book V of the IPC). The legislative part of the code is mainly the result of the Order No. 2001-670 of 25 July 2001 and the Law of 29 October 2007 on 'the fight against counterfeiting', which respectively transpose into French law the Directive No. 98/71/EC of 13 October 1998 and the Directive No. 2004/83/EC of 29 April 2004 on the enforcement of intellectual property rights.

§7.03 SUBJECT MATTER

The purpose of the law of designs, which has been for a long time difficult to determine because of its closeness to copyright, has been clearly defined by the 2001 reform. The requirements for protection under the law of design rights are '*novelty*' and '*individual character*', as set out in the Article L. 511-2 of the IPC. These requirements are described in Articles L. 511-3 and L. 511-4, with the Article L. 511-5 providing further detail in the case of products consisting of a component part of a complex product.

[A] Definition/What Constitutes a Protectable Design?

The Article L. 511-1 of the IPC first specifies that the appearance of a product, or of part of a product, may be protected as a design, and that any industrial or handicraft item, packaging, get-up, graphic symbols and typographic typefaces, but excluding computer programs, are regarded as '*products*'.

[1] Shape, Configuration

The French law of designs protects the appearance of a product or part of a product which are characterized in particular by its lines, its colours, its outlines, its shape but also its texture or materials. These features can be those of the product itself or of its ornamentation.[3] As a consequence the notion of appearance is broadly defined because the list of elements likely to characterize it is only indicative.

The part of a complex product, however, must be visible during normal use of the product by the end user to be protectable. It excludes only visible parts during maintenance, service or repair.[4] This provision tends to promote competition between the manufacturers of spare parts especially in the automotive sector. It is not limited to this sector since the complex product within the meaning of the directive is a '*product composed of multiple parts that can be replaced to allow disassembly and reassembly of the product*'.

3. Article L. 511-1 of the IPC.
4. Article L. 511-5 of the IPC.

Compared to copyright, the scope of the design rights is also limited by the product definition of Article L. 511-1 of the IPC: '*any industrial or handicraft item, packaging, get-up, graphic symbols and typographic typefaces, but excluding computer programs, are regarded as products*'.

The products or parts of products must therefore have utility and functionality and are likely to be produced in several copies. The appearance of purely artistic creations with a unique copy is not protected. Moreover, the protection must cover the practical application of the design or model to a product: the protection is not about an idea, a '*genre*' (which consists of an ensemble or group of objects which possess common characteristics), a style or a product category. Registration may thus be cancelled or refused if the deposit does not show the clear appearance of the product.

Computer programs are not considered as products: so the user interfaces cannot be protected as industrial design and may be only protected by copyright. Conversely, the case law includes in the definition of the product the interior design of commercial premises.[5]

Finally, the designs that are contrary to public order and morality (appearance of objects of a pornographic nature or relating to the use of illegal substances) are not protected.[6]

[2] Ornamental or Functional

The ornamental or aesthetic character is not required by law. A registered design may be protected provided that it does not fall under the exclusions set out in Article L. 511-8 of the IPC: appearance whose characteristics are dictated solely by a technical function.

The main objective of this exclusion is to prevent to the protection for exclusively technical functions for a longer period than the one provided by patent law. The criterion used by case law is the multiplicity of possible forms: for the exclusion to be retained, it must be the only form capable of exercising this technical function. Conversely, the form that includes one or more features non-imposed by the technical function is protectable. The shape of a product incorporating a patented invention may thus constitute a protectable model if it is separable from the technical result.

The Paris Court of Appeal has held in a recent judgment concerning awnings models that '*the square form and the existence of several awnings are not dictated by solely technical or functional conditions; they result from arbitrary choices left to the discretion of the user who could opt for other forms such as rectangular or triangular*'.[7]

The appearance of a product whose exact form and dimension must necessarily be reproduced to be mechanically associated to another product under conditions enabling each product to perform its function ('assembly parts').

5. CA Paris, 23 May 2008 (Court of Appeal).
6. Article L. 511-7 of the IPC.
7. CA Paris, 5 pôle, Ch. 2nd, 11 Jan. 2013, No. 12/01304 (Court of Appeal).

The mechanical association between the products will be a connection, a contact or a placement inside or outside. This exclusion tends mainly to promote competition in the market of spare or assembly parts and therefore results in lower prices for consumers.

Article L. 511-8 provides, however, that a design or a model developed to allow multiple assembly or connections to products that are interchangeable in a modular set can be protected. The product part that is used for assembly may thus be protected if it meets the substantive requirements of the protection. This is the '*Lego clause*' which allows the protection of the shape of the brick of a building set.

[B] Originality/Novelty

The substantive conditions of protection depend on three criteria: novelty, individual character and disclosure.

Article L. 511-3 of the IPC defines the requirement of novelty:

> A design shall be considered to be new if, on the date of the filing of the application for registration or on the date of priority claimed, no identical design has been disclosed. Designs shall be deemed to be identical if their features differ only in immaterial details.

Article L. 511-4 of the IPC defines the requirement of individual character:

> A design has individual character if the overall visual impression it produces on the informed observer differs from that produced by any design disclosed before the date of the filing of the application for registration or before the date of priority claimed. In assessing individual character, the degree of freedom of the creator in developing the design shall be taken into consideration.

Article L. 511-6 of the IPC defines disclosure, as referred to in the preceding two articles:

> A design shall be deemed to have been disclosed if it has been made available to the public through publication, use, or by any other means. No disclosure has taken place if the design could not reasonably have become known, according to the normal course of business in the sector concerned, by professionals operating in the European Community, before the date of filing of an application for registration or before the date of priority claimed.

The above articles show that the novelty and individual character of a design will be the subject of a relative examination since the entire prior art cannot be usefully relied upon.

Moreover, in assessing individual character, the courts will have to assume the perspective of the *informed observer* and take into consideration the degree of freedom left to the creator in developing the design.

These conditions from the 2001 reform only apply to drawings filed after its entry into force.

[1] Test for Originality Generally

The provisions of Article L. 511-4 of the IPC define the individual character which is an entirely new condition as a result of the 2001 reform.

In fact, the French legislature adopted an *'own character'* condition while the European Directive 98/71 of 13 October 1998 speaks of *'individual character'*. With this difference in terminology, the French legislature has just wished to avoid confusion with the subjective condition of originality for copyright. Thus it has been held that the requirement of a creative effort is not required by Article L. 511-2 of the IPC.[8] However, the condition of *'own character'* and the condition of *'individual character'* convey the same meaning.

A design has its individual character, when *'the overall visual impression on the informed observer differs from that produced by any design disclosed before the filing date of the application for registration or the date of claimed priority'*.

The assessment is performed *globally* and disregards detail points. It involves comparing the design or model for which protection is sought with the design and model disclosed before the filing date of the application or the date of claimed priority (*for a definition of disclosure,* please see *below* §*7.03[B][2] Novelty*). Recital 13 of the EU Directive 98/71 of 13 October 1998 specifies that a clear difference for the informed user must exist between the design or model for which protection is sought and those previously disclosed. The impression of *'déjà vu'* destroys the individual character.

This overall visual impression must be limited to the perspective of an informed observer (*'informed use*r' within the meaning of the Directive). It was ruled that the informed observer means an observer with special vigilance, either because of his personal experience or his extensive knowledge of the sector. It can be identified depending on the class of product mentioned in the deposit act (see *below* §7.07[C]).

The condition of individual character is quite flexible since the degree of freedom left to the creator in achieving the design or model must also be taken into consideration.[9] This allows to take into account the constraints facing the creator (technical, commercial, regulatory and abundance of creation in the sector). A little margin for manoeuvre of the creator allows to retain that only slight differences are sufficient to create a different overall visual impression on the informed observer.

Lastly, the overall visual impression does not take into account the elements of the design unlikely to be protected because they are exclusively functional or not visible under normal conditions of use.[10]

In summary, the European case law describes a method in four steps for assessing the individual character that should be applied by the French judge because of the principle of interpretation in the light of the Directive:

- To determine the sector of products into which the design is intended to be incorporated.

8. CA Paris, 10 Apr. 2009, No. 07/08596 (Court of Appeal).
9. Article L. 511-4 subpara. 2 of the IPC.
10. Cass. Com, 28 Nov. 2006, No. 04-18154 (French Supreme Court).

- To define the informed user of such products according to their purpose and determine, in reference to this informed user, his degree of knowledge of the prior art and the level of attention he is likely to display.
- To determine the degree of freedom of the creator in developing the design.
- To determine the result of the comparison of the designs in question in the light of the concerned sector, the degree of freedom of the creator and overall impression produced on the informed user by the contested design and any prior design disclosed to the public.

[2] *Novelty*

With the 2001 reform, an objective novelty requirement was created that clearly differs from the criterion of originality (necessarily subjective) applicable only to copyright. Pursuant to Article L. 511-3 of the IPC, a design is regarded as new if on the date of filing of the application for registration or the date of the claimed priority, no identical design has been disclosed. It follows that a design or model may be new without having its own character.

One must therefore compare the design or model, the protection of which is sought with those disclosed previously and which form a state of the art similar to patent law. Two drawings and two industrial models are considered identical if their features differ only in immaterial details. The judge's discretion must be applied '*in a comprehensive manner, depending on the overall appearance produced by the combination of the characteristic elements and not by examining each of them separately*'.[11] Therefore, the appearance of a product consisting of the combination of elements which cannot be individually protected since disclosed or functional can be new.

For example, the Paris High Court of First Instance ('Tribunal de Grande Instance') stated about an umbrella, that '*the particular combination of thin horizontal stripes going around the umbrella and the design of a small zipper on one of the sides, which is not an immaterial detail, is not found on any of the previously designed umbrellas so that the model (...) actually appears as new*'.[12]

There will be disclosure destroying novelty when the design or model has been made available to the public '*by publication, use or any other means*',[13] which includes:

- Prior Use.
- Prior Publication.
- Prior Application or Registration.

[a] *Prior Use*

Prior use of the design or model can be considered as disclosure and therefore shall destroy novelty.

11. CA Paris, 8 Nov. 2006, No. 05/20938 (Court of Appeal).
12. TGI Paris, 10 Oct. 2013 (High Court of First Instance).
13. Article L. 511-6 of the IPC.

[b] Prior Publication

This may be the disclosure of the design or model through publication of a design previously created by a third party.

Public disclosure is specific. Indeed, disclosure is only destructive of novelty if, prior to the date of filing for registration or the date of the claimed priority, design or model could have been reasonably known in the normal course of business in the sector concerned, by professionals operating in the European Community. These concepts are relatively vague and are an important source of litigation. The Office for the Harmonisation of the Internal Market (OHIM) has retained the disclosure for the official publication of an American model available on online databases since '*in the normal course of business, we should expect that the circles specialized in the field of motors and acting in the EC will keep themselves abreast of designs registered by their competitors, at least in the most important countries, like the United States*'.[14] Evidence of a disclosure localized in the European Community implies presumption of knowledge of the latter by professionals.[15] Article L. 511-6 of the IPC thus reduces the scope of the prior state of applied art opposable to registration: a disclosure which is little-known or forgotten by the circles specialized in the sector operating in the European Union (EU) cannot be enforceable against a registration in France.

The disclosure to third parties expressly or impliedly held secret will not be taken into consideration.[16] A grace period is also planned. A disclosure that occurred within twelve months before the date of application filing or the date of claimed priority will not be taken into account:

- When the disclosures are attributable to the creator, his successor or a third party based on information provided or acts performed by them.
- Where the disclosure took place following abusive behaviour towards the designer or his successor.

[c] Prior Application or Registration

This may be the disclosure of the design for which protection is sought or of a design previously created by a third party.

§7.04 DURATION OF PROTECTION

The protection by design rights becomes effective as from the filing of the registration application, for a period of *five years*, which may be extended by periods of five years up to a *maximum of twenty-five years*.[17]

14. OHIM, 30 Aug. 2006.
15. OHIM, 7 Jul. 2008.
16. Article L. 511-6 subpara. 2 of the IPC.
17. Article L. 513-1 of the IPC.

Prior to the entry into force of Order No. 2001-670 of 25 July 2001, the period of protection conferred by a design was twenty-five years from filing, which could be extended once for an identical period. To resolve the problems of transition in the law posed by the reduction of the period of protection, paragraph 2 of Article L. 513-1 of the IPC provides that designs that were filed before 1 October 2001 '*shall remain protected, without any extension being possible, for a period of twenty-five years from their date of filing*', specifying that those whose protection was, prior to 1 October 2001, given an extension of twenty-five years, shall remain protected until the expiry of this period.

Any extension shall take effect from the day following the expiry of the registration date. The first extension may be requested during the filing. Otherwise, the extension is the result of a statement with the French National Institute of Intellectual Property (INPI) presented during a six-month period ending on the last day of the month in which ends the protection period. The extension declaration must include evidence of fee payment[18] and must come from the owner entered in the Public register (known as National Designs Register) on the day of the statement or his representative.[19] As a consequence, if the sale of designs was not recorded on the public register, the transferee will not be able to request the extension.

However, against the payment of an additional fee, statement may be made or the fee paid within a further period of six months from the day following the last day of protection expiry. In addition, Article L. 512-3 provides a *notice of lapse* if the holder can provide a legitimate excuse. No inadmissibility of the extension request can be pronounced without the holder of the right to be invited to submit observations.

At European level, Article 12 of the Council Regulation No. 6/2002 on the Community design provides a period of protection for registered designs which is identical to that provided for French designs. In the terms of Article 11 paragraph 1 of the above-mentioned Council Regulation, the unregistered Community design is protected for a period of three years starting from the date on which it was first made available to the public.

§7.05 INFRINGEMENT

The right on a registered design is defended by an action in civil or criminal infringement. The legal provisions are similar to those provided for trademark and patent since the infringement actions system has been unified by the Law of 29 October 2007.

[A] Scope of Infringement Action

The registration of a design shall confer on the right holder the exclusive right to exploit it.[20] Indeed, the rights attached to the filing shall be binding on third parties starting

18. Article R. 513-2 of the IPC.
19. Article R. 513-1 2° of the IPC.
20. Article L. 513-2 of the IPC.

from publication. However, the right holder of the design or model is not obliged to use it as in trademark matters (no revocation for non-use), and has not to pay fees either to the French National Institute of Intellectual Property as in patent matters.

The rules on the scope of protection were amended by the 2001 reform and apply when the facts at issue were committed after its entry into force.

Article L. 513-4 of the IPC lists the acts, which are deemed to be an infringement. Except with the consent of the holder, the following acts are prohibited when they concern a product incorporating a registered design:

- Manufacturing.
- Offering.
- Placing on the market.
- Import and export.
- Use.
- Holding for these purposes.

The holder of the design may object to these acts when they concern an identical design. In addition, this protection extends to *any design which does not produce a different overall visual impression on the informed observer*[21] (see *above* §*7.03[B][2]*). The objective is to protect the essential function of the design rights that is to protect the appearance against unauthorized use.

Regarding internal transit of goods (goods in transit originate from one EU Member State and thus put into free circulation), it appears that the solution adopted for trademarks[22] could be transposed to the design matter. The transit will be an act of infringement of the design registered in the State of transit if the goods are intended to be marketed in the EU. It will not be the case if the Community goods in transit are intended to be marketed in a Third State.

Regarding the external transit (the goods come from a Third State and are not put into free circulation within the EU), the goods must be considered as not imported into the State of transit, although the manufacturing and marketing are illegal respectively, in the Third State of origin and in the State of destination (EU Member or Third State).

Concerning acts of offering for sale and marketing made on or from a foreign website, they are likely to constitute infringement of a French design only if they can be regarded as located in France. Therefore the site must target the public in France within the meaning of the Hugo Boss ruling on trademarks[23] the solution of which has been implemented into the law of designs. It has been considered concerning sale offers on a Monaco website that:

> the disputed site contained no warning on the impossibility of buying the displayed products in France, it was written in French and the product price was given in French francs; that granting access to the site in English, the plaintiff intention was probably to interact with consumers established in several states, without excluding French consumers; that under these conditions, the website

21. Article L. 513-5 of the IPC.
22. CJCE, 18 Oct. 2005, Class International (ECJ).
23. Cass. Com., 11 Jan. 2005 (French Supreme Court).

was targeting the French public; that the counterfeiting acts in France are characterized.[24]

By way of exception, a number of acts are permitted by Article L. 513-6 of the IPC:

- Acts done privately and for non-commercial purposes:
  Acts performed 'off trade' thus avoid the complaint of infringement on two grounds. First, if the conditions of Article L. 513-6 of the Code of intellectual property are met and secondly, pursuant to Article 26 section 1 of the TRIPS which provide there can be no right infringement if the acts are not 'undertaken for commercial purposes.'
- Acts done for experimental purposes.
- Acts of reproduction for purposes of illustration or teaching on condition that the filing and the name of the rights holder are mentioned, conform to fair trade practices[25] and does not impact the normal design or model commercial operation.[26]

The Paris High Court of First Instance (TGI) has recently commented on this exception concerning the use of design as a visual in an advertising campaign dedicated to hair products.[27] According to the Court, the actual use of the design did not answer the definition of acts done privately and in the case of an advertisement, the acts could not be considered as made '*for non-commercial purposes*'.

Article L. 513-7 of the IPC also provides exceptions whose practical relevance is limited:

- The rights to the designs are not enforceable on the equipment installed on vessels and aircraft registered in another country when these temporarily enter French territory.
- These rights are not enforceable on spare parts and accessories imported on French territory for the repair of these vessels and aircraft.

Article L. 513-8 of the IPC finally recalls the principle of exhaustion of rights on products to which the design is applied or that incorporate a model. The national monopoly conferred by a design or model cannot be opposed to the free movement of property, goods and services within the EU, provided that the right holder has agreed to the marketing of its products in the European Economic Area.[28]

If the design meets the originality requirement of copyright, Article L. 122-5 of the IPC also permits, once the work is disclosed:

- Copies and reproductions strictly reserved for private use and not intended for public use.

24. CA Paris, 23 Mar. 2007 (Court of Appeal).
25. Requirement borrowed from the concepts of *fair use* and *fair dealing* in Anglo Saxon law.
26. The design may not be invoked if the acts category of the disputed act may cause serious harm to the economic interests of the incumbents.
27. TGI Paris, Ch. 3rd, s. 4th, 10 Oct. 2013 (High Court of First Instance).
28. See the ECJ case law on the exhaustion of rights.

- Analysis and short quotations justified by the critical, polemic, educational, scientific or information of the work in which they are incorporated.
- Representation or reproduction of excerpts from works for the sole purpose of representation for teaching and research purposes to an audience of concerned students or researchers; provided the name of the author and the source are clearly indicated (such representation or reproduction entails the payment of a lump sum).

Infringement proceedings are also limited by the period during which the infringement action may be exercised:

- Only acts after the publication of the registration can be sued for infringement. However, acts subsequent to the notification of the application for registration can be prosecuted.
- Since the 'Yung' Law of 11 March 2014, statute of limitations for civil and criminal proceedings is within five years from the disputed facts (instead of three years).

[B] Test for Infringement/Criteria to Constitute Infringement

The right holder is protected against exploitation of products incorporating either the identical design or a similar design. Article L. 513-5 of the IPC adopts the definition of the condition of individual character: under L. 513-4 of the IPC, acts constitute infringement when they concern a product incorporating a design or a model that produces on the informed observer a similar overall visual impression.

In the civil action, the owner does not have to show that the defendant copied the design but only that the overall visual impression generated by the disputed design is the same as the registered design. Thus, the infringer can be held liable even if he acted with carelessness or negligence. For the criminal action, the material element consists of the acts defined in Articles L. 513-4 to L. 513-8 of the IPC. However, an element of intent must be demonstrated that is to say that the act is done knowingly.[29]

The use of the definition of individual character to qualify infringement involves reference to comments already made about it (*Please* see *above* §7.03[B][1]).

In practice, the judge must only appreciate infringement by taking the perspective of an informed observer to compare the overall visual impressions produced by the conflict designs. He must thus identify the protected characteristics and seek whether they appear in the allegedly infringing product such as to create the same overall visual impression to the informed observer, whatever the differences between the products may be. The high degree of freedom left to the creator in the sector concerned will make it easier for the judge to admit an identity of overall visual impressions and thus infringement.

29. Article L. 521-11 of the IPC.

The Paris Court of Appeal recently restated that infringement is assessed by the similarities.[30] In this case, the court held that the only difference between the two models consisted in a detail difference that did not allow the complaint of infringement to be dismissed.

Moreover, the criterion of confusion likelihood in the area of Trademark Law was clearly excluded by design law in a judgment of 26 March 2008[31] holding that '*having found that the contested goods reproduced the essential features of the registered design and created the same overall impression, the Court of Appeal, which was not bound to conduct an unnecessary research on the potential likelihood of confusion, has, on the sole basis of these findings, accurately deduced the existence of an infringement*'. In addition, the reference to the average consumer of Trademark Law has also been ruled out since '*the criteria for infringement by imitation of model, which does not include the risk of confusion, is appreciated in view of the informed observer and not of the average consumer*'.[32]

Finally, the scope of the exclusive right is determined solely by reference to the drawing or photograph contained in the deposit and annexed to the identity certificate of the design. Therefore, the shape features that do not appear or are not sufficiently visible in the reproduction integrated to the identity certificate of the design cannot be enforced against third parties.

[C] Standing to Sue

For the civil action before a civil court or attached to a criminal action, person who may take action for infringement:

- The right holder of the filing including the assignee for acts subsequent to its registration in the National Designs Register.
- The holder of an exclusive license, if the contract does not prohibit it and after a formal demand that remained unsuccessful to the right holder.

The other licensees may join infringement proceedings to seek compensation for their prejudices, even if the license has not been entered in the National Designs Register.

For the criminal action: the public prosecutor or the victim exercising civil action under the terms of the Code of Criminal Procedure may act.

Pursuant to Article L. 511-9, the author of the application for registration is regarded prima facie as the beneficiary of the protection, that is to say, as the creator or his successor. The filing creates a rebuttable presumption of right ownership. A third party may therefore reverse the presumption by showing that he had already made the creation at that time, even if it was not disclosed.

30. CA Paris, 5 pôle, Ch. 1st, 10 Apr. 2013, No. 11/12675 (Court of Appeal).
31. Cass Com, 26 Mar. 2008, No. 06-22013 (French Supreme Court).
32. Cass Com, 23 Sep. 2008, No. 07-15053 (French Supreme Court).

The issue of ownership of rights for foreigners is provided for under Article L. 511-11: foreigners, nationals, domiciled or established in a Member State of the Paris Convention or the WTO have the same rights as the French (national treatment). It is the same for foreigners domiciled or established in a Member State of the EU or the European Economic Area. Other foreigners may also protect their industrial design provided that their country grants reciprocity to the French.

Otherwise, Book V does not contain any provision on employees' creations. One should reason by analogy with copyright, which provides that the work does not belong to the employer simply because of the existence of an employment contract.[33] The employer must therefore acquire the work to the employee after creation. However, if the work is collective, it belongs to the natural or legal person under whose name it was disclosed. Nevertheless, French courts admit that without a claim by the individual who did the drawing or model, a company that has disclosed and exploited them enjoys a presumption of acquisition.

In terms of subject matter jurisdiction, civil action for infringement falls within the jurisdiction of the Regional High Courts ('TGI'),[34] while the criminal action falls within the jurisdiction of the local criminal courts. For territorial jurisdiction, the court is determined by the ordinary rules of competence: the applicant has the choice between the courts of the place where the defendant is domiciled or place of infringement or the place where the damage occurred. In civil cases, the competent TGI are exhaustively designated by regulation: Bordeaux, Fort de France, Lille, Lyon, Marseille, Nancy Nanterre, Paris and Rennes, but also Strasbourg since the 'Yung' Law of 11 March 2014. They are also responsible for related unfair competition actions (Please see *below* §7.09[C]).[35]

[D] Remedies for Infringement

As a result of recent legislation dated 11 March 2014, the remedies in respect of infringement have been increased. The IPC law provides for the penalties as detailed below.

Penalties differ depending on whether acting before civil or criminal courts (a. and b.). Investigative and interim measures can also be taken to prove acts of infringement (c.).

[1] The Sanctions of Civil Action for Infringement

Damages[36]

The assessment of damages is at the discretion of the judge.

33. Article L. 111-1 subpara. 3 of the IPC.
34. Article L. 521-3-1 of the IPC.
35. Article 46 of the French Civil Procedure Code.
36. Article L. 521-7 of the IPC.

The court must consider the negative economic impact of infringement for the injured party including the shortfall, the material and non-material loss suffered by the victim and the profits made by the infringer (cost savings of intellectual, material or promotional investments resulting from infringement).

As an alternative, on request of the injured party, the court may award a lump sum which may not be less than the fees that the infringer would have paid if he had asked permission to use the design or the model. It does not exclude compensation for non-material loss.

Withdrawal of Counterfeited Products and Counterfeit Instruments[37]

At the request of the injured party, the court may order the removal from distribution channels of infringing products, materials and instruments that were mainly used for their creation or manufacture. The court may also order their destruction or forfeiture in favour of the injured party.

The infringer is bound to pay the costs of confiscating the infringing articles and the costs of any equipment installed with a view to the illegal reproduction of the article and all copies. The confiscated objects must be given back to the victim of the infringement.

Publication

The court is empowered to order any appropriate measure to advertise the judgment (display of the judgment or an extract notably on the infringer's website, publications in newspapers ...). This measure is ordered at the expense of the infringer.

[2] The Sanctions of Criminal Action for Infringement

Fine and Imprisonment[38]

> The infringer shall incur a prison sentence of up to three years and a EUR 300,000 fine. The penalties are increased to five years and EUR 500,000 when the offence is committed by an organized group or where the facts relate to dangerous goods for health, human or animal safety. The penalties are doubled in the event of a repeat offence or if the infringer is or has been bound by an agreement with the injured party.[39]

Additional Penalties

If a company is held criminally liable, the fines are five times greater. Both the civil and criminal courts can further order that the establishment which committed the

37. Article L. 521-8 of the IPC.
38. Article L. 521-10 of the IPC.
39. Article L. 521-13 of the IPC.

infringement be closed, either definitively or temporarily for a duration not exceeding five years.

When the closure is final, it results in the payment of additional compensation for dismissed employees. Temporary closure may not result in the termination or suspension of the employment contract, nor any financial loss to employees.

Natural or legal persons criminally responsible can be sentenced, at their expense, to remove the infringing trade channels and all objects that were used to commit the offence. The court may order the destruction or delivery to the injured part of these objects[40] and order the publication or display of judgment.[41]

[3] Proving Infringement

Counterfeiting is proved by any means (Article L. 521-4 of the IPC). Many measures can be taken to prove acts of infringement, as elaborated below.

The Procedure for Seizure

A counterfeit seizure procedure allows any person entitled to exercise the infringement action to proceed to seizure of allegedly infringing articles and related documents.[42] It is authorized by an order issued on request of a competent jurisdiction and performed by a bailiff assisted by experts appointed by the applicant. The seizure can be an actual or a 'descriptive' seizure; the latter being a procedure whereby the goods seized are left in the hands of the infringer but their existence and nature are described in an official statement by the bailiff. In the case of an actual seizure, the bailiff describes the goods in his or her written statement and also leaves the goods with the clerk of the court. The infringer is therefore stripped of the infringing goods.

The implementation of these measures may be subject to the provision of guarantees to ensure compensation of the defendant if the infringement is not subsequently recognized or if the infringement seizure is cancelled. The applicant has a period of twenty working days or thirty-one days (whichever is longer) from seizure to refer the matter to the courts. If this time period is exceeded, the seizure is deemed null and void. The order of the President of the Court is subject to appeal.

Cancellation is declared at the request of the seized who does not need to justify itself. The 'Yung' Law of 11 March 2014 enhances the effectiveness of the counterfeit seizure procedure by allowing the bailiff to seize documents in the absence of allegedly infringing products at the place of seizure, where there is sufficient evidence.

The case law specifies when the infringement seizure must take place.[43] The Paris Court of Appeal requires the bailiff to indicate in the minutes of the infringement seizure the time at which the notification of the order took place, and the time at which seizure operations began. If these data are missing, the minutes do not establish if the

40. Articles L. 521-11 and L. 521-12 subparas 1 and 2 of the IPC.
41. Article L. 521-4 of the IPC.
42. Article L. 521-4 of the IPC.
43. CA Paris, pôle 5-1, 13 Feb. 2013, no. 11/14124 (Court of Appeal).

seized party could 'have a reasonable time to understand the scope of the powers of the bailiff and scope of his rights'.

The Paris Court of Appeal has also clarified the notion of 'reasonable time'.[44] In this case of infringement seizure in a shop, the minutes of notification of the order were written at 16h25, while the minutes of infringement seizure mentioned the start of entry operations at 16h35. The seized company disputes the validity of the seizure of counterfeited goods, arguing that the ten minutes that the saleswoman (on which the act was notified) disposed was not sufficient. But the Court considers that a period of ten minutes between the notification of the order and the entry operation is 'exclusive of the notion of immediacy invoked and constitutes a reasonable time to do this'. The court also notes that the infringement seizure is valid when it is served on a person that is not representative of the company but who declared being empowered, since it is accepted that the bailiff has no obligation to verify the accuracy of this statement.

Right to Be Informed

To determine the origin and distribution channels of the infringing products, it is possible to ask the court to order, possibly subject to a fine, the communication of information by the defendant, by persons found in possession of infringing goods, by suppliers of service used in the infringing activity and any person involved in the production or distribution of these products.[45] Nevertheless, they can claim a lawful impediment (especially banking secrecy, medical confidentiality but excluding trade secret).

Since the Law of 11 March 2014, the communication of documents or information held by the defendant or by third parties may be requested as well an action on the merits or a summary hearing. This reform has also extended the scope of information and documents whose disclosure is likely to be ordered by the judge by repealing the provisions of the Intellectual Property Code which listed the information and documents concerned (names and addresses of the producers, manufacturers, distributors, wholesalers, quantities produced and sold the goods in question and the price ...).

The Supreme Court had the opportunity to specify the information procedure, describing it as the adversarial procedure '*to order the defendant to produce information and elements, of commercial or accounting nature, which could enable the right holder, which in parallel supplied evidence of the alleged infringement, to determine the origin and extent of infringement and improve his demands*'.[46]

Interim Measures[47]

The applicant for such measures can be anyone who can sue for infringement and must apply to the competent court in a summary manner. However, the measures can

44. CA Paris, pôle 5-2, 8 Feb. 2013, no. 11/02407 (Court of Appeal).
45. Article L. 521-5 of the IPC.
46. Cass. com., 8 Oct. 2013, no. 12-23.349 (French Supreme Court).
47. Article L. 521-6 of the IPC.

be ordered on request when circumstances require that they are not taken contradictorily (e.g., A delay would cause irreparable loss to the applicant). The applicant must provide evidence that are reasonably available to him, demonstrating that his right is being infringed or that such infringement is imminent. The court may order, possibly subject to a fine, all measures to prevent an imminent or actual infringement of the rights conferred by the registration. Thus, the court may prohibit alleged infringing acts or make such continuation subject to a provision of guarantees to compensate the plaintiff.

To prevent the introduction or movement of allegedly infringing products, the court may also order the seizure and placement in the custody of a third party. If the applicant demonstrates that the circumstances are such as to jeopardize the potential damages recovery, the court may order the preventive seizure measures of the property of the applicant, including his bank accounts (it may require the disclosure of information necessary to operate them). The court may grant a provision to the applicant if his prejudice does not appear seriously questionable. The implementation of these measures may be subject to the provision of security by the applicant for indemnification of the defendant if the infringement was not recognized or were cancelled. If the measures are ordered before the exercise of the action on merits, the applicant must take civil or criminal infringement proceedings, within the same period as for the infringement seizure. Otherwise, the action will be cancelled at the request of the defendant, without the need to justify his request (damages can be also awarded to the defendant).

Seizure by Customs' Authorities

The proprietor of a registered design or the beneficiary of an exclusive right of exploitation can, in writing, require the customs' authorities to retain under their control, any goods which appear to constitute an infringement. The authorities then immediately inform the head of the prosecution department of the 'Tribunal de Grande Instance', the plaintiff and the owner of the goods that such goods are being held. Within ten days from the date that the goods are seized, the plaintiff must formalize the seizure of the infringed goods on application to the President of the Tribunal de Grande Instance or request that the President fix the amount of bail intended to cover any liability of the applicant in the event of the infringement not being proven. Obviously, the applicant may obtain from the Customs authorities the names and addresses of the exporter, importer and destinee of the goods. If within this ten-day period no authorization has been given by the President of the Tribunal de Grande Instance, the goods will be directly returned to their destinee.

Apart from any application, the customs authorities may also detain goods likely to infringe a protected design. The detention will be waived if within three working days from its notification, the right holder or beneficiary of the exclusive right of exploitation does not file a request for it.

The last reform of 11 March 2014 reinforces the powers of the customs' authorities especially in view of these detentions. Indeed, this law extends customs competence to carry out infiltrations and 'coups d'achat' to the whole of the infringing goods (acquisition of a certain amount of products suspected of being counterfeit to see if infringement is proven or not).

Intervention by the Judicial Police Officers[48]

Judicial police officers finding an infringement can seize products manufactured, imported, held, sold, delivered or supplied illegally as well as equipment or instruments specially installed for the purpose of such acts.

§7.06 DEFENCES TO INFRINGEMENT

The exception of nullity is the classic defence against the infringement action. The 2001 reform inserted in the IPC a new section called '*invalidity of a registration*' that brings together the grounds for invalidity of a protected design. These provisions apply to designs filed after its entry into force.

[A] Validity

The defendant for infringement, who is seeking the cancellation of the registered design as a counterclaim, has to demonstrate the existence of one of the causes of invalidity referred to in Article L. 512-4 of the IPC. The request shall be within the exclusive jurisdiction of the Regional High Court (TGI) referred by decree (Please see *above §7.05[C] 'Standing to sue').*

The first ground of cancellation provided for in Article L. 512-4 subparagraph (a) concerns the design that does not comply with the requirements of Articles L. 511-1 to L. 511-8 of the Intellectual Property Code specifying the substantive requirements of protection for designs. Any interested party may request cancellation for lack of novelty, individual character, compliance with public order or morality, the invisibility of the spare part or the exclusively functional feature (Please see *above §7.03 Subject matter*).

In other cases referred to in Article L. 512-4 (subparagraphs (b) to (e)), only the holder of the right opposed can act. However, the Public Prosecutor may initiate ex officio a request for cancellation whatever the reason may be.

Cancellation may be partial since the 2001 reform provided that the appearance of the product is not changed significantly. Article L. 512-5 stipulates that:

> if the grounds for cancellation affect the design or a model only in part, the registration can be maintained in a modified form provided that in this form, drawing or model meets the criteria for granting protection and that its identity is preserved.

Article L. 512-6 provides that the judicial decision of nullity shall be absolute, that is to say, it is enforceable against third parties.

48. Article L. 521-9 of the IPC.

Where a third party succeeds in defending an action brought against it on the basis of a filing by demonstrating that the filed design does not in fact fulfil the requisite criteria to be protected, such a decision has both an effect on third parties and on the filing itself.

The decision is listed in the National Designs Register at the request of the clerk or one of the parties[49] and is published in the Official Bulletin of Industrial Property ('Bulletin Officiel de la Propriété Intellectuelle' BOPI).

[1] Not Novel/Original

It is up to the defendant for infringement contesting novelty to prove the contents and date of disclosure it invokes. Case law requires '*precise prior art with precise dates*' for the novelty destroying disclosure to be retained. It is also up to the person challenging the validity to demonstrate that the conditions of Article L. 511-5 are met and justify the protection, namely: that the part is always incorporated into a complex product and is invisible in normal use of the product by the end user.

Conversely, it is for the holder of the right whose validity is challenged, to establish that the prior art that is opposed to it could not reasonably be known, according to the normal business practice in the area concerned by professionals acting in the EU ('*informed user*' see above).

[2] Functional/Not Ornamental

The registered design being presumed valid, it belongs to the third party requesting the nullity to demonstrate the existence of an exclusively functional feature. It is therefore necessary to apply the criterion of multiple forms (Please see *above* §*7.03[A]2]*).

[3] Other Possible Criteria, Such as Non-obviousness

Article L. 512-4 (c) provides that registration may be cancelled if it infringes the rights attached to a drawing or national or Community design registered beforehand, but disclosed by publication after the date of filing or date of priority of the model whose validity is challenged.

Registration is also void if the design infringes the copyright of a third party (L. 512-4 (d)). This provision is justified by the fact that the unity of art provides only for possible accumulation of protections. Whether the work has been disclosed or not, it may be opposed by the rights holder to a later filing. If the work has been disclosed so as to be made reasonably available to industry professionals operating in the EU, the deposit will be met with a lack of novelty or specific feature likely to be raised by any interested party (L. 512- 4 (a)).

49. Article R. 512-4 of the IPC.

[4] Faults in Application

Article L. 512-4 (b) covers cases where the owner could not benefit from the expected protection because he had no right to registration. Indeed, Article L. 511-9 provides that the right to protection is granted to the designer or his successor and only a rebuttable presumption is established in favour of the applicant. The creator or his successor, as victims of abuse, may thus apply for cancellation of a registration obtained by a third party (*fraus omnia corrumpit*). Therefore the infringement defendant cannot pretend, except by claiming to be the actual creator, that the registration is invalid on the ground that the holder is not the creator or his successor.

In addition to this ability to obtain the cancellation, the law provides for an opportunity to claim the property attached to a registration. Article L. 511-10 provides that '*if a design has been filed in fraud of a third party or in violation of a statutory or contractual obligation, the person who claims an interest in the design may claim legal ownership in judicial proceedings*'. This action is prescribed three years after the publication of the design registration. However, the limitation starting point is the expiry of the protection period in case of bad faith at the time of the registration publication or design acquisition.

[B] Marking

Article L. 511-4 (e) provides that registration is void if the design includes a prior protected distinctive sign, without the authorization of the holder. (Please see *below* §*7.09[A]*)

[C] Non-infringement

To demonstrate the non-infringement, several arguments can be brought: the lack of identity of overall visual impressions of disputed drawings or models, the acts of exploitation by the alleged infringer are not included in the list Article L. 513-4 of the IPC or are excluded from protection since they are not undertaken for commercial purposes, statute of limitations (five years).

[D] Other Defences

The infringement defendant may invoke his good faith, namely his ignorance as to infringing an exclusive right. The defendant will thus avoid criminal penalties, but will incur civil liability once the counterfeit material element is established.

§7.07 PROCEDURES FOR FILING APPLICATION FOR REGISTRATION

To enjoy design right protection in France, it is necessary to register the design (except in the specific case of the unregistered Community design). Indeed, Article L. 511-9 of the IPC states that only designs and models which have been duly filed enjoy the benefit of the Book V of the IPC from the filing date of the application or from the priority claimed.

There is no time period within which a filing must be made, where the design has not already been commercially exploited by the author. However, the author of the design or model should make his or her entire filing as soon as possible after the work has been created to avoid novelty bars.

It is no longer possible to consider a deposit after a market testing of the creation because disclosure would be destructive of novelty. In the absence of deposit, it will however be possible to invoke the grace periods, the protection by copyright or the protection by unregistered Community design.

The filing must be carried out either at the French National Institute of Intellectual Property, for authors domiciled in Paris or abroad or at the Clerks' Office of the Commercial Court within the territorial jurisdiction in which the author resides.[50]

The deposit may result from sending a registered letter with acknowledgment of receipt or use of any means of remote transmission defined by the director of the INPI. The filing date is then the date of receipt by the INPI (last paragraph Article R. 512-1 of the IPC).

A valid filing must state the identity of the beneficiary of the filing, the number of designs or models filed, a description of the designs or models and whether or not confidentiality is requested. It must also state whether or not the application includes any claim of priority, include a graphic or photographic reproduction of the design, the payment of a filing fee and a power of attorney if the filing is not made directly by the author.

The agent must have domicile, seat or establishment in the EU or in a country belonging to the European Economic Area (EEA). People who do not have their home or office in a State of the EU or EEA shall present such a representative within a period specified by the INPI (Article R. 512-1 and R. 512-2 of the IPC). The creation of a common agent, meeting the same criteria, is mandatory in case of multiple applicants.

The registers are national and all matters which, after the filing, may affect the ownership or the existence of the filing and the identity of the holder (in other words, any transfer, license, total or partial waiver, change of name or address and so forth) are entered on them.

Once the design has been filed the contents of the filing are published by the INPI in the Official Bulletin of Industrial Property (Bulletin Officiel de la Propriété Industrielle: BOPI) (R. 512-10 of the IPC) after a short period of technical preparation, unless the applicant requests that the contents remain confidential. This means that while the existence of the filing will be public, its contents will remain confidential both from the public and the public authorities. A request for confidentiality may not be made unless it is in respect of all the objects which are the subject of the filing. In addition,

50. Article L. 512-1 of the IPC.

publication of a filing is obligatory in order to avail oneself of its protection against a third party and bring an action for infringement. In this respect it is possible for the holder of the filing to subsequently request that it is made public by the release to the general public of a copy of the object in question. The contents of the filing may not remain confidential for more than three years (*compare below* §*7.08 Varying Registration*).

Once published, the deposit is entered in the National Design Register mentioning the identity of the holder (R. 512-13 of the IPC) and is presumed valid.

The INPI does not perform a substantive examination and the registration cancellation is obtained *ex post* by courts.[51]

The director of the INPI only has the power to reject any filing which is not presented in the appropriate form and manner. This power of rejection can be extended on moral grounds or where the publication would be contrary to public policy (R. 512-8 of the IPC). Such a rejection may only be made however, after the party making the filing has been given the opportunity to rectify the filing or make any comments on it that he feels appropriate (R. 512-9 of the IPC).

If he supplies a legitimate excuse,[52] the applicant's revocation which has incurred for missing a deadline may be lifted. The application must be filed within two months from the end of the impediment. The omitted act must be completed within this period. However, the revocation lifting request is not allowed after the lapse of a six-month set period from the expiry of the period that was not observed (R. 512-12). Any decision may be challenged on the merits before a competent Court of Appeal within a period of one month from the notification (L. 411-4 and R. 411-19 of the IPC).

[A] Priority

Article R. 512-5 deals with claims of priority at the time of filing and is the result of Article 4 of the Paris Union Convention which provides a six-month priority right from the deposit in an EU Member State to seek protection in other Member States. The international deposit also triggers the priority period. This priority right has a strong practical advantage: prior disclosures prior to filing and in principle destructive of novelty are neutralized within this six-month period.

[B] Drawings

The drawing or photograph contained in the deposit is annexed to the identity certificate of the design or model. It determines the scope of the exclusive right in case of infringement. This requires that all shape or colour features appear and are sufficiently noticeable.

51. Article L. 512-4 of the IPC.
52. Article L. 512-3 of the IPC.

[C] Design Part of Another Article/Environment of Design

Designs are not subject to the principle of specialty: the appearance is protected regardless of the product that incorporates it. However, Article R. 512-3 (e) requires the filing to specify the common name of the product in which the design is intended to be applied. The class of products covered in the deposit thus allows the court to determine the *informed observer* which serves as an assessment criterion of the conditions for protection.

[D] Description of Design (Where Required)

The description or drawing or photograph caption contained in the filing is optional and carries no legal value. It has been held that '*the filing of a model protects the model as filed and not just what is said in its caption, which is merely an informative complement*'.[53]

§7.08 EXPUNGING/CANCELLING OR VARYING REGISTRATION

[A] Expunging/Cancelling

There is no revocation action for failure to use designs. The title cancellation and removal from the National Register of Designs will necessarily be the result of:

- A court decision on cancellation.
- A full or partial waiver of his rights by the holder at any time upon a simple declaration to the French National Institute of Intellectual Property (INPI).
- Expiry of the validity period (twenty-five years) or non-renewal.

[B] Varying Registration

There are also two methods of filing in order to obtain the registration.

(1) The *standard filing* with publication or (optional) deferral of the publication for up to thirty-six months[54] (in this case, publication must be requested before the thirtieth month in order to preserve a design right).
(2) The *simplified filing*, provided by Article L. 512-2 section 5, in which case publication is deferred automatically (in this case, publication must be requested before the thirtieth month in order to preserve a design right). This form of filing applies to industries which frequently change the shape and presentation of their products, such as clothes, shoes and textiles used in the

53. CA Paris, 10 Dec. 2004, No. 04/480 (Court of Appeal).
54. Articles R. 512-3 f. and R. 512-10.

fashion and accessories industry and to packaging and bottling predominantly used in the perfume industry. There are fewer formalities for a simplified filing than for a full filing. The simplified filing primarily deals with the presentation of the reproductions. This no longer includes the requirements for height and form. As well, the simplified filing allows a reduced fee. This only includes the filing fee which is independent of the number of designs and reproductions provided. This particular filing necessitates the postponement of publication for a nonrenewable period of three years so that the designs or models remain confidential during that period. The applicant who wishes to extend the duration of protection must request renewal six months prior to the expiration of the three-year period.[55]

§7.09 OTHER

[A] Interface between Industrial Design and Trademarks

The purpose of Trademark Law is to protect distinctive signs which guarantee the identity or origin of the goods or services. This consideration is irrelevant to the law on industrial designs and the criteria of '*likelihood of confusion*' and the '*average consumer*' are clearly inapplicable to designs (Please see *above §7.05[B] Test for infringement*). The combination of the two protections may however be possible subject to compliance with the respective conditions for protection.

A design's registration can however be cancelled if the design includes a prior protected distinctive sign, without the authorization of its owner.[56]

The lack of novelty and individual character may be constituted by an earlier figurative (or three-dimensional) mark, which will then allow the holder to request the cancellation of the design's registration on that basis. In a decision of the European Court of Justice, the judges decided that the contested design had no individual character and that '*the international trademark registration No. 369072 was proof of the existence and the prior disclosure to the public of the design or model against the contested design.*'[57]

[B] Relationship between Industrial Design and Copyright/Design Copyright

It is settled French law that the same industrial product may be protected by a design right (Book V of the IPC) and by copyright (Book I of the IPC). Article L. 513-2 provides that the protection of registered designs is '*without prejudice to the rights resulting from the application of other laws, including Books I and III*'.

55. Article R. 512-11 of the IPC.
56. Article L. 511-4 (e) of the IPC.
57. Trib. UE, Ch. 1st, 7 Nov. 2013, aff. T-666/11.

By virtue of this theory of Unity of Art, cumulative protection is permitted subject to the requirements specific to each type of protection being fulfilled. For example, the judges point out that '*the protection of copyright applies only to works meeting the requirement of originality which is understood as a reflection of the personality of its creator*'[58] and that '*concepts of novelty and individual character are foreign to copyright*'.[59] Consequently, the conditions are truly distinct and simultaneous protection is obviously not automatic.

The copyright provisions apply to all original intellectual works '*irrespective of their type, form, artistic merit or purpose*'.[60] Both case law and legal academic thinking have long considered that a judge's role is not that of an art critic and that it is not the judge's job to determine whether those features present in a work allow it to be defined as 'art'. Thus, the provisions relating to intellectual property in Book I of the IPC, protect without distinction, pure art and applied art subject only to the condition that the work in question is original.

Originality may be summarized as the reflection of the creator's personality. In recent years, a new definition of '*originality*' emerged from case law: it is the existence of an '*aesthetic bias*' of the author.

The Paris Court of Appeal recently stated that:

> if the company rightly points out that this box does not reproduce the overall characteristics of the claimed box in the same combination because of the material used and the same applies to the Chinese box because of the poor quality of its representation, it should be remembered that the concept of prior rights is irrelevant to copyright since only proof of the original character is required as condition for protection under Book I of the Intellectual Property Code; it is therefore for the one who relies on these provisions to justify, not the novelty of the claimed design, but that it has an individual appearance which reflects an aesthetic bias and the personality of its creator.[61]

Moreover, no formalities need be completed for one to benefit from the protection of these particular provisions. The author of a work enjoys, by virtue of the mere act of its creation, an exclusive incorporeal property right opposable to the rest of the world.[62] Therefore, original works of applied art are considered as intellectual works within the meaning of Book I of the IPC.[63]

French judges show great rigour in analysing the conditions for protection under copyright and designs, whilst clearly distinguishing them for the respective protection sought and imposing a higher threshold in their assessment of originality. For instance, the Paris Court of Appeal has retained the protection of a shirt model under designs law whilst refusing it under Copyright Law because the mere addition of buttons on the shirt pattern, which exactly reproduced an ornamental configuration already present

58. CA Caen, Ch. 2nd, civ., 6 Dec. 2012 (Court of Appeal).
59. CA Paris, 5 pôle, 2nd ch., 19 Apr. 2013 (Court of Appeal).
60. Article L. 112-1 of the IPC.
61. CA Paris, 5 pôle, Ch. 2nd, 17 Jan. 2014, No. 13/03710 (Court of Appeal).
62. Article L. 111-1 of the IPC.
63. Article L. 112-1 of the IPC.

on competing shirt models, was neither original nor the reflection of the author's personality.[64]

Nonetheless, case law does regularly recognize cumulative protection. For example, the Paris Court of Appeal has accepted the protection of a shoe model on the basis of both an unregistered Community design and of copyright.[65] The court held that the defendant in infringement proceedings had not provided evidence of prior art with precise dates in the EU which would be capable of showing a lack of novelty. The defendant did not demonstrate that this model of footwear could be different from previous forms for the informed observer. Moreover, the court also admitted that copyright subsisted in the shoe model because there was a real creative effort which reflects the author's personality.

Consequently, there is always an interest for companies to apply for cumulative protection of their designs. However, this double protection should be invoked at the first instance level because no new arguments may be raised on appeal pursuant to Article 564 of the Civil Procedure Code.[66]

The duration of protection under Book I of the IPC (seventy years after the author's death), which is longer than that under Book V, may extend to registered designs.

[C] Relationship between Industrial Design and Unfair Competition

The practice of the courts is to apply Article 1382 of the Civil Code to punish acts which are contrary to standards of fair competition and business ethics. The action for unfair competition is a classic tort action for which the evidence of the three elements of Article 1382 of the Civil Code are required: a fault, a prejudice and a causal link. This action aims to remedy the loss suffered because of the wrongful act resulting from a competitor's activities. The holder of a registered design and/or copyright and the licensee (if the agreement is registered in the National Designs Register) will be able to bring an action for unfair competition.

As an action for unfair competition is based on classic tort law, it is therefore assumed that the plaintiff has to rely on separate facts from those used to prove the infringement. In such cases, the plaintiff can act both on the basis of infringement and on the basis of unfair competition.

The imitation of a design or a model used by a competitor may constitute misconduct, regardless of any design infringement. Indeed, the similarity can create a damaging confusion in the public's mind. In a judgment dated 8 April 2014,[67] the Court of Cassation upheld the dismissal of an action by Ferrari based on both copyright and designs law, covering the alleged infringement of its model 'Modena' by a vehicle called 'Turismo' used in a video game. It was alleged that the Court of Appeal had ruled on the differences and not on similarities, but the Supreme Court found that the trial

64. CA Paris, 5 pôle, Ch. 2nd, 31 May 2013, No. 12/13618 (Court of Appeal).
65. CA Paris, 5 pôle, Ch. 2nd, 8 Feb. 2013, No. 11/02407 (Court of Appeal).
66. CA Paris, 5 pôle, Ch. 2nd, 19 Apr. 2013, No. 12/01742 (Court of Appeal).
67. Cass. com. 8 Apr. 2014 no. 13-10689 (French Supreme Court).

judges had conducted a considered factual appreciation and that they were not required to decide in light of the overall visual impression. However, the Court of Cassation upheld Ferrari's argument on the issue of unfair competition. The Judges of Appeal had examined each complaint separately (choice of name, logo and typography) while they should really have considered these complaints as a whole that was likely to generate confusion in the mind of the consumers.

Beyond the likelihood of confusion, parasitism or commercial free-riding is often invoked. The parasitic behaviour can be defined as the behaviour of a company that interferes in the wake of another to take advantage of its investments without incurring any expenses. The scope of this action is wide because companies do not have to be in direct competition. Thus, one may invoke a design infringement and parasitism as well because a third party has benefited unlawfully from advertising and financial efforts. The defendant can be ordered pay compensation for the entire prejudice resulting from such acts but the assessment of the amount of damages is quite difficult.

CHAPTER 8

Industrial Design Rights: Germany

*Uwe Lüken**

§8.01 INTRODUCTION

There have been substantial developments in design law since the last edition. The European Design Law has been harmonised by way of a Directive[1] as well as a Regulation.[2] Consequently, designs can be protected by either a Community Design (registered or unregistered) or by way of a German national design for the territory of Germany. After the trademark, the design is the second right that is fully harmonised in the EU. Like with the trademark, the harmonisation has pushed the relevance of the right substantially. Again, like with the harmonised trademark, this also led to a higher economic importance of design rights. There are a number of reasons for this, as elaborated below:

(1) There is a European Wide Registered Design Right
The European Union consists of twenty-eight jurisdictions at the moment.[3] The territory of the Union has a substantial economic importance.[4] To acquire a European wide right with one application is a very substantial advantage. The applicant has to invest relatively less both financially as well as administratively to achieve a right in such a broad territory.[5] At a later stage, the

* The author thanks Ms. Julia Präger for her substantial support in preparing this contribution.
1. Directive 98/71/EC of the European Parliament and of the Council of Oct. 13, 1998 on the legal protection of designs.
2. Council Regulation (EC) No. 6/2002 of Dec. 12, 2001 on Community designs.
3. *See* <http://europa.eu/about-eu/countries/member-countries/> Last accessed Dec. 12, 2014.
4. The EU's GDP in 2012 was EUR 12,945,402 million, thus the EU's economy is now bigger than the U.S.'s. <http://europa.eu/about-eu/facts-figures/economy/index_en.htm> Last accessed Dec.17, 2014.
5. Münchner Anwalts Handbuch, Gewerblicher Rechtschutz, 4th ed. 2012, s. 48 para. 3.

European wide rights allows the owner also European wide remedies, i.e., European wide injunctions.[6]

(2) The Harmonised Design is a Very Strong Right
The harmonised design right is a strong right. While there is no examination of the design application, there is a presumption of validity, section 39 Design Act. The burden of proof for invalidity is on the defendant. In addition, given that the design right is not examined the official fees are low and the costs for applications are also usually quite moderate.

(3) There is Also an Unregistered European Design Right
The harmonised European design right provides for an EU wide unregistered design right, Article 11 of the European Design Directive.[7] This is a very interesting right as it provides for a lot of the advantages of the registered right. The idea behind the non-registered right was *inter alia* to allow fast moving industries like the fashion industry to get protection without the need to file an application because of the high number of short-lived designs they produce (see Recital 18 and 25 of the Regulation). The main difference is that: (a) there is limitation to three years (non-extendable) and (b) an additional requirement of copying.[8]

The Geschmacksmustergesetz of 2004[9] has brought broad change to the German design law by implementing the Design Directive into German law. With the new law, the design has become economically much more important. A good example is the Apple and Samsung design dispute which has led to numerous German court decisions including a Preliminary Injunction valid for the whole of the European Union which had been granted ex parte.[10] In 2014, there has been a further amendment to the design law which has changed the name of the right from "Geschmacksmuster" to "Design".[11] The new German Design Act (Designgesetz)[12] has come into effect on January 1, 2014.

§8.02 CONVENTIONS AND LEGISLATION

In addition to the EU framework the following international treaties apply.

6. David Stone, "European Union Design Law, A Practitioners' Guide", 2012, para. 20.02.
7. Council Regulation (EC) No. 6/2002 of 12 Dec. 12, 2001 on Community Designs.
8. For the requirements of copying *see* i.a. David Stone, "European Union Design Law, A Practitioners' Guide", 2012, para. 17.35.
9. Gesetz über den Schutz von Mustern und Modellen (Geschmacksmustergesetz - GeschmG), published Mar. 12, 2004 (BGBl. I S. 390); it has been subsequently amended various times.
10. Landgericht (Regional Court) Düsseldorf, Sep. 9, 2011, 14c O 194/11.
11. A description of the changes with the amendment of 2014 can be found at Kappl GRUR 2014, 326–331.
12. Designgesetz, published Feb. 24, 2014 (BGBl. I S. 122); full text only available in German at <http://www.gesetze-im-internet.de/geschmmg_2004/index.html#BJNR039010004BJNE0001 06360> Last accessed Dec. 17, 2014.

[A] Paris Convention[13]

The Paris Convention contains only rudimentary regulations concerning design. It names designs as industrial property in Article 1(2), but it does not offer a definition. Furthermore in Article 5quinquies it declares designs protected in the Member States[14] but again does not specify what kind of protection that is.

However Article 4 grants a six-month priority period in other Member States for a national filing in one of the Member States that being an enormous advantage.

At last Article 19 allows for Member States to sign treaties concerning industrial property with each other - one of those being the Hague Agreement (see below).

[B] TRIPS[15]

Members of TRIPS coincide with members of the World Trade Organization (WTO). According to Article 25(I) TRIPS Member States have to protect designs which are new and original. Although the members are able to exclude those not being significantly different from already known designs or those dictated by technical or functional aspects. TRIPS leaves the decision whether design protection is provided through industrial design law or copyright law to its members.

According to Article 26, the owner of the design can prevent the making of copies and more importantly substantial copies. Further the exceptions to protection should be limited and protection should be extended for a minimum duration of ten years.

Finally it should be borne in mind when assessing European or German design law that the interpretation of European and German courts has to be compliant with the before mentioned provisions.[16]

[C] Hague Agreement[17]

The Hague Agreement is administered by WIPO and provides a system for international application - up to 100 designs can be registered in its contracting parties[18] with one single application. Such an application can be filed directly with the International Bureau at WIPO or with the German Patent and Trade Mark Office (DPMA). The latter has to forward the application immediately to the International Bureau without examination.[19]

---

13. The Paris Convention on the Protection of Industrial Property of Mar. 20, 1883.
14. A list of the members can be found at <http://www.wipo.int/treaties/en/ip/paris/> Last accessed Dec. 17, 2014.
15. Agreement on Trade-Related aspects of Intellectual Property Rights of Apr. 15, 1994.
16. ECJ, C-431/05, paras 34 et seq. - Merck Genéricos-Produtos Farmacêuticos Lda.
17. Hague Agreement Concerning the International Registration of Industrial Designs of Nov. 6, 1925.
18. For a list of the parties *see* <http://www.wipo.int/hague/en/members/> Last accessed Dec. 17, 2014.
19. *See* s. 67 and s. 68 Design Act.

The application leads to a bundle of national rights subject to national design laws in each country.[20] Or in case of the European Union to a registered Community Design.

[D] Locarno Agreement[21]

Members of the Locarno Agreement[22] agreed to use the same classification system for the registration of designs. This system consists of classes and subclasses grouping industrial products in which designs are incorporated. Within the classes and subclasses products are listed alphabetically. Additionally there is an alphabetical list containing every product indicating the class it belongs to.[23]

When applying for a German registered design the applicant has to choose the products he seeks protection for from a list based on the Locarno Classification.[24]

§8.03 SUBJECT MATTER

[A] Definition/What Constitutes a Protectable Design

The subject matter of the design is defined in section 1 No. 1 Design Act. According to this a design can constitute of the two or three dimensional appearance of a product. According to section 1 No. 2 Design Act a product is an industrial or handcraft item but not a computer program. The overview here does not allow to go into too many details. However, it is no exaggeration that the scope of these definitions is very broad. Even separate parts that can be put together to form a complex product meet the requirement. Included are also *inter alia* the following:

- Packaging.
- Get-up of the product.
- Graphic symbols.
- Typographical typefaces.

These four terms are expressly mentioned in the respective articles. At the same time there are designs excluded from protection:

- Those violating the public order or mores.[25]
- Consisting of a sign mentioned in Article 6ter Paris Convention.[26]
- Consisting of an official coat of arms, emblem or insignia.[27]

20. David Stone, "European Union Design Law, A Practitioners' Guide", 2012, para. 15.01.
21. Locarno Agreement Establishing an International Classification for Industrial Designs of Oct. 8, 1968, as amended on Sep. 28, 1979.
22. *See* all members at <http://www.wipo.int/treaties/en/classification/locarno/> Last accessed Dec. 17, 2014.
23. *See* full current Locarno Classification at <http://www.wipo.int/classifications/nivilo/locarno10/index.htm?lang=EN> (status: Dec. 17, 2014).
24. Eichmann/v. Falckenstein, Geschmacksmustergesetz, 4th ed. 2010, s. 11 para. 64.
25. *See* s. 3 (1) No. 3 Design Act.
26. *See* s. 3 (1) No. 4 Design Act; for Paris Convention *see* above s. 8.02[A].
27. *See* s. 3 (1) No. 4 Design Act.

[1] *Shape, Configuration*

An appearance protected by the Design Act can be part of the product itself such as its lines, contours, shapes, texture or material.[28] This enumeration is not final - every visible appearance can be protected.[29] Of course a material itself cannot be registered but the way it is used and the effect it creates can.[30]

Although the definition is very broad there are limitations. The following cannot be protected:

- Thoughts or ideas, i.e., something that has not yet achieved a visible form.
- Sound, smell or taste as they are not visible.
- Design concepts.

However, if the exact shape or dimensions are dictated by the necessity to fit into another product or to be applied to another product the appearance is excluded from design protection.[31] These are the so-called must-fit parts. This provision secures the availability of accessories and spare parts.[32] Nevertheless, if the appearance matching the before mentioned criteria has the purpose to enable the assembly or connection in a modular system - a so-called must-match part - the exclusion does not apply.[33] The economical important question of spare parts especially in the car industry has been of high political importance during the legislative process.[34]

[2] *Ornamental or Functional*

Also expressly mentioned in section 1 No. 1 Design Act and a protected appearance are ornaments. These decorative features may very well be achieved by using the products' lines, shape, contours etc. The listed appearances do not exclude each other.[35]

But an appearance solely dictated by a technical function is not amenable to design protection according to section 3 (1) No. 1 Design Act. Otherwise technical innovation might be hindered.[36] This exclusion may apply to one of a product's features or to the whole design if the appearance as such is due to a technical function.[37] But that does not mean that a design only has to contain aesthetical features - single technical features will just not be taken into consideration when assessing the design's originality and scope of protection.[38]

28. *See* s. 1 No. 1 Design Act.
29. Fischoeder in: Stöckel, Handbuch Marken- und Designrecht, 3rd ed., p. 583.
30. Eichmann/v. Falckenstein, Geschmacksmustergesetz, 4th ed. 2010, s. 1 para. 11.
31. Section 3 (1) No. 2 Design Act.
32. Fischoeder in: Stöckel, Handbuch Marken- und Designrecht, 3rd ed., p. 604 f.
33. Section 3 (2) Design Act.
34. For more information on this *see* Fischoeder in: Stöckel, Handbuch Marken- und Designrecht, 3rd ed., p. 607 ff.
35. Eichmann/v. Falckenstein, Geschmacksmustergesetz, 4th ed. 2010, s. 1 para. 13.
36. Eichmann, GRUR 2000, 751, 758.
37. OHIM, R-690/2007-3, para. 38 - Chaff cutters.
38. Landgericht München (Regional Court of Munich), GRUR-RR 2007, 266, 267.

[B] Originality/Novelty

In general a registered design requires originality (the legal term under EU and hence German law is "individual character" - nevertheless this text uses the term "originality" as this is the term used internationally and consequently in other parts of this international publication) and novelty,[39] but a peculiarity applies to a design being part of complex product: the design protection is sought for has to be visible when incorporated in the complex product and its visible characteristics themselves have to be original and novel.[40]

[1] Test for Originality Generally

According to section 2 (2) Design Act a design is original if its overall impression on an informed user differs from the overall impression created by another already disclosed design. Even if all known designs have to be taken into consideration the product in question has to be assessed in comparison to one other design at a time - it is improper to pick features of different known designs and compare them to the design in question.[41] This means a design combining appearances of several disclosed other design can still be original.[42]

Furthermore, when assessing originality the degree of the designer's freedom in developing the design has to be taken into consideration.[43] The result of which is that if there is only a small margin of design freedom - due to technical reasons or high number of disclosed designs - a slight difference compared to the closest known design might already create originality.[44]

The first step in examining the design's overall impression is determining what kind of impression the design creates and in a second step what the features causing this impression are.[45] Then the features have to be evaluated with regard to their contribution to the overall impression.[46] That means identity in details whilst having differences in characteristics does not hinder originality.[47]

Which features take part in the assessment depend on the user's perception: A feature he does not perceive whilst using it nevertheless perceiving it when purchasing the product are being taken into consideration.[48] However, criteria outside the design itself like a design trend or the economic success of a design are not standards for assessing the overall impression and will not be taken into account.[49]

39. Section 2 (1) Design Act.
40. Section 4 Design Act.
41. Oberlandesgericht (Higher Regional Court) Frankfurt/Main, GRUR-RR 2009, 16, 17 - Plastik-Untersetzer.
42. Fischoeder in: Stöckel, Handbuch Marken- und Designrecht, 3rd ed., p. 593.
43. Section 2 (3) Design Act.
44. Fischoeder in: Stöckel, Handbuch Marken- und Designrecht, 3rd ed., p. 593.
45. German Federal Supreme Court (BGH), GRUR 2001, 503, 505 - Sitz-Liegemöbel.
46. Eichmann/v. Falckenstein, Geschmacksmustergesetz, 4th ed. 2010, s. 2 para. 19.
47. Fischoeder in: Stöckel, Handbuch Marken- und Designrecht, 3rd ed., p. 594.
48. Oberlandesgericht (Higher Regional Court) Dusseldorf, 20 U 128/06, para. II.2.
49. EC, T-153/08, para. 58 - Fernmeldegerät.

The reference point for assessing the difference between two designs is the informed user. First of all an informed user is not a real person[50] and second an informed user is not a design professional such as a critic or expert.[51] It rather he is a fictitious person using the product in which the design is incorporated and showing a relatively high degree of attention when he uses them, as well as having knowledge of the various existing designs in the concerned sector and of the features which those designs normally include.[52]

In principle, the informed user will when possible make a direct comparison between the designs at issue but an indirect comparison based on an imperfect recollection is not out of the question if a direct one is impracticable or uncommon in the sector concerned, because of specific circumstances or the characteristics of the products.[53]

[2] Novelty

A design is regarded as new according to section 2 (2) Design Act if no identical or insignificantly differing design has been disclosed prior to the application day. So contrary to originality when examining a design's novelty not all disclosed designs are taken into account, but only those designs which are identical or those having insignificantly different characteristics from the design at issue. Whether a difference is insignificant has to be assessed with regard to the design's overall impression.[54]

But the process of comparing the designs itself is the same: one design after the other is compared to the one at issue.[55] Only if all features of the design at issue are combined in a prior design, is it not novel.[56]

The key element for the criteria of novelty is disclosure. A design has been disclosed if it was publicised, exhibited, used or made available to the public in another manner. Unless people in the sector concerned could not have known the design in the ordinary course of business prior to its filing or unless it was only disclosed to a third party under the condition of confidentiality.[57] A disclosure outside of Germany is sufficient as long as specialists in the sector concerned get to know designs in foreign countries in the ordinary course of business.[58]

If a design has been made available to the public by the designer himself or his successor twelve month prior to the application it does not constitute a disclosure according to section 5 Design Act.[59] This constitutes a grace period to the benefit of the

50. Oberlandesgericht (Higher Regional Court) Frankfurt/Main, GRUR-RR 2009, 16, 17 - Plastik-Untersetzer.
51. Oberlandesgericht München (Higher Regional Court Munich), GRUR-RR 2010, 166, 168 - Geländewagen.
52. EC, T-153/08, paras 46 et seq. - Fernmeldegerät.
53. ECJ, C- 281/10P, paras 55 et seq. - Grupo Pomer.
54. Eichmann/v. Falckenstein, Geschmacksmustergesetz, 4th ed. 2010, s. 2 para. 6.
55. *See* above s. 8.03[B][1].
56. German Federal Supreme Court (BGH), GRUR 1980, 235, 236 - Play-family.
57. *See* s. 5 Design Act.
58. German Federal Supreme Court (BGH), GRUR 2009, 79 para. 22 - Gebäckpresse.
59. Section 6 Design Act.

designer. The same applies if the design got disclosed through improper actions of a third party.[60]

[a] Prior Use

Prior use requires that an existing product enter the market place – import or export actions are only relevant if they are perceptible to the public.[61]

[b] Prior Publication

A publication in accordance with section 20 Design Act by the German Patent and Trade Mark Office meets this requirement as well as the publications of other IP Offices such as OHIM or WIPO as long as it is the ordinary course of business in the relevant sector[62] (see above).

[c] Prior Application or Registration

A prior application or registration does not automatically disclose a design. First the applicant can request that the publication of the design's drawing will be pushed back by thirty months,[63] so the design is not available to the public yet. However, a design might be available to the public without registration – that being the case if a design is expunged.[64] The owner of the prior design does not need to be a third person. A prior design by the same owner will also destroy the novelty (or even more importantly, the originality). This can be relevant if a company revises a prior design for e.g., a new model. In other words, self-collision is relevant.

§8.04 DURATION OF PROTECTION

According to section 27 (1) Design Act protection begins with the design's registration. However, section 27 (2) Design Act states that the period of protection is a maximum of twenty-five years and starts with its filing. This means the beginning of protection differs from the duration period's beginning. So there is no protection between filing and registration but the twenty-five-year period is already running out. However, given that there is no substantial examination the period between application and registration is usually short.

In order to maintain protection after the first five years the right holder has to pay a maintenance fee every five years and can repeat this every five years until the total

60. *See* above.
61. Eichmann/v. Falckenstein, Geschmacksmustergesetz, 4th ed. 2010, s. 5 para. 3.
62. Eichmann/v. Falckenstein, Geschmacksmustergesetz, 4th ed. 2010, s. 5 para. 6.
63. Section 21 (1) Design Act.
64. Eichmann/v. Falckenstein, Geschmacksmustergesetz, 4th ed. 2010, s. 5 para. 6.

period of twenty-five years is reached.[65] This means there are separate payments for years six to ten, eleven to fifteen, sixteen to twenty and twenty-one to twenty-five – it is not possible to pay for the remaining twenty years at once in advance.[66] But a special request for maintaining the registered design is not required.

The fee is due two months after the last month within the protected five year period ends. But after this initial two month period there is another four month period within the right holder can uphold the registered design by paying the maintenance and an additional late fee.[67]

If the maintenance fee (and late fee) is not paid within the these periods the design is no longer protected under the Design Act and will be removed from the register according to section 36 (1) No. 1 Design Act.

§8.05 INFRINGEMENT

[A] Scope of Infringement Action

The design provides the owner with an exclusive right to use the design, section 38 (1) (1) Design Act. What is more important, the same provision provides the owner with a right to prevent others from the use of the design. There are examples for relevant use in section 38 (1) (2):

- making;
- offering;
- putting on the market;
- importing;
- exporting;
- using of a product in which the design is incorporated;
- stocking such a product for those purposes.

As these are examples it is clear that further ways of infringing use are clearly possible.

This is a very strong right as the law provides for an assumption of validity of the design, section 39 Design Act. Such a presumption is not alien to IP laws.[68] However, it has to be noted that the design is not examined in the registration process. The only prerequisite for the presumption is registration. This is very important in a jurisdiction that is not afraid to grant far reaching preliminary injunctions on an ex parte basis.[69]

65. *See* s. 28 Design Act.
66. Eichmann/v. Falckenstein, Geschmacksmustergesetz, 4th ed. 2010, s. 28 para. 3.
67. Eichmann/v. Falckenstein, Geschmacksmustergesetz, 4th ed. 2010, s. 28 paras 5 et seq.
68. In relation to trademark law this is settled case law, *see* e.g., German Federal Supreme Court (BGH), GRUR 2010, 1103, 1104 – Pralinenform II, *see also* Hacker in Ströbele/Hacker, Markengesetz, 11th ed. 2014, s. 14 para. 18.
69. In relation to this *see* Dissmann in: Stöckel, Handbuch Marken- und Designrecht, 3rd ed., p. 696. Critical to that is David Stone, "European Union Design Law, A Practitioners' Guide", 2012, para. 18.07 because of the not-examined nature of the right.

[B] Test for Infringement/Criteria to Constitute Infringement

The owner can prohibit any identical design or any design that does not produce on the informed user a different overall impression, section 38 (2) Design Act. The test for infringement does not differ substantially from the question whether a prior design has originality.[70]

In summary the following aspects are relevant:

(1) The comparison has to be made between the registered design and the potentially infringing design. The two have to be examined directly vis-à-vis each other. Different from trademark law it is not relevant whether a customer would have a chance to see both vis-à-vis each other in the marketplace.[71] It is also not relevant whether there is a danger of confusion. However, if there is evidence of actual confusion that can indicate a design infringement.[72]
(2) The overall impression is decisive. There is a focus on the elements of the registered design that provide for the originality. This means that e.g., functional elements might have to be taken out of the comparison.
(3) The yardstick for the comparison is the informed user.[73]

While it is clear that the infringement test has to be conducted based on the overall impression it is accepted that a design cannot only be infringed in full but also in part.

[C] Standing to Sue

The design's owner that is the person registered as the right's holder, as well as rightful claimants are entitled to take legal action.[74]

Under certain circumstances a licensee can be a rightful claimant according to section 31 Design Act. In general a licensee can only take legal action against infringement if the owner approves.[75] However, if an exclusive licence is at issue, the licensee can sue if he asked the owner to take legal action and he didn't do so within an appropriate period of time.[76]

Another rightful claimant is the owner's legal successor.[77] A design is transferable in general according to section 29 (1) Design Act in two possible ways: contract or inheritance.

70. *See* s. 8.03[B] above.
71. Eichmann/v. Falckenstein, Geschmacksmustergesetz, 4th ed. 2010, s. 38 para. 18.
72. Oberlandesgericht (Higher Regional Court) Hamburg, NJOZ 2007, 3055.
73. *See* s. 8.03[B][1] above.
74. Section 42 (1) Design Act.
75. Section 31 (3) Design Act.
76. Section 31 (3) Design Act.
77. Eichmann/v. Falckenstein, Geschmacksmustergesetz, 4th ed. 2010, s. 42 para. 9.

[D] Remedies for Infringement

German design law provides for a wide range of remedies.

[1] The Most Important Remedy by Far Is the Cease and Desist Order, section 42 (1) Design Act

The prerequisite to such a claim is that there either has been an actual infringement already (so-called Wiederholungsgefahr) or a first infringement is imminent (so-called Erstbegehungsgefahr). This is not so much a question of design law as a question of general requirements for cease and desist claims under German law. The question of how far or how limited a claim and consequently an order can be worded is one of the most complex questions under German Law. This cannot be discussed more in-depth here. However, it is worth to pointing out that a German order might be more limited than an order in other jurisdictions.

[2] Damages, section 42 (2) Design Act

The second most relevant claim is the claim for damages. In addition to an infringement the prerequisite for such a claim is default including any form of negligence. This requirement does not provide a very strict test. Given the nature of the design as an exclusive right, it is accepted that there is a requirement to search for prior rights before introducing a new product.[78] Consequently, it is rather unlikely for a defendant to successfully argue a lack of negligence.

However, the claimant has to calculate the amount of damages he claims. In line with the practice relating to other IP rights German law provides three ways to calculate damages:

(1) loss of design owner;
(2) gain of infringer;
(3) so-called licence analogy.

The two first ways to calculate damage have substantial disadvantages. The design owner will not want to give insights into his calculation. However, that would be necessary to calculate his damage. Even more problematic is that the design owner can only ask for the loss that he has incurred because of the infringement. In most cases there can be many reasons for loss that are not limited to the infringement.

In relation to the gain of the infringer, the main problem had been that the infringer included high overall costs in the calculation. The German Federal Supreme

78. Eichmann/v. Falckenstein, Geschmacksmustergesetz, 4th ed. 2010, s. 42 para. 16 with references to case law.

had dealt with this problem and has limited the costs that the infringer can include into the calculation to the production and distribution cost of the infringing products.[79]

A rather popular method to calculate damages is called licence analogy. The infringer has to pay as damage what he would have to pay as a licence had the parties agreed on such in advance. Of course, the problem with this method is the determination of the royalty rate. It has to be taken into account that German law does not provide for punitive damages.

[3] Request for Information

German Law provides in relation to damages a claim for information. The right owner does not have a basis for his damage calculation if he does not have the relevant information. The infringer has to provide the owner with information on the extent of the infringement. This includes the amount of products sold, the price, the profit etc. While there is no specific provision for this claim in the Design Act, it is settled practice that this claim exists.[80] The extent of the information to be provided depends on the calculation method for the damage.

In general, it can be said that it is quite burdensome for the owner to enforce a damage claim. Often the enforcement is achieved by way of the following stages:

- The owner initiates an infringement proceeding in which he asks for information in addition to his cease and desist claim.
- In a second stage, the infringer provides the owner with the information. Usually, this leads to discussions between the parties on the accuracy and/or the required level of detail of the information. In complicated cases the parties might need the courts to resolve conflicts in relation to the information claim as well.
- Finally, the owner has to sue the infringer for a concrete amount of damages if the parties cannot agree on an appropriate amount.

It is apparent from the number of different steps that have to be taken in relation to a damage claim that this is a costly and time consuming process. It is not unusual that the owner does not pursue his damage claim because of the burden involved. It is also not seldom that the parties agree on an amount without giving information as the infringer often does not want to give away information.

Section 46 Design Act provides for an additional claim of information. The owner can ask for information on source and the distribution channel of the design. This claim allows the owner of a design to try to get to the source of an infringement. This claim can procedurally be enforced by way of a preliminary injunction, section 46 (7) Design Act.

79. German Federal Supreme Court (BGH), GRUR 2001, 329, 332 - Gemeinkostenanteil; *see also* Eichmann/v. Falckenstein, Geschmacksmustergesetz, 4th ed. 2010, s. 42 para. 25 with further citations.
80. German Federal Supreme Court (BGH), GRUR 1980, 227, 232 - Monumenta Germania Historia.

[4] *Destruction, section 43 (1) Design Act*

The owner has a claim for destruction of infringing products in the possession of the infringer. This includes any appliances that have primarily been used for the production of the infringing products. The claim is limited by the requirement of proportionality, section 43 (4) Design Act.

[5] *Recall, section 43 (2) Design Act*

In addition to the destruction of products in the infringer's possession he has to recall products still within the distribution chain. This does not include end users of the products as they are not part of the distribution chain. This claim is also limited by the requirement of proportionality, section 43 (4) Design Act.

§8.06 DEFENSES TO INFRINGEMENT

[A] Validity

Given that the design right is a non-examined right, it is obvious that validity questions have a high importance as a defense. Technically, it is not correct to call this a defense. Validity questions need to be brought into the proceedings by counterattack. Section 52 a Design Act expressly states that invalidity in relation to a registered design can only be claimed by way of a counterclaim or a request for cancellation in front of the German Patent and Trademark Office (Deutsches Patent - und Markenamt - DPMA).

The German Design Act has been amended in that regard with effect from 2014 onwards. It will be standard procedure that a defendant will bring that counterclaim. The counterclaim is governed by section 52 b (1) Design Act. The court can deal with the counterclaim itself. Alternatively, it can order the defendant to request a cancellation with the DPMA and set aside the proceedings until the cancellation request has been decided, section 52 a (3) Design Act. This has two consequences. There is no exclusivity of the DPMA to cancel designs. The court can decide whether it wants to find on validity itself.

The relationship between an infringement proceeding and a cancellation proceeding relating to the same design is further defined in section 34 b Design Act. The court can set aside the infringement proceedings in its own discretion. Traditionally, German civil courts are reluctant to set aside infringement proceedings. The reason is that cancellation proceedings can last years. The infringement would continue during that time. However, the court has to set aside the proceedings if it is of the opinion that the design has to be cancelled.

The relationship between the proceedings is further stipulated as follows:

- A decision by the DPMA and by the court in which the design is cancelled leads to the cancellation and, hence, is valid for everybody (*inter omnes*), section 36 Design Act.

- A decision that a design will not be cancelled is only valid between the parties to the proceedings (*inter partes*). Consequently, a counterclaim relating to validity is not admissible if the DPMA had already decided against such a cancellation, section 52 b (2).

In addition to the grounds for invalidity discussed below, German Law provides for the following additional grounds for invalidity:

- not a design, section 33 (1) No. 1 and section 1 No. 1 Design Act;[81]
- design not protectable according to section 3 Design Act (especially functionality - see below §8.06[A][2]), section 33 (1) No. 2 Design Act; and
- interference with other rights like copyright, prior design or trademark, section 33 (1) No. 3 Design Act.

[1] Not Novel/Original

This is the main reason to attack validity. Novelty and originality have been discussed above.[82]

[2] Functional/Not Ornamental

According to section 3 (1) No. 1 Design Act functional elements cannot be protected by design. Hence, those elements have to be excluded when accessing the overall impression.[83] While section 3 (1) No. 1 Design Act primarily deals with validity its main practical importance is in relation to infringement. The reason is that a lot of designs have numerous technical elements but rarely *only* consist of technical elements. A design with substantial technical elements might still be valid but its scope of protection is limited to the non-technical parts. This is one of the more complex questions of German design law.[84]

[3] Other Possible Criteria, Such as Non-obviousness

The following further defenses are applicable:

- Design not registered or not disclosed.
  A design can only be enforced if it is registered. Secondly, it is necessary that it has been disclosed. The disclosure of the design can be postponed by thirty months. During the time of the postponement an infringement claim is only granted if there is copying, section 38 (3) Design Act.
- No German representative.

81. *See* s. 8.03[A] above.
82. *See* s. 8.03[B] above.
83. Eichmann/v. Falckenstein, Geschmacksmustergesetz, 4th ed. 2010, s. 38 para. 13.
84. Fischoeder in: Stöckel, Handbuch Marken- und Designrecht, 3rd ed., p. 603 f.

Section 58 Design Act requires that owners without a place of business in Germany have to appoint a German representative which is mostly a German attorney or patent attorney.

[4] Faults in Application

Faults in the application are only relevant for cancellation proceedings.

[B] Marking

Marking is not relevant in Germany.

[C] Non-infringement

Section 40 Design Act provides for a number of reasons for non-infringement:

- No. 1 acts done privately and for non-commercial purposes;
- No. 2 acts done for experimental purposes;
- No. 3 acts of reproduction for the purpose of making citations or teaching; and
- No. 4 equipment on non-EU ships and aircraft.

[D] Other Defenses

[1] Prior Use, section 41 Design Act

There is an exception for infringement for designs that have been used before the filing date of the registered design. The practical relevance of this defense is fairly limited. Every prior use that has been disclosed is an obstacle for novelty/originality. The defense is not limited to actual use but also available for serious and effective preparations.

[2] Exhaustion, section 48 Design Act

Exhaustion is a common defense in IP rights. Exhaustion is not limited to Germany but any putting on the market within the European Union/European Economic Area (EEA) leads to exhaustion.

An interesting case in relation to exhaustion has been decided by the Federal Court of Justice (Bundesgerichtshof – BGH). Car maker Daimler had taken action against a company which stretched genuine Daimler cars. According to the BGH exhaustion was not applicable as the stretched cars differed from the ones put on the market by Daimler.[85]

[3] Statute of Limitation, section 49 Design Act

Section 49 Design Act refers to the German Civil Act (Bürgerliches Gesetzbuch - BGB) and more specifically to section 5 (Abschnitt 5). In essence there is a limitation of three years to the end of the year.

§8.07 PROCEDURES FOR FILING APPLICATION FOR REGISTRATION

Designs have to be filed at the German Patent and Trade mark Office (Deutsches Patent- und Markenamt - DPMA). The application has to contain, section 11 (2) Design Act:

- an application for registration, No. 1;
- the identity of the owner, No. 2;
- and a representation of the design suitable for reproduction, No. 3.

Additionally, the application has to contain an indication of the products on which it is intended to be incorporated, section 11 (3) Design Act. However, this specification does not have an effect on the scope of protection, section 11 (6). There are further possible elements which can be made in the application, section 11 (5) Design Act. This includes a description.

The office has a fairly limited scope of examination. This basically mirrors the prerequisites under section 11 (1) No. 1-3 Design Act and the payment of the official fee, section 16 Design Act.

[A] Priority

The design has the priority from the date that the application is received by the DPMA, section 13 (1) Design Act. There is the possibility to rely upon a foreign priority, section 14 Design Act. The foreign priority needs to be in a Member State of the Paris Convention or TRIPS.[86] Furthermore, there is a priority for certain exhibition, section 15 Design Act.

[B] Drawings

There has to be a representation. The formats typically used are either drawings or photographs. One design can be depicted by more than one representation. The

85. BGH GRUR 2010, 718 - Verlängerte Limousinen.
86. For both *see* §8.02 above.

maximum number of representations per one design is ten, section 7 (1) (1) Design Regulation.[87]

It is a question of application strategy whether to use drawings or photographs.

[C] Design Part of Another Article/Environment of Design

Generally, the design should be limited to the actual design and not include elements for which no protection is sought. However, it is accepted that there are instances in which it is necessary or at least helpful to present the design in a surrounding that allows someone to understand the design better. DPMA does not have any rules on how to deal with this. However, OHIM (Office for Harmonisation in the Internal Market - which is responsible for the registration of community designs) has guidance in the Examination Guidelines of OHIM.[88] Paragraph 11.4 reads *inter alia*:

> The representation of a design should be limited to the features for which protection is sought. However, the representations may comprise other elements that help to identify the features of a design for which protection is sought. In an application for registration of a Community design the following identifiers will be allowed:
>
> 1. Dotted lines may be used in a view either to indicate the elements for which no protection is sought or to indicate portions of the design which are not visible in that particular view, i.e., non-visible lines. Therefore, dotted lines identify elements which are not part of the view in which they are used.
> 2. Boundaries may be used to surround features of the design for which protection is sought.
> 3. Colouring may be used on a black and white drawing to highlight the features of the design for which protection is sought only. Where the design concerns ornamentation, a product to which it is applied may be identified by dotted lines or the ornamentation may be surrounded by boundaries. It is the applicant's responsibility to use the dotted lines, boundaries and coloring in such a way to make clear for which features protection is sought and for which not, since the examiner will not conduct any examination other than verifying the suitability of the reproduction for publication.

This has been accepted by the General Court.[89]

[D] Description of Design (Where Required)

Section 11 (4) No. 1 Design Act allows for a description of the design. Different from other jurisdictions descriptions are not common in Germany. The description is limited to 100 words, section 10 (2) (1) Design Regulation. The description can only include words and cannot consist of graphical or other non text elements, section 10 (2) (2)

87. Verordnung zur Ausführung des Designgesetzes, DesignV, published Jan. 2, 2014 (BGBl 2014, 18).
88. OHIM Examination Guidelines for Design can be found on OHIM's website at www.oami.eu at http://oami.europa.eu/en/office/aspects/pdf/ExamGuidelines.pdf (status: Dec. 22, 2014).
89. EC T-68/10 para. 60 ff.

Design Regulation. The description can only relate to elements visible in the representation, section 10 (1) (1) Design Regulation.

A description can be useful to explain the relationship between different representations. The description can also describe the relevance of different visible elements in the representations.

§8.08 EXPUNGING/CANCELLING OR VARYING REGISTRATION

A design cannot be amended after it has been applied for. The owner can waive its rights in the design, section 36 Design Act. Interestingly, the owner can also cancel parts of a design, section 36 (2) design Act. It is fairly unclear what that could mean as it seems to be impossible to amend without changing the representation.[90]

The Community Design regulation provides for a possibility to amend a design in Article 25(6) of the Regulation. However, according to that stipulation the amendment must be in a form that retains the identity of the design. As it seems to be hard for the design to remain identical while amending there does not seem to be much room for partial cancellation.[91]

§8.09 OTHER

[A] Interface between Industrial Design and Trademarks

Intellectual property rights are independent and coexist. This means requirements for protection and the scope of protection vary for the different IP rights and are solely governed by the respective laws.[92] Therefore it is possible that an appearance can be protected under the German Trade Mark Law (Markengesetz) and Design Law at the same time or consecutively. This is likely to happen in the case of graphic symbols as they can be registered as figurative marks.

Three dimensional designs, such as the shape of goods or of their packaging, as well as other wrapping are eligible for trademark protection, too[93] which matches section 1 No. 2 Design Act. However, as stated by section 3 (2) No. 3 Trade Mark Act signs consisting exclusively of a shape which gives substantial value to the goods are not amenable to protection as a Trademark. This paragraph is fulfilled if the sign's aesthetic content is the only factor creating value.[94] So the aesthetics based Design Law falls exactly under this provision which was created to differentiate between IP rights.[95] Nevertheless under certain circumstances it is possible to register such an aesthetics

90. Similar David Stone, "European Union Design Law, A Practitioners' Guide," 2012, para. 12.148 in relation to Article 26(1) of the regulation: "Community designs are either valid or they are not."
91. For further information on the Community level *see* David Stone, "European Union Design Law, A Practitioners' Guide," 2012, para. 12.149 ff.
92. Ingerl/Rohnke, Markengesetz, 3rd ed. 2010, s. 2 para. 14.
93. *See* section 3 (1) Trade Mark Act.
94. German Federal Supreme Court (BGH), GRUR 2010, 138 – ROCHER-Kugel.
95. Ströbele/Hacker, Markengesetz, 11th ed. 2014, section 3 para. 131.

based three dimensional design as a Trademark. Because the exclusion does not apply to trademarks being established in trade circles according to section 8 (3) Trade Mark Act. So if the design is well known to a degree that it is perceived by the public as an indication of origin – a trademark's essential function – it overcomes the exclusion and can be protected as a Trademark.[96]

It most certainly is advantageous to seek trademark registration for a successful design because trademarks are protected without time limit because they can be renewed every ten years according to section 47 (2) Trade Mark Act.

[B] Relationship between Industrial Design and Copyright/Design Copyright

According to section 2 (2) Copyright Act (Urhebergesetz) it protects intellectual creations whereas the Design Act protects originality.[97] These two aspects meet in applied art. Contrary to visual art whose sole purpose is the art itself applied art does have a purpose of use – just like designs do.[98] For years German Courts had declared copyright paramount to design protection thus requiring higher standards for copyright protection of applied arts.[99] But because of legislative amendments designs have become an independent Intellectual Property right.[100]

Therefore design and copyright protection coexist[101] as they have different reasons for protection, prerequisites and different legal consequences.[102] Accordingly designs are per se eligible for copyright if they meet the general criteria of an intellectual creation. In the particular case of applied arts, an intellectual creation is given if the design's aesthetic effect is not dictated by the purpose of its use but due to an artistic accomplishment.[103] This requires the possibility and space to implement an idea thus permitting the expression of creativity.[104] But a commodity's purpose of use often dictates certain features leaving less freedom for creativity.[105] So it seems that copyright for designs will not be an easy path but seeking its protection may very well be worth it:

- Duration of copyright:
  it expires seventy years after the creator's death according to section 64 Copyright Act. This is especially attractive for older designs whose lack of novelty prevents a Design registration.
- Remuneration:

96. German Federal Supreme Court (BGH), GRUR 2006, 679 – Porsche Boxter.
97. *See* section 2 (3) Design Act.
98. German Federal Supreme Court (BGH), GRUR 2012, 58 para. 17, 34 – Seilzirkus.
99. German Federal Supreme Court (BGH), GRUR 2009, 856 para. 45 – TrippTrapp-Stuhl.
100. German Federal Supreme Court (BGH), GRUR 2014, 175 para. 35 – Geburtstagszug.
101. ECJ, C-168/09 paras 35 et seq. – Flos/Semerano.
102. German Federal Supreme Court (BGH), GRUR 2014, 175 para. 39 – Geburtstagszug.
103. German Federal Supreme Court (BGH), GRUR 2012, 58 para. 36 – Seilzirkus.
104. ECJ, C-393/09, paras 49 et seq. – BSA/Kultusministerium.
105. German Federal Supreme Court (BGH), GRUR 2014, 175 para. 41 – Geburtstagszug.

The creator has the right to remuneration for the permission to exploit his creation. As the judicial amendment leading to the coexistence of Design and Copyright Law came into effect on June 1, 2004 the creator can seek payment as of this date.[106]

[C] Relationship between Industrial Design and Unfair Competition

Unfair Competition Law sanctions deception and fraudulent behavior whereas Design Law protects a particular accomplishment.[107] Due to these differing reasons behind Unfair Competition and Design Law, as well as different prerequisites and legal consequences[108] protection under the Design and Unfair Competition Act coexists if the tort of the unfair action goes beyond the characteristics protected through the Design Act.[109]

This applies in case of section 4 No. 9 lit. a and lit. b Unfair Competition Act (Gesetz gegen den unlauteren Wettbewerb) when a person offers goods that are replicas of a competitors' goods and that causes avoidable deception of the purchaser regarding their commercial origin or unreasonably exploits or impairs the assessment of the replicated goods.[110] An avoidable deception is only possible if the copied product is known by the public[111] - without the knowledge of the original's existence there can be no deception about the origin.[112] Furthermore the public is deceived about the commercial origin of goods when they assume the replica's manufacturer is the same as the original's or an associated company.[113] This deception is avoidable if adequate and reasonable measures are able to prevent it.[114]

The second possibility of actionable exploitation or impairment refers to the good reputation of a product. So potential buyers need a positive image of the original product.[115] Section 4 No. 9 lit. b Unfair Competition Act protects against the transfer of that image to the replica[116] or the reputation's damage i.e., through inferior quality.[117]

106. German Federal Supreme Court (BGH), GRUR 2014, 175 para. 42 - Geburtstagszug.
107. German Federal Supreme Court (BGH), GRUR 2006, 79 para. 18 - Jeans I.
108. German Federal Supreme Court (BGH), GRUR 2013, 951, para. 20 - Regalsystem.
109. German Federal Supreme Court (BGH), GRUR 2003, 359, 360 - Pflegebett.
110. German Federal Supreme Court (BGH), GRUR 2008, 790 para. 35 - Baugruppe.
111. German Federal Supreme Court (BGH), GRUR 2009, 79 para. 35 - Gebäckpresse.
112. Köhler/Bornkamm, UWG, 32rd ed. 2014, para. 9.41a.
113. German Federal Supreme Court (BGH), GRUR 2001, 251 - Messerkennzeichnung.
114. German Federal Supreme Court (BGH), GRUR 2007, 339, para. 43 - Stufenleitern.
115. Köhler/Bornkamm, UWG, 32rd ed. 2014, para. 9.52.
116. German Federal Supreme Court (BGH), GRUR 2009, 500 para. 22 - Beta Layout.
117. German Federal Supreme Court (BGH), GRUR 2000, 521 - Modulgerüst.

CHAPTER 9

Industrial Design Rights: India

Sudhir D. Ahuja

§9.01 INTRODUCTION

[A] Highlights of New Developments in This Jurisdiction since Last Edition

The Designs (Amendment) Rules, 2008, was notified on 17 June 2008, and provided *inter alia*, a more detail classification of design conforming to international standards.

§9.02 CONVENTIONS AND LEGISLATION

Design registration in India is granted under the Designs Act, 2000. There is no utility model law in India. The formalities to be observed for securing design registration including physical requirements, prescribed forms and fees are governed by the Designs Rules, 2001 and Designs (Amendment) Rules, 2008.

India is a member of the Paris Convention and priority can be claimed from a design application filed within six months from the date of first filing of a design application in a convention country. India is not a member of the Hague Convention. However priority can be claimed from a design application filed under Hague Convention designating a Paris Convention member country.

§9.03 SUBJECT MATTER

[A] Definition/What Constitutes a Protectable Design?

[1] Shape, Configuration

'Design' means only the features of shape, configuration, pattern, ornament or composition of lines or colours applied to any article, whether in two dimensions or three dimensions or in both forms, by any industrial process or means, whether manual, mechanical or chemical, separate or combined, which, in the finished article appeal to and are judged solely by the eye, but does not include any mode or principle of construction or anything which is in substance a mere mechanical device, and does not include any trademark as defined in section 2(1)(v) of the Trade and Merchandise Marks Act, 1958, or property mark as defined in section 497 of the Indian Penal Code or any artistic work as defined in section 2(c) of the Copyright Act, 1957.

[a] Essential Requirements for the Registration of a 'Design'

In order for a design to be registered under the Designs Act, 2000, certain requirements must be met. First, the design shall be new or original and not previously published in India or in any other country by publication in tangible form or by use in any other way prior to the filing date, or where applicable, the priority date of the application for registration. The design shall be significantly distinguishable from known designs or combinations of known designs. The design shall not comprise or contain scandalous or obscene matter.

Publication of design occurs when the public is put in possession of the design either by making it available to the public in documents such as books or catalogue or when the design is disclosed to an individual member of public who is under no obligation to keep it secret.

Second, the design shall relate to features of shape, configuration, pattern or ornamentation of an article. Thus, designs of industrial plans, layouts and installations are not registrable under the Act.

Third, the design shall be applied to any article by any industrial process. Thus, designs of an artistic nature such as paintings, sculptures and the like which are not produced in bulk by any industrial process are excluded from registration under the Act.

Fourth, the features of the design in the finished article should appeal to and are judged solely by the eye. This implies that the design must appear and should be visible on the finished article for which it is intended. Thus, any design in the inside arrangement of a box, money purse or almirah cannot be registered by showing such articles in the open stage, as those articles are generally put on the market in the closed state.

Lastly, any mode or principle of construction or operation, or anything which is in substance a mere mechanical device, is not a registrable design. For instance, a key which is novel only in the shape of its corrugation or bend at the portion intended to engage the levers inside the lock it is associated with cannot be registered as a design under the Act. However, when a design suggests any mode or principle of construction, mechanical or other action of a mechanism, a suitable disclaimer in respect thereof is required on each representation, provided there are other registrable features in the design.

[b] Design Applied to Part of an Article

A design applied to part of an article is registrable under the Designs Act, 2000 only if that part is made and sold separately.

[c] What Is an Article?

'Article' means any article of manufacture and any substance, artificial, or partly artificial and partly natural; and includes any part of an article capable of being made and sold separately.

Stamps, labels, tokens and cards are not considered articles for the purpose of registration of a design because once the alleged design, i.e., ornamentation is removed, only a piece of paper, metal or like material remains and the article referred to ceases to exist. The article must have its existence independent of the designs applied to it. However, novel shapes of labels, tokens, card, etc., can be registered as designs.

[2] Ornamental or Functional

The design should relate to features of shape, configuration, pattern or *ornamentation* applied or applicable to an article. 'Design' under the Designs Act *does not include any mode or principle or construction or any thing which is in substance a mere mechanical device.* When any design suggests any mode or principle of construction or mechanical or other action of a mechanism, a suitable disclaimer in respect there of is required to be inserted on its representation, *provided there are other registrable features in the design.*

[B] Originality/Novelty

[1] Test for Originality Generally

'Original' in relation to design means:

(a) originating from the author of design;
(b) include cases which though old in themselves are yet new in their application.

[2] Novelty

Novelty in respect of designs is absolute. In order to meet novelty requirements the article should not have been published in India or in any other country. Novelty in India in respect of designs is lost if the product is published, sold or used within India or in any other country prior to the filing date, or where applicable the priority date, of the application for registration.

[C] Unobviousness (if Appropriate)

Not applicable

§9.04 DURATION OF PROTECTION

The duration of registration of a design is initially ten years from the *date of registration* (priority date or filing date, if priority is not claimed). The initial period of registration may be extended for a final period/term of five years on an application to the Controller before the expiry of the copyright, accompanied by the prescribed fee.

Where a design has ceased to have effect by reason of failure to pay the prescribed fee for extension of copyright, the proprietor of the design or his legal representative or one or more of joint holders without joining the others, may, with the leave of the Controller, make an application for restoration of the design in the prescribed manner with the prescribed fee, within one year from the date on which the design ceased to have effect.

§9.05 INFRINGEMENT

[A] Scope of Infringement Action

See §9.05[B] - Test for Infringement/Criteria to Constitute Infringement.

[B] Test for Infringement/Criteria to Constitute Infringement

The application of a design, or its fraudulent or obvious imitation, to any article belonging to the particular class of goods in which the design has been registered for the purpose of sale or importation of such articles without the consent of the registered proprietor constitutes infringement of the design. Publishing such articles or exposing terms for sale with knowledge of the unauthorized application of the design also amounts to infringement of the design.

[1] Legal Proceedings

Under section 22 of the Designs Act, 2000, during the existence of copyright in any design it shall not be lawful for any person:

(1) for the purpose of sale to apply or cause to be applied to any article in any class of goods in which the design is registered, the design or any fraudulent or obvious imitation thereof, except with the licence or written consent of the registered proprietor, or to do anything with a view to enable the design to be so applied;
(2) to import for the purpose of sale, without the consent of the registered proprietor, any article belonging to the class in which the design has been registered, and having applied to it the design or any fraudulent or obvious imitation thereof; or
(3) knowing that the design or any fraudulent or obvious limitation thereof has been applied to any article in any class of goods in which the design is registered without the consent of the registered proprietor, to publish or expose or cause to be published or exposed for sale of that article.

If any person acts in contravention of this section, he or she shall be liable:

(1) to pay to the registered proprietor of the design a sum not exceeding INR 25,000 (USD 417 approx.) recoverable as a contract debt; or
(2) if the proprietor elects to bring a suit for the recovery of damages for any such contravention, and for an injunction against the repetition thereof, to pay such damages as may be awarded and to be restrained by injunction accordingly, provided that the total sum recoverable in respect of any one design under clause: (i) shall not exceed INR 50,000 (USD 834 approx). When the Court grants a decree in a suit under the above provisions, it shall send a copy of the decree to the Controller, who shall cause an entry thereof to be made in the Register of Designs.

Suit for infringement of registered design cannot be instituted in any court lower than that of a District Judge, and, in the event of defence being taken by the defendant or alleged infringer on the grounds available for cancellation of registration of the concerned design, the proceedings shall be transferred to the High Court for decision.

[C] Standing to Sue

A proprietor shall have no right to institute a suit or proceeding in respect of piracy of design or infringement copyright, which has been committed between the date on which the design ceased to have effect and the date of restoration of the design.

[D] Remedies for Infringement

[1] Penalty for Infringement of a Registered Design

Anyone contravening the copyright in a design is liable to pay for every offence a sum not exceeding INR 25,000 to the registered proprietor, subject to a maximum of INR 50,000 in respect of any one design. The registered proprietor may bring a suit for the recovery of the damages for any such contravention and for an injunction against repetition of the same.

§9.06 DEFENCES TO INFRINGEMENT

[A] Validity

There are four defences to allegations of infringement of a design:

(1) The design has been previously registered in India.
(2) The design has been published in India or in any other country prior to the date of registration.
(3) The design is not a new or original design.
(4) The design is not registrable under the provisions of the Designs Act, 2000.

[B] Marking

[1] Marking of an Article

It is not compulsory that an article to which a registered design has been applied be marked by the registered proprietor. However, it is advantageous to the registered proprietor to mark the articles so as to indicate the registered design number (except in the case of textile designs and in respect of articles made of charcoal dust which are brittle). Otherwise, as is the case with patents, a registered proprietor would not be entitled to claim damages from any infringer unless it can be shown that all proper steps to ensure the marking of the article were taken, or unless it can be shown that the infringement took place after the infringer knew or had received notice of the existence of the copyright in the design.

An infringer who has proceeded on the basis of a report from the Patent Office that no registration subsists in respect of a particular design and is later held as infringing copyright for the said design can plead as an innocent infringer and get relief from damages.

[C] Non-infringement

Not applicable

[D] Other Defences

Not applicable

§9.07 PROCEDURES FOR FILING APPLICATION FOR REGISTRATION

[A] Who Is Entitled to Register a 'Design'?

Any person claiming to be proprietor of a new or original design not previously published in any country, and which is not contrary to public order or morality, may apply for registration of such design to the Head Office of the Patent Office in Calcutta and in the Branch Offices of Indian Patent Office in Mumbai, Delhi and Chennai. However, design applications are examined, prosecuted and registered only in Calcutta (Kolkata) Head Office. In case of an application claiming priority, the application should be made by a person who made the application in the United Kingdom or any of the other convention countries or groups of countries that are members of inter-governmental organizations, or by his or her legal representative or assignee, within six months from the date of the convention application. Convention countries include members of the Paris Convention and the World Trade Organization.

[B] Procedure for Filing an Application for the Registration of a Design

In India, there is no provision to file a multiple-design application. Separate applications have to be filed for designs which are similar but not the same. Protection can be obtained for a design applied to an article in a particular configuration/state/position. Separate applications have to be filed for designs illustrating the article in different configurations/states.

In India, divisional applications cannot be filed for designs. It is recommended to file the important variations in the first instance.

Application for registration of a design (Convention and non-Convention) under the new Designs Act, 2000, shall be filed on the prescribed form (Form 1 under the Designs Rules, 2001) accompanied by the prescribed fee.

A design may be registered in more than one class by making separate applications, and in respect of any or all of the articles in a prescribed class of articles.

Requirements for filing a design application remain as follows unless notified otherwise:

(1) Four copies of drawings, photographs, tracings or specimens suitable for record, duly prepared or pasted on durable paper of A4 size should be filed in the Calcutta Patent Office, or Branch Offices of the Indian Patent Office in Mumbai, Chennai and Delhi, accompanied by the prescribed fee.
(2) The design should be shown as applied to the article and if one view of the article is not sufficient to properly show the design, different views such as side elevation, top view, bottom view, front view, rear view, or perspective

views should also be filed. In case of textile designs with repeated surface pattern, each specimen should be of sufficient portion of the repeat in length and width and shall not be less than 5 inches by 4 inches or 13 centimetres by 10 centimetres, so as to clearly show at least one full repetition of the design.

(3) The name, address and nationality of the applicant must be provided. A power of authority can be filed any time within six months of filing the application (no legalization is necessary). One general power of authority is sufficient for all applications by a particular applicant.

India became a member of the Paris Convention with effect from 7 December 1998. Priority can be claimed by anyone who first filed applications in the United Kingdom or any of the other convention countries or groups of countries which are members of inter-governmental organizations. The list of convention countries including members of the Paris Convention and the World Trade Organization will be notified separately. Priority cannot be claimed from an EPO/PCT application even if it designates one of the convention countries. Certified copy of the design first filed in a convention country is required for a priority application. It is preferable that a brief statement of novelty claimed for the design be endorsed on each copy of the design.

Where a design has been registered in respect of an article contained in a class, the application of the proprietor of the design to register it in respect of other article(s) in that class shall not be refused, nor the earlier registration invalidated, either on the ground of novelty, or on the ground of publication, by reason only that it was previously registered, provided that such subsequent registration shall not extend the period of copyright in the design beyond that arising from the previous registration.

Further, if any person applies to register in respect of an article, a design not substantially different from a design previously registered by another person in respect of the same or some other article, and if at any time while the application is pending the applicant becomes the registered proprietor of the previously registered design, then the applicant shall be treated as the registered proprietor of the previously registered design from the time of making the application.

The Designs Rules, 2000, has notified the adoption of the International Classification for Goods.

In case of any doubt regarding the class to which any particular description of article belongs, it shall be determined by the Controller, whose decision shall be final.

[C] Post-Application Stage

Following an application for registration of a design, the application is numbered, dated and examined and defects (objections), if any, are communicated to the applicant.

The applicant should meet the official objections and put the application for the design in order for acceptance within six months from the date of application. The said period of six months may be extended for maximum three months under Rule 18 of the Designs Rules, 2001, as amended in 2008, with the prescribed fee before the expiry of the six months. When an application for registration of a design is in order, it will be accepted; on acceptance, a certificate of registration is issued to the applicant and the registration is recorded in the Register of Designs.

The Designs Rules, 2000, provides for the applicant or his agent to be 'heard' by the Controller before the Controller takes the final decision for grant or refusal of a design. Appeal from the Controller's decision will lie before the High Court.

The Controller has been empowered to make orders regarding substitution of applications at any time before a design has been registered on satisfactory submission, in the prescribed manner, of claims of joint applicants or rights of surviving applicant(s) by common consent or by operation of law.

On acceptance of the design in respect of an application, the Controller has also been empowered to cause publication of the particulars and registration of the design, and as notified in the Designs Rules, 2000, one or more view(s) of the design will be selected by the Controller for its publication in the Gazette, which in his opinion would best depict the design.

[D] Register of Designs

The Register of Designs is maintained by the Calcutta Patent Office as a statutory requirement. It contains the design number, the class number, date of filing (in India) and reciprocity date (if any), the name and address of the proprietor, and such other matters as would affect the validity or proprietorship of the design.

The Register of Designs is open to public inspection on payment of the prescribed fees, soon after the design is registered.

The registration of a design confers upon the registered proprietor copyright in the design for the period of registration. 'Copyright' means the exclusive right to apply a design to any article belonging to the class in which it is registered.

§9.08 EXPUNGING/CANCELLING REGISTRATION

The registration of a design may be cancelled on any of the following grounds:

(1) the design has been previously registered in India;
(2) the design has been published in India or in any other country prior to the date of registration;
(3) the design is not a new or original design; or
(4) the design is not registrable under the provisions of the Designs Act, 2000.

A petition to the Controller for cancellation of the registration of a design may be made at any time after the registration of the design on the basis of the above grounds. Appeal lies before the High Court against any order of the Controller, in the above matter.

§9.09 OTHER

[A] Amendments in the Copyright Law

The Copyright Act, 1957, amended in 1994, offers some measure of protection to the owners of copyright in two-dimensional drawings of three-dimensional works. The 1994 amendments to the Copyright Act, 1957, which came into force on 10 March 1995, expand the purview of section 14 (the Meaning of Copyright) by defining copyright as 'the exclusive right to do or authorize the doing of, among other acts in respect of a work or any substantial part thereof, the three dimensional reproduction of a two dimensional work in case of artistic works such as industrial designs'.

Engineering and product drawings qualify as works of art under Copyright Law and the new amendments would effectively bar 'drawing' to 'product' copying, i.e., a three-dimensional production of an article from a two-dimensional drawing/illustration in which copyright subsists would be considered contrary to the law.

[B] Procedure for Litigation

As the law stands now, the best way to secure authenticity of an industrial design is by obtaining registration for the design under the Designs Act, 2000.

If no such registration is possible or available, the other option is to bring an action for passing off. To sustain such an action, the plaintiff is required to establish exclusive association of the design in question with its proprietor in the eyes of consumers. This may be proved by extensive publicity given to the product by its owners/manufacturers, not only by its name, but through other means. The design must also be distinctive of the plaintiff, and not a mere copy of the other products in the market performing the same or similar function.

In exceptional cases, it could also be possible to advance a case of copyright infringement to check the piracy of an industrial design, but in that case one should note the limitations imposed under section 15(2) of the Copyright Act, 1957 which states:

> Copyright in any design, which is capable of being registered under the Designs Act, 1911, but which has not been registered, shall cease as soon as any article to which the design has been applied has been reproduced more than fifty times by an industrial process by the owner of the copyright, or with his licence, by any other person.

[C] Non-prejudicial Disclosure

Section 21 of the Designs Act, 2000, relating to the provisions as to exhibitions, states that the exhibition of a design, or of any article to which a design is applied, at an industrial or other exhibition to which the provisions of this section have been extended by the Central Government by a notification in the Official Gazette, or the publication of a description of the design before or after the period of the holding of the exhibition, or the exhibition of the design or the article, or the publication of a description of the design by any person elsewhere during or after the period of the holding of the exhibition, without the privity or the consent of the proprietor, shall not prevent the design from being registered or invalidate the registration thereof provided that:

(1) the exhibitor exhibiting the design or article, or publishing the description of the design, gives prior notice to the Controller in the prescribed form; and
(2) the application for registration is made within six months from the date of first exhibiting the design or article or publishing a description of the design.

[D] Unauthorized Disclosure

An unauthorized disclosure of the subject design by a third party shall not affect the novelty of the design for an application for design registration in India under section 16 of the Designs Act 2000. The proprietor of the design has to establish that such disclosure was made by the third party in breach of good faith, when the registration is challenged on the ground of novelty for such prior disclosure.

Although there is no specific time limit prescribed under this section, it is recommendable to file the design application as soon as unauthorized disclosure of the design comes to the knowledge of the proprietor of the design.

[E] Searches and Copies of Designs

Searches can be conducted and a constant watch maintained in respect of new designs registered in any class of articles. Copies of registered designs can be obtained at any time after the publication of a design in the Register of Designs. Under the Designs Act, 2000, publication of particulars of registered designs will take place immediately after registration.

CHAPTER 10

Industrial Design Rights: Italy

Luigi Pavanello

§10.01 INTRODUCTION

[A] Highlights of New Developments in This Jurisdiction since Last Edition

Italian law has been substantially reorganised and some important changes have been introduced in respect of industrial design since the last edition of this book.

Now, under Italian Law,[1] the protection of industrial design is generally obtained by means of a model patent. The exclusivity rights last for five years renewable for additional five-year terms up to a maximum term of twenty-five years. The validity requirements, the procedure for the issuance, the formalities to be complied with, and the protection afforded to a design patent are modeled on those of "regular" patent. In appropriate cases, other forms of protection, such as the ones afforded by Italian Copyright Law or the Trademark Law is and may be sought and even combined, when appropriate, with the protection afforded by a design patent. Copyright was rarely used, or rather was rarely permitted to protect industrial design, since: (i) industrial design could not be contemporaneously protected pursuant to a design patent and a copyright and (ii) the use by the Italian courts of the *scindibilità* test (separation test), i.e., the possibility of dividing the aesthetic/artistic portion from the functional/industrial character of the article of manufacture, set forth in Article 2, paragraph 4 of the Copyright Law,[2] severely curtailed, if not almost always prevented any attempt to protect industrial design under the Copyright Law. The Copyright Law

1. Legislative Decree No. 30 dated Feb. 30, 2005.
2. Law No. 633 dated Apr. 22, 1941 (the "Copyright Law").

grants protection and exclusivity rights for a long period of time, that is, as a general rule, until seventy years after the death of the author of the work. It should be noted though, that Article 27 of Law No. 266 of August 7, 1996 had expressly limited the copyright protection of industrial design works to only fifteen years, i.e., for the same period of time as a design patent as then in force. Such term was then extended to twenty-five years (i.e., the maximum protection afforded to model patents) by Legislative Decree 95/2001 until the enactment of Article 44 of Legislative Decree 131/2010, which has finally equated the term of the protection afforded to industrial design to the one granted to the other works protected by copyright (i.e., until seventy years after the death of the author of the work).

The issue, which made the legislative life of industrial design so troublesome in Italy was whether the same could be also protected under copyright laws in Italy. The Copyright Law provides paragraph number 4 of Article 2 of this law provided that sculpture, painting, design and drawing, engraving and similar figurative arts, works are all protected by the law, provided that their artistic value/portion may be separated from the industrial character of the article of manufacture to which they are associated. This paragraph of Article 2 of the Copyright Law set forth the so-called *scindibilità* test (separation test), which, as interpreted by the Italian courts and the majority of the Italian scholars,[3] provided that in case of art applied to industrial products, such as industrial design, said products may be copyrighted only in the event that the aesthetical/ artistic portion of the work may be separated and distinguished from the industrial/functional character of the same. The majority of the Italian courts interpreted this provision very narrowly and, in the past, have consistently denied that industrial design may be protected by means of the Copyright Law,[4] so that in Italy, it has been maintained that industrial design could be protected under the Copyright Law only if the aesthetic portion of the industrial design could be independently enjoyed from the resulting industrial product, as any other work of art.[5] This position of the majority of the Italian courts and scholars was perfectly summarised by the *Corte di Cassazione*, the Italian Supreme Court in Civil and Criminal matters, which in denying copyright protection to a piece of furniture held:

> the original cut of a dress or suit, the new design of the body of a car, the texture of a piece of fabric, the shape of a dish, or, as in this case, the peculiar structure of

3. *See*, for example, Corte di Cassazione, Jul. 5, 1990, no. 7077, *Foro Italiano*, Giuffrè, Milano, Rep. 1990, item "Diritto d'Autore" no. 68; Corte di Cassazione, Dec. 5, 1994, *Poro Italiano*, Giuffrè, Milano, 1995, I, 811; for a detailed summary of the opinion of the Italian scholars on point, *see* G. Bergomi, *op. cit.*, endnote no. 1; and M. Franzosi, "La protezione de! disegno industriale (Unmittelbare Leistungsübernahme quale base di protezione)", *Contratto ed Impresa*, CEDAM, Padova, 1991, I, 97.
4. *See* for example, Corte di Cassazione, Jan. 25, 1933, no. 359, *Il Diritto d'Autore*, Giuffrè, Milano, 1933, 51; Corte di Cassazione, Jan. 27, 1977, no. 3238, *Giun:Sprudenza Annotata di Diritto Industriale*, Giuffrè, Milano, 1978, no. 993, 3; and Tribunal of Milan, Jun. 12, 1980, *Giurisprudenza Annotata di Din'tto Industriale*, Giuffrè, Milano, 1980, no. 1321, 492.
5. *See* for example, Corte di Cassazione, Jan. 27, 1977, no. 3238, *Giurisprudenza Annotata di Diritto Industriale*, Giuffrè, Milano, 1978, no. 993, 3; Appeal Court of Florence, Jan. 9, 1978, *Giurisprudenza Annotata di Diritto Industriale*, Giuffrè, Milano, 1978, 202; and Appeal Court Milan, Sep. 28, 1976, *Giurisprudenza Annotata di Diritto Industriale*, Giuffrè, Milano, 1976, 701.

a piece of furniture, can never be separated from the industrial product to which they are associated with.[6]

For the foregoing reasons, in the past the Italian courts had consistently refused to grant copyright protection to industrial design products, including, for example, a stove,[7] a lamp,[8] a bed,[9] a bag,[10] an ashtray, a basket and an umbrella stand declared to be of a high artistic value,[11] and even to chairs designed by Le Corbousier (Chaisse-long),[12] Breuer (Wassily),[13] Rietwield of the Der Stijl artistic movement (red and blue).[14]

The Italian Parliament tried to substantially reform the protection afforded to industrial design. In fact, with the passing of subsection 58 of Article 1 of Law No. 650 of December 23, 1996 for the first time, it had, expressly allowed the protection of industrial design pursuant to the provisions of the Copyright Law, then in force. The exact limits, the procedure and the requirements for, and manner of this protection were left to regulations that should have been issued by the Italian government. The issuance of such regulations was subject to the enactment of a European Union ("EU") Directive covering this matter, so that the same would comply and implement the provisions of such proposed Directive. Unfortunately, EU Directive 98/71/CE on Industrial Design has been issued only on October 13, 1998 and, in the meantime, the Italian Parliament repealed subsection 58 Article 1 of Law 650/1996 reinstating the *status quo ante* (paragraph 1 of Article 27 of Law 266/97), until the implementation of the EU Directive on Industrial Design in Italy. Thereafter, the Italian Parliament and government decided to reorganise the intellectual sector by uniting in a single document, restating and revising most of the legislation (which originally dates back to the early 1940s) in order to take into account developments in the law, EU legislation and international treatises, and enacted Legislative Decree No. 30 dated February 10, 2005, also known as the Industrial Property Code, as subsequently amended ("IPC").

In addition to the Italian national legislation, EU legislation is also available to protect industrial design, with one major difference, contrary to EU law, under Italian law no protection is afforded to "non registered" industrial design, so that to protect "non registered" industrial design a party can only rely upon EU law, which is part of Italian Law and is duly applied by Italian courts.

After the aforesaid reforms Italian industrial design law and protection are basically in line with the major industrialised countries.

---

6. Corte di Cassazione, Jan. 27, 1997, no. 3238, *Giurisprudenza Annotata di Diritto Industriale,* Giuffrè, Milano, 1978, no. 993, 3.
7. Corte di Cassazione, Jan. 20, 1932, no. 317, *Foro Italiano,* Giuffrè, Milano, 1932, I, 869.
8. Appeal Court of Milan, Sep. 17, 1914, unpublished.
9. Appeal Court of Milan, Dec. 31, 1937, unpublished.
10. Appeal Court of Milan, Feb. 16, 1962, unpublished.
11. Tribunal of Milan, Oct. 3, 1973, *Foro Italiano,* Giuffrè, Milano, Rep. 1974, item "Diritto d'Autore", no. 13.
12. Corte di Cassazione, Dec. 7, 1994, no. 10516, *Foro Italiano,* Giuffrè, Milano, 1995, I, 811.
13. Corte di Cassazione, Jun. 26, 1980, *Giurisprudenza Annotata di Diritto Industriale,* Giuffrè, Milano, 1980, no. 1258, 61.
14. Corte di Cassazione, Oct. 22, 1977, no. 3238 (criminal case).

§10.02 CONVENTIONS AND LEGISLATION

The main legislation is represented by the IPC, which basically contains the whole of the Italian legislation in respect of intellectual property including industrial design, except copyright, which is still regulated by Copyright Law, as amended.

From a civil procedure law, an important piece of legislation is Legislative Decree No. 168 dated June 27, 2003 which has created an intellectual property specialised section in each *Tribunale* (trial court) sitting where twelve Appellate Courts[15] have their seat. All of the intellectual property litigation in the circuit of any of these Appellate Courts must be filed with and will be decided by the locally competent intellectual property specialised section and no longer by a random judge (who may or may not be an expert in intellectual property law).

EU Directives, when implemented, and EU Regulations, upon enactment, are also part of Italian law, including:

- EU Directive 98/71/CE on Industrial Design dated October 13, 1998.
- EU Regulation 207/2009 on EU trademark dated February 26, 2009.
- EU Regulation 6/2002 dated December 12, 2001 of EU design and models (the "EU Design and Models Regulation").
- EU Regulation 2245/2002 dated October 21, 2002 on rules of procedure of the Board of Appeals of the Office for the harmonisation in the Internal Market.

Italy is also a party to the following multiparty international conventions:

- Paris Convention on industrial property dated March 20, 1883 (the "Paris Convention").
- Madrid Agreement on the international registration of trademarks dated April 14, 1891, as amended.
- Nice Agreement regarding the international classification of goods and services for the purposes of registration of marks dated June 15, 1957, as amended.
- Locarno Agreement establishing an international classification of Industrial Designs, as amended.
- Patent Cooperation Treaty dated June 19, 1970.
- Strasbourg Agreement on the international classification of patents dated March 24, 1971 as amended.
- European Patent Convention dated 5 October 1973.
- Marrakesh TRIP agreement dated April 15, 1994.
- Protocol relating to the Madrid Agreement regarding the international registration of trademarks.
- Trademark Law Treaty dated October 27, 1994.
- Patent Law Treaty dated June 1, 2000.
- Singapore Treaty on trademarks dated March 27, 2006.

15. Bari, Bononia, Catania, Florence, Genua, Milan, Naples, Palermo, Roma, Turin, Trieste, Venice.

- Common regulations under the Madrid Agreement regarding the international registration of trademarks and the Protocol relating to that Agreement.

In addition, Italy has entered into bilateral agreements concerning INTELLECTUAL PROPERTY including:

- Joint Declaration with the United States Patent and Trademark Office dated June 17, 2003.
- Cooperation Agreement and several memorandum of understanding with the People Republic of China Patent and Trademarks Office.

§10.03 SUBJECT MATTER

Under Italian Law, industrial design is generally defined as "the design of the shape of mass production goods for the purpose of harmonizing aesthetic and functional character of the article [of manufacture],"[16] and, as outlined in detail below, the protection of same is generally based upon model patents which grant exclusivity rights and consequent protection from infringement for up to twenty-five years (the protection is for five years and may extended for additional five year terms up to a maximum of twenty-five years).[17] In the event that copyright protection is also applicable and obtained then the exclusivity rights and consequent protection from infringement until seventy years after the death of the author.

[A] Definition/What Constitutes a Protectable Design?

Design and Model

Under Italian Law, the protection of industrial design is now generally based upon the provisions on "models" and "designs" set forth in Articles 31 through 44 of the IPC.

[1] Shape, Configuration

According to Article 31 of the IPC design and models may be registered as "models" in the event they provide the shape or configuration of certain industrial products or a part thereof, as it results from a peculiar combination of lines, contours, colors, shape, surface structure or the material of the product or its ornament provided that they are new and have an individual distinctive character.

16. S. Auteri, Industrial Design, in *Dizionario di Diritto Commerciale ed Industriale,* Giuffrè, Milano, 1981, 585; *see also,* G. Bergomi, "Industrial Design", *Contratto ed Impresa,* CEDAM, Padova, 1987, II, 966.
17. Article 37 of the IPC.

According to the same Article 31 the term "product" means any industrial or artisanal object, the components to be assembled, the packaging, the presentations and the graphic symbols and fonts.

[2] Ornamental or Functional

It should be noted that the characteristics of the shape of the product that are mandated solely by the technical function of the product cannot be validly registered as a model or design.[18]

[B] Originality/Novelty

[1] Test for Originality Generally

A design or a model is deemed to have an individual distinctive character, if the general impression to an informed observer of the design or model differs from the general impression to the same observer by any other design or model published before the date of filing or the application or, in case priority rights are claimed, before the date of the latter. In ascertaining the individual distinctive character the freedom enjoyed by the author in developing the design or model is taken into account.[19]

A design or a model is deemed published if it has been made available to the public by the filing of the application or otherwise, or, if it has been exposed, marketed or otherwise made publicly available, unless such events could reasonably be known by specialised operators in the relevant field, operating in the European Union, during normal operations prior to the filing of the application and or, in case priority rights are claimed, before the date of the latter. Disclosure to a party subject to a confidentiality or non-disclosure obligation does not amount to publication, nor: (i) it is deemed publication any action or disclosure of any information by the author or any third party in the twelve months preceding the filing of the application and or, in case priority rights are claimed, before the date of the latte, nor (ii) it is deemed publication the mere fact that the design or model has been made public by means of, directly or indirectly, any infringement of or violation of the author or any third party rights, nor (iii) the publication requirement is met by the mere exposition of the model or design during trade fairs.[20]

[2] Novelty

A design or a model is deemed new if no identical model or design has been published prior before the fling of the application[21] or, whenever priority rights are asserted

18. Article 36 of IPC.
19. Article 33 of IPC.
20. Article 34 of IPC.
21. Article 31 of IPC.

before the filing of the application. Designs or models are deemed identical when their characteristics differ only for irrelevant minor details.[22]

It should be noted that the characteristics of the shape of the product that are mandated solely by the technical function of the same cannot be validly registered as a model or design.[23]

Finally, the protection and rights afforded by the law to a design or a model may be combined, in the appropriate cases, to ones granted by a utility patent and the copyright.[24]

[a] Utility Patent

Although industrial design was generally protected by means of a model or a design, in the appropriate case, even a utility patent may be used to protect industrial design.

Utility patents are defined as "the models which enhance the effectiveness, ease of application or use, of machinery, parts thereof, equipment, tools or any other article [of manufacture] normally used, such as new models having new shapes, disposition or combination or parts."[25] For example, if a certain article is protected only by a utility patent, but the utility of such patent is in whole or in part related to the particular shape of such article, then in practice, for as long as the utility patent is in force, such article is indirectly protected as if a model or design application had been filed. In fact, no one could copy or counterfeit such shape, which theoretically is subject to a patent in respect of its functional/industrial character, but not in respect of its aesthetic/artistic character, since the shape to be used is always the same and cannot be used without violating the exclusivity rights of the holder of the utility patent.

Moreover, it should also be noted that Article 40 of IPC expressly allows the simultaneous filing of an application for both a utility patent and a design patent, in the event that the shape or design of an article of manufacture provides such article not only with a new ornament, but also enhances the utility of the article. However, the same article specifies that the protection afforded cannot be cumulated in a single application, but two separate applications should be obtained.

This means that in Italy, under certain circumstances, industrial design articles may be protected in their technical/industrial portion, as well as in their aesthetic/artistic portion, for the time periods respectively provided by the law.

An issue concerning industrial design that has had a particular development and interest in Italy is whether the body of a car must be patented (with a design patent) as a whole or whether a separate design patent can also be obtained for single parts of the same. The importance of this issue is clear. If a car producer is entitled to patent each single part of the body of a car (e.g., doors, lights), no one could manufacture spare parts or replacements except for the owner of the design patent on the car body parts.

22. Article 32 of IPC.
23. Article 36 of IPC.
24. Articles 40 and 44 of IPC.
25. Article 82 of IPC.

Both the Italian courts and scholars appear to be divided;[26] although it should be noted that in the most recently published decision,[27] the Court of Appeals of Turin, in confirming the trial court decision[28] on different grounds, has held that a part of the body of a car (in the specific case, the lights) cannot be separately patented with a design patent separate from the one protecting the car body to which it belongs. The court held that a separate design patent cannot be obtained since a single part cannot be separately used by the consumer because it is an integral and substantial part of the body of the car and therefore lacks the "utility" requirement required under Italian Law for the issuance of valid design patent. Second, the part loses its separate patentability once it becomes part of the car body that is patented as a whole (in practice, a prohibition against double patenting).

The term of the protection starts from the date of filing of the application and lasts for ten years.

[b] Copyright

In the past, alternatively and in addition to the traditional protection afforded by design patents, an attempt was often made in Italy to protect industrial design by means of a copyright. However, in most instances such attempts were frustrated by the narrow and strict interpretation of the provisions of the Copyright Law by the majority of the Italian courts.

As stated above, Article 44 of the IPC expressly provides for Copyright Law rights to apply to industrial design. Of course, they would apply if Copyright Law applies to the specific product.

Article 2, paragraph 1 No. 10 lists industrial designs among the works that may be protected under Copyright Law. However, in order to be afforded the protection granted by Copyright Law, the requirements provided by such law must be met.

In a recent decision, the intellectual property section of the Court of Rome[29] has clearly outlined them in recently issued order. First the Court clarified and confirmed that: (i) industrial design works (i.e., products) are among the artistic works protected by the Copyright Law and (ii) neither the mass production nor the fact that the work may be reproduced or that it is aimed to the consumer market are valid grounds to deny

26. *See*, N. Zorzi, *Diritti di Privativa su pezzi di ricambio e disciplina de/la concorrenza Gee,* Contratto ed Impresa, *CEDAM v. l., Padova*, 1989, I, 419; G. Floridia, *Ancora sui pezzi di ricambio per macchina*, Rivista di Diritto Industriale, Giuffra, Milano, 1985, II, 511; Tribunal of Turin, Jun. 19, 1989, *Foro Italiano,* Giuffrè, Milano, Repertorio 1990, Giuffrè, Milano, item "Patents" nos. 75–77 and *Rivista di Diritto Industriale,* Giuffrè, J'vl.ilano, 1989, II, 337, note Franceschetti; and Tribunal of Milan, Nov. 10, 1986, *Foro Italiano,* Repertorio 1988, Giuffrè, Milano, item "Patents" no. 39.
27. Appeal Court Turin, Jan. 18, 1993, *Foro Italiano,* Giuffrè, Milano, I, 1994, 2232.
28. Tribunal of Turin, Apr. 21, 1990, *Rivista di Diritto Industriale,* Giuffrè, Milano, II, 13, note Vignali.
29. Tribunal of Rome, IP Section, no. 16642/2014 dated Sep. 8, 2014, *see also* Tribunalof Milan no. 9173 dated May 3, 2012.

protection under the Copyright Law. With respect to the requirement for the granting of the copyright protection the Court has stated that the work (i.e., the industrial design product) must present and have a distinctive and creative character expressing some kind of artistic value. The court further added that: (i) the artistic vale is no per se excluded by the fact the work has also a "practical" function and (ii) the artistic value of the work may deemed ascertained if the public generally acknowledges such value to the work (e.g., though exhibitions, notoriety, expert valuations); though, such an acknowledgement by the public requires the underlying fact that the work had been presented to the public as the expression of the personality and creativity of the author and that the latter has maintained that it was his/her intention to create a new and creative work.

In short, copyright protection is available to industrial design products only if the product has a recognised artistic value and such protection cannot be denied even if the product may be mass produced.

As a final note, it is worth remembering that even before the recent changes in the law, the Italian courts have occasionally declared copyright applicable to industrial design in the case of bi-dimensional products, such as an artistic and innovative drawing applied to a product[30] provided, of course, that such bi-dimensional product is artistic, that is, creative, new, and original. This different treatment has been justified by the Italian courts on the ground that, contrary to an industrial design tridimensional product, a bi-dimensional product may be enjoyed as art separately from the industrial product on which it is applied, since the industrial product may be sold even without the art work applied on it, and the bi-dimensional art work may be recognised and enjoyed as art regardless of the type of support on which it is applied.[31]

[c] Trademark

Industrial design may also be protected with a trademark; provided that, certain conditions are met; though, this happened rarely and has not be much used lately because of the difficulty of obtaining a valid one.

Trademark protection is not alternative, but may be combined with the protection granted by a design patent and/or a utility patent so that the industrial design product may be fully protected for as long as the trademark (valid for ten years and indefinitely renewable for additional ten year periods) and the design/model registration and/or a utility patent (valid for twenty-five and ten years respectively) are valid and in force.

Clearly, once the design and/or utility patents have expired, the only protection available will be the one provided by the trademark in respect of the distinctiveness of the mark.

30. Corte di Cassazione, Oct. 22, 1956, no. 3806, *Foro Italiano*, Giuffrè, Milano, 1957, I, 5.
31. Needless to say that in the past there were only a handful of cases in which the Italian courts have permitted industrial designs to be protected by means of copyright. *See*, for example, most recently, Corte di Cassazione, Dec. 7, 1994, no. 10516, *Foro Italiano*, Giuffrè, Milano, I, 811.

In Italy, in particular after the amendments to the provisions of law concerning *"Marchio di Forma"*, i.e., a tridimensional trademark, have been carefully redrafted in order to avoid the situation where a trademark can be used to circumvent the public policy underlying design, models, inventions and utility patents, namely, requiring that the same will become part of the public domain after a certain period of time. In this manner, industrial design or another patent cannot be granted the theoretically indefinite exclusive rights and protection granted by law to a valid trademark.

For the foregoing reason, Article 9 of the IPC provides that:

> the marks resulting solely by the shape imposed by the same nature of the products, the shape of the product required to obtain a certain technical result, or the shape which gives the product its value, cannot be registered as trademarks.

Although it has been originally introduced for the purposes of implementing an EU Directive, the above-reported amended text of Article 9 of the IPC clarifies and incorporates into a statutory provision the conclusions that have already been reached by the Italian courts and scholars.

In fact, after having for a long time denied the validity of a tridimensional trademark, the Italian courts have slowly changed their position and admitted the validity of such trademarks, provided that the distinctive character of the mark can be separated from its utility (i.e., the functional character) and shape (i.e., the aesthetic/artistic character).[32] Nevertheless, according to the Italian courts, such separation occurs whenever the distinctiveness of a drawing or shape on one side and its aesthetic/artistic value and/or technical utility on the other side, may actually be deemed autonomous from one another in light of the cultural and technical framework in which they have been created.[33] For the foregoing reason, the Italian courts have upheld the validity of the tridimensional trademarks represented, for example, by the bottle of the *"Campari Soda"* aperitif[34] or the *"San Pellegrino"* soft drinks,[35] or the peculiar container for yogurt of a dairy producer.[36]

In short, under Italian Law a certain shape may be validly registered as a tridimensional trademark both if such shape is new, distinctive and lawful and secondly if such shape also has an aesthetical/artistic value and/or a utility, provided however, in the latter case that the distinctiveness of the mark may be separated for its aesthetic/artistic value and/or utility, otherwise the trademark would be null and void.

---

32. *See*, for example, Corte di Cassazione. Apr. 7, 1975, no. 1233, *Giurisprudenza Annotata di Diritto Industriale*, Giuffrè, Milano, 1975, 32; Corte di Casszione, May 21, 1981, no. 3333, *Giustizia Civile*, Giuffrè, Milano, 1981, I, 2960; and Appeal Court of Milan, May 26, 1978, *Rivista di Diritto Industriale*, Giuffrè, Milano, 1980, II, 113.
33. *See*, Tribunal of Milan Milan, Feb. 5, 1976, *Rivista di Diritto Industriale*, Giuffrè, Milano, II, 1977, 47; and Tribunal of Milan, Feb. 26, 1976, *Foro Padano*, Foro Padano di Mario Franzosi & C. S.a.s., Milano, 1976, I, 190.
34. Tribunal of Rome, Oct. 9, 1972, *Giurisprudenza Annotata di Diritto Industriale*, Giuffrè, Milano, 1972, 1329.
35. Tribunal of Milan, Jan. 19, 1961, *Foro Italiano*, Giuffrè, Milano, I, 1254; and Corte di Cassazione, Apr. 7, 1975, no. 1233, *Giurirprudenza Annotata di Diritto Industriale*, Giuffrè, Milano, 1975, 32.
36. Tribunal of Milan, Feb. 26, 1976, *Fora Padano*, Foro Padano di Mario Franzosi & C. S.a.s., Milano, 1976, I, 190.

[d] Patent Law

In the appropriate cases, that is, provided that the industrial design meets the requirements set by the law for the granting of valid patent, namely, novelty, originality (non-obviousness), utility and lawfulness, industrial design may also be protected by means of a regular patent.

A regular patent is distinguished from a design or model in that the latter only refers to aesthetical/artistic character of the product and does not represent an advancement of the state of the art in that specific field. In fact, industrial design most often refers to commonly used products or objects, which are presented to the consumer in new and unusual shapes.

The distinction between a regular patent and a utility patent is often, in practice, a little bit more complicated. The Italian courts are consistent in holding that a regular patent is distinguished from a utility patent in that the former represents an original improvement of the state of the art and may concern both a material thing (machinery, equipment, tools etc.) and formula or proceeding; while a utility patent may only refer to a material thing and does not represent an original improvement of the state of the art, but only enhances the effectiveness, the ease of application of use of the existing state of the art.[37]

In short, according to the Italian courts, the difference between a regular patent and a utility patent is basically a qualitative one, based upon the fact that only a regular patent represents an effective and original improvement of the state of the art.

§10.04 DURATION OF THE PROTECTION

Models/Design: The protection of the model or design starts from the filing of the application for registration and lasts for five years and it is renewable for additional five year terms up to maximum term of protection of twenty-five years.[38] Non-registered industrial design is protected only for three years from its first publication in the European Union.[39]

Utility Patent: The protection of the model or design starts from the filing of the application for registration and lasts for ten years and it is not renewable.

Copyright: The protection granted is now the same as the other works of art that until seventy years after the death of the author.

Trademark: The protection of the model or design starts from the filing of the application for registration and lasts for ten years and it is renewable for additional ten year terms without limitation.

Patent: The protection of the model or design starts from the filing of the application for registration and lasts for twenty years and it is not renewable.

37. *See*, for example, Corte di Cassazione, Sep. 22, 1978, no. 4245, *Giurisprndenza Annotata di Diritto Industriale*, Giuffrè, Milano, 1978, 80; Tribunal of Milan, Apr. 23, 1979, *Giurisprndenza Annotata di Diritto Industriale*, Giuffrè, Milano, 1979, 515.
38. Article 36 of IPC.
39. Article 11 of the EU Design and Models Regulation.

§10.05 INFRINGEMENT

The IPC has placed great attention to the litigation and procedural matters regarding intellectual property and it has dedicated a whole chapter and substantial number of articles[40] to rationalising and harmonising the litigation process of all kind of intellectual property regulated by the same.

[A] Scope of Infringement Action

The registration of the industrial design as a design or a model, or the granting of utility patent or the fact that the industrial property is a work protected by copyright, all grant exclusive rights to author/owner. Any violation of these rights grants to the author or owner certain remedies to enforce his/her/its rights.

[B] Test for Infringement/Criteria to Constitute Infringement

The lack of a firm infringement/no infringement test leads to the fact that sometimes it not so easy to determine whether or not an infringement has taken place. In some cases, the law itself gives some help and provides an indication when a certain behavior is deemed or not deemed an infringement. For example, Article 42 of the IPC expressly states that the following actions do not constitute infringement of a model or design:

- actions made in a strictly personal/private situation or for non-commercial purposes;
- actions made for experimental purposes;
- action made for study or teaching purposes; provided that the same are performed within the limits of a correct professional behavior, do not improperly limit the normal use of the model or design, and the source is always specified.

Italian court have adopted a case by case approach, in particular in respect to the so-called preparatory actions, that is, those actions that if fully implemented could lead to an infringement. A review of the criteria upon which the Italian courts declare the infringement of a design patent is useful. In general, according to the majority of jurisprudence of the Italian courts, a design patent is infringed when the counterfeiter has copied the main and distinctive characteristics of the product, since minor and/or secondary modifications to the original product are not sufficient to avoid infringement.[41] However, according to the Italian courts, no infringement occurs when the "infringing" product may be easily distinguished from the original one, even by a

40. Articles 177–146 of the IPC.
41. Appeal Court of Milan, Oct. 10, 1972, *Giurisprudenza Annotata di Diritto Industriale*, Giuffrè, Milano, 1977, 435; and Tribunal of Milan, Dec. 22, 1977, *Giurisprudenza Annotata di Diritto Industriale*, Giuffrè, Milano, 1977, 435.

layperson[42] or there are substantial differences from the original design, for example, in the manner in which the product is manufactured and in the colors used to decorate it.[43]

With respect to the quantity of counterfeited product to be placed in the market, the Italian courts have adopted a tough and narrow position and have held that it is not necessary to have substantial amount of sales or a substantial diffusion of the counterfeited products in the market. It is sufficient that the mere potential and possibility exists through the commencement of the counterfeited products distribution.[44] On the basis of this narrow position, it has been held that the manufacturing of a limited number of products[45] or even mere display of a few samples at a fair without the intention to sell them,[46] amounts to the infringement of a design patent.

[C] Standing to Sue

Standing to sue to enforce the exclusivity rights deriving from the registration of the industrial design as a design or a model, or the granting of utility patent or the fact that the industrial property is a work protected by copyright, is vested with the owner of the industrial property rights and/or the author when appropriate (an author always has the right to be recognised as the author of the work or the utility patent. Author of copyrighted works also have other rights as better described below).

Instead, standing to challenge any registration of a model or design for industrial design or the granting of a utility patent (or even a regular patent) is vested in anyone who has an interest in doing so.[47]

[D] Remedies for Infringement

The remedies available under Italian law in case of infringement of intellectual property rights (including copyright) are the same for every kind of intellectual property since the IPC has tired harmonise and rationalise as much as possible the available remedies.

They generally consist of provisional remedies such as description, attachment and temporary injunction and an action for damages and/or a permanent injunction, as well as other relief such as, for example, the destruction or assignment of the of the counterfeited goods and tools, and the publication of the decision in the newspapers.

---

42. Appeal Court of Brescia, Jun. 23, 1973, *Giurisprudenza Annotata di Diritto Industriale*, Giuffrè, Milano, 1973, 870.
43. Tribunal of Padua, May 12, 1973, *Giurisprudenza Annotata di Diritto Industriale*, Giuffrè, Milano, 1973, 623.
44. Appeal Court of Milan, Jun. 27, 1980, *Giurisprudenza Annotata di Diritto Industriale*, Giuffrè, Milano, 1980, 502.
45. Tribunal of Milan, Jan. 10, 1980, App. Milan, Jun. 27, 1980, *Giurisprudenza Annotata di Diritto Industriale*, Giuffrè, Milano, 1980, 502.
46. Tribunal of Turin, Oct. 12, 1981, *Giurisprudenza Annotata di Diritto Industriale*, Giuffrè, Milano, 1981, 606.
47. Article 122 of the IPC.

[1] Provisional Remedies

Provisional remedies may be obtained by motions filed before or during the litigation also on an ex parte (or *inaudita altera parte* as said in Italy) basis. Although Italian courts are not very keen to grant ex parte motions, in intellectual property cases they are frequently granted probably due to the very nature of the cases. Provisional remedies may be granted even in respect of applications for the registration of a model or a design, or an application for a utility patent. The applicable provisional remedies are:

(a) *Attachment:*[48] It is one of the most common provisional remedies. Attachments may be granted, even on ex parte motions, provided that the moving party (only a patent owner, applicant and licensee have standing for this motion) provides the court with the so-called *fumus bonis juris* (likelihood of success on the merits) and *periculum in mora* (irreparable harm), as required by Italian practice for the granting of attachments. In its order granting the motion, the court may impose the posting of a bond upon the moving party.
(b) *Preliminary Injunction*: It is one of the most common provisional remedies. As in the case of attachment, may be granted, even on ex parte motions, provided that the moving party (only a patent owner, applicant and licensee have standing for this motion) provides the court with the "*fumus bonis juris*" (likelihood of success on the merits) and "*periculum in mora*" (irreparable harm), as required by Italian practice for the granting of provisional remedies. In its order granting the motion, the court may impose the posting of a bond upon the moving party.
(c) *Description:*[49] This is a peculiar provisional remedy typical of intellectual property litigation. The purpose of this remedy is on the one side to a have a clearer picture of the infringement and on the other side to prevent the counterparty to hide evidence of its infringement. Description is performed by an "*Ufficiale Giudiziario*" (bailiff/sheriff) consists into the listing and description of the goods, tools and machinery allegedly used to effect the infringement. Also in the case of description the court may, and most often, grant it on an ex parte basis and, as in case of any provisional remedy, Italian practice requires the moving party to provide proof of both "*fumus bonis juris*" (likelihood of success on the merits) and "*periculum in mora*" (irreparable harm).

[2] Declaratory Judgment

Another remedy used is the declaratory judgment, that is, the plaintiff requests the court to issue a judgment declaring the plaintiffs model, design, utility patent or

48. Article 129 of the IPC.
49. Article 128 of the IPC.

trademark etc. valid or alternatively, the defendant's model, design, utility patent or trademark etc. null and void. Needless to say, this remedy may be, and is, more often used in counterclaims.

[3] Action for Damages

With respect to the ordinary action, the same may be commenced prior to a motion for a provisional remedy, or in the event one of these motions is made, within the mandatory terms provided by law.

In an ordinary action the remedies normally requested are damages and a permanent injunction (e.g., an order not to sell, use and/or manufacture the products infringing the model, design, utility patent or trademark). In determining the quantum of damages, the court will follow the common law rules since the IPC does not provide for any special kind of damages (there are no treble damages. Moreover, Italian Law does not know and grant punitive damages).[50] In particular, according to the prevailing opinion of the Italian courts, the damages paid in case of infringement have the nature of lost profits.[51] However, in some cases the Italian courts have identified such lost profits with the fair compensation that the counterfeiter or infringer would have paid to lawfully obtain a licence from the owner of the patent in the case of the infringement of a patent not yet granted,[52] or with the enrichment of the counterfeiter/infringer[53] derived from his or her tortious actions. Finally, the IPC grants to the plaintiff the right to plead the court to calculate damages in a lump sum based upon the pleading and the presumptions deriving therefrom.

Since the actual determination of damages suffered by the plaintiff is often quite difficult, the law expressly allows the plaintiff to request the court to determine damages on the basis of the pleadings and other documentation filed with it, as well as the presumptions arising therefrom. In other words, the determination of damages is left at the prudent discretion of the court, which may also determine the quantum of damages on equitable grounds.[54]

Moreover, as mentioned above, the plaintiff may also request a declaratory judgment concerning the validity of its model/design registration or utility patent, as the case may be.

50. Article 125 of the IPC.
51. *See* for example, Tribunal of Milan, May 29, 1976, *Giurisprudenza Annotata di Diritto Industriale*, Giuffrè, Milano, 1976, 513.
52. *See* for example, Tribunal of Rome Rome, Mar. 10, 1977, *Giurisprudenza Annotata di Diritto Industriale*, Giuffrè, Milano, 1978, 165.
53. Tribunal of Milan Milan, Dec. 13, 1978, *Giurisprudenza Annotata di Diritto Industriale*, Giufffre, Milano, 1978, 682; and Appeal Court of Bononia, Jul. 21, 1981, *Giurisprudenza Annotata di Diritto Industriale*, Giuffrè, Milano, 1981, 515.
54. Article 125, para. 2 of the IPC. For previous jurisprudence, *See* for example, Tribunal of Milan, Oct. 18, 1973, *Giurisprudenza Annotata di Diritto Industriale*, Giufre, Milano, 1973, 1203; Appeal Court of Milan, Feb. 4, 1972, *Giun'sprudenza Annotata di Diritto Industriale*, Giuffrè, Milano, 1972, 442; Tribunal of Bononia Bononia, Apr. 20, 1979, *Giurisprudenza Annotata di Diritto Industriale*, Giuffrè, Milano, 1979, 505; and Appeal Court of Bononia, Jul. 21, 1981, *Giurisprudenza Annotata di Diritto Industriale*, Giuffrè, Milano, 1981, 515.

Finally, with respect to a permanent injunction, it should be noted that Italian courts will not grant one if the registration of the model or design or the utility patent expires prior to the issuance of the decision.[55]

[4] Other Relief

In addition to damages and a permanent injunction, the court may, upon request from the plaintiff, but at its own discretion, provide other kinds of relief. As stated above, these additional forms of relief are represented by:

(1) the publication of the judgment in one or more newspapers at the expense of the losing party;[56] although it should be noted that Italian courts have consistently denied such remedy in the event the patent expires prior to the issuance of the decision;[57]
(2) the assignment of the counterfeited goods and/or the tools and machinery used to manufacture them, to the plaintiff;[58]
(3) the setting of a penalty payable to the plaintiff in case of any further infringement or delay in executing the court decision on the part of the defendant;[59]
(4) the attachment of the counterfeited goods until the expiration of the patent;[60]
(5) the destruction of the counterfeited goods.[61]

[5] Criminal and Administrative Sanctions

Finally, it should be noted that protection of industrial design is also achieved through the administrative and criminal sanctions provided by the IPC,[62] the Copyright Law,[63] and the Criminal Code[64] respectively relating to the criminal forgery of intellectual property.

55. *See* Tribunal of Reggio Emilia, Jul. 27, 1978, *Giurisprudenza Annotata di Din.tto lndustriale,* Giuffrè, Milano, 1978, 480.
56. Article 126 of the IPC. According to the majority of Italian courts and scholars, the publication of the judgment has both a redress and a preventive purpose. *See* for example, Appeal Court of Catania, Sep. 12, 1984, *Giurisprudenza Annotata di Diritto Industriale,* Giuffrè, Milano, 1984, 606.
57. *See* for example, in respect of the former design patents, Tribunal of Modena, Apr. 1, 1978, *Giurisprudenza Annotata di Diritto lndustriale,* Giuffrè, Milano, 1978, 337; and Appeal Court of Milan, May 28, 1974, *Giurisprudenza Annotata di Diritto Industriale,* Giuffrè, Milano, 1974, 727.
58. Article 124 of the IPC.
59. Article 124 of the IPC.
60. Article 124 of the IPC.
61. Article 124 of the IPC.
62. Article 121 of the IPC.
63. Articles 171 through 174 of the Copyright Law.
64. Article 127 of the IPC and Articles 473, 474, 514, and 515 of the Criminal Code (Royal Decree No. 1398 of Oct. 19, 1930).

[6] Customs Filing

In Italy, the owner of a design or a model may prevent the importation of goods that might infringe its rights pursuant to EU Regulation 1383/2003, which calls for the seizure and stop at the EU frontiers of any suspect goods. In order to start the procedure (which is valid for one year and may be renewed for additional one year terms) an application must be filed with the *Agenzia delle Dogane* (Italian Customs Service) located in Rome, specifying which designs or model are to be controlled attaching a copy of the registration certificate and filing a "Declaration" where as many as possible characteristics of the original product that may help distinguish it from the counterfeited one. In the event, the Customs suspect that a certain goods may be counterfeited, it will get in touch whit the owner who will have ten business days to confirm whether the same are counterfeited or not. If the answer is positive, the Customs will seize the goods and a criminal action will commence. The owner of the model or design has the right to be informed of the sender and final destination of the counterfeited goods and may commence a civil action against the intended consignee of the goods.

§10.06 DEFENSES TO INFRINGEMENT

[A] Validity

[1] Not Novel/Original

The first and one of most common defenses is the lack of novelty of the registered design as provided for by Article 31 of the IPC.

[2] Functional/Not Ornamental

Another fairly common defense is that the characteristics of the shape of the product are mandated solely by the technical function of the product. This is a violation of Article 36 of the IPC.

[3] Other Possible Criteria, Such as Non-obviousness

Since under the Italian law: (a) an "inventive step" (non-obviousness in U.S. terms) must be provided in order to a valid industrial design registration and patent and (b) this inventive step: (i) may consist of the shape or configuration of the product or a part thereof, as it results from a peculiar combination of lines, contours, colors, shape, surface structure or the material of the product or its ornament and (ii) must have a new and individual distinctive character, this defense fairly commonly used to defeat and oppose an infringement claim.

[4] Faults in Application

This is not a very common defense for since the IPTO review of the application is concentrated on the compliance of the formalities and requirements set forth by the law and faults in the application are generally detected. Moreover, also upon demand by the IPTO, the applicant has time (two months) to integrate and supplement the application in the event any formality has not been or has only partially complied with.[65]

The most used and important affirmative defenses or counterclaims used in infringement cases involving industrial design and intellectual property in general, are that the design, model, utility patent, or trademark is null and void, it has lapsed or it has expired.

As a general rule, the party who asserts the nullity or the lapse of a patent (or trademark) bears the burden of proof of such nullity.[66]

Since industrial design will be generally protected by means of the registration of a model or a design, the grounds for nullity of the registration are provided by Article 43 of the IPC and, namely:

(i) the design or model could not be registered since it does not meet the requirements set forth by Articles 31, 32, 33, 34, 35 and 36 of the IPC, i.e., the subject matter, novelty, originality, publication lawfulness, and technical function requirements of the registration;
(ii) the design or model is contrary to public order or is immoral;
(iii) if the party registering the model or the design had no right to obtain it and the inventor did not claim the authorship and ownership rights as provided by Article 118 of the IPC;
(iv) the use of the design or the model would violate a mark or copyrighted work;
(v) the use of the design or the model would violate Article 6*ter* of the Paris Convention or sign, emblems or coat of arms of specific interest for the Government; and
(vi) another model or design has priority.

In case of industrial design based upon a utility patent then the grounds for nullity - the same as the ones of a regular patent[67] - will be quite similar to the ones stated above, and namely:

(i) the patent cannot be granted since it does not meet the requirements set forth by Articles 45, 46, 48, 49, and 50 of the IPC, i.e., the subject matter, novelty, non-obviousness, capacity to be industrialised, and lawfulness;
(ii) the invention is not clearly described so that a party cannot actually implement it;

65. Article 148 of the IPC.
66. Article 121 of the IPC.
67. Article 76 of the IPC.

(iii) the subject matter of the patent goes beyond the one specified in the original application;
(iv) the use of the design or the model would violate a mark or copyrighted work;
(v) if the party registering the model or the design had no right to obtain it and the inventor did not claim the authorship and ownership rights as provided by Article 118 of the IPC;
(vi) another patent has priority.

[B] Marking

Not applicable

[C] Non-infringement

Article 42 of the IPC expressly states that the following actions do not constitute infringement of a model or design:

- actions made in a strictly personal/private situation or for non-commercial purposes;
- actions made for experimental purposes;
- action made for study or teaching purposes; provided that the same are performed within the limits of a correct professional behavior, do not improperly limit the normal use of the model or design, and the source is always specified.

[D] Other Defenses

Other defenses are: (i) the industrial design is contrary to public order or is immoral,[68] (ii) the industrial design infringes a copyright,[69] (iii) the design in improperly uses any of the elements provided for by Article 3*ter* of the Paris Convention or signs, elements or badges that are deemed of public interest buy the Government,[70] or (iv) the party filing the application had no right to file and obtain the registration.

§10.07 PROCEDURES FOR FILING APPLICATION FOR REGISTRATION

[A] Priority

Priority is granted affective the day when the application is made available to the public, i.e., the date of the filing of the application for registration, and for a period of

68. Article 33*bis* of the IPC.
69. Article 43 of the IPC.
70. Article 43 of the IPC.

six months (twelve months if the regular patent is used to protect the industrial design).[71] Whenever priority is claimed the relative documentation must be attached to the application for registration.[72] If the documentation is in a foreign language, a translation must be provided.

[B] Drawings

The drawings of the design are a mandatory part of the application.[73] In case of failure to do so, even partially, the IPTO may request the applicant to supplement or amend the application by filing drawings or additional/supplemental drawings[74] within a reasonable term (no less than two months). In the event the request is not duly and timely complied with, the IPTO will dismiss the application.[75]

[C] Design Part of Another Article/Environment of Design

Designs that are part of another article/environment of design may also be protected provided that they meet the novelty, originality, and distinctive character (i.e., non-obviousness/inventive step) requirements, as well as has per se an individual and distinctive character. The procedure is the same as the one for a product not a part of another article/environment of design.[76] The design is protect whether the same is visible or partially visible during the use of the product of which it is a part. In this second phase, the above-described requirements must be met by the visible portion of the design.[77]

[D] Description of Design (Where Required)

Description of the design is mandatory part of the application whenever it is necessary for a complete and correct filing.[78] In case of failure to do so, even partially, the IPTO may request the applicant to supplement or amend the application to file the description or to amend or supplement the description.[79] In the event the request is not duly and timely complied with, the IPTO will dismiss the application.[80]

71. Articles 4 and 169 of the IPC.
72. Article 167 of the IPC.
73. Article 167 of the IPC.
74. Article 172, para. 3 of the IPC.
75. Article 173 of the IPC.
76. Articles 31 and 35 of the IPC.
77. *See* Tribunal of Venice, Oct. 10, 2005, Sezioni. Specializzate. Proprietà Intellettuale, 2005, 2, 189, Tribunal of Turin, May 22, 2008.
78. Article 167 of the IPC.
79. Article 172 para. 3 of the IPC.
80. Article 173 of the IPC.

Again the IPC has, whenever possible, streamlined and simplified the procedure and formalities for the filing of the registration for designs and models, and the application for a utility patent.[81]

[1] Models and Designs

The filing is quite simple.[82] The application for registration is filed with Italian Patent and Trademarks Office ("IPTO") or the locally competent Chamber of Commerce who will then forward the application to the IPTO. Filing can be in the form of either paper or electronic filing. The application must be supported with accurate description, including precise drawings and eventually a sample, of the design or the model, of the design or the model to be registered, as well as any other useful and necessary documentation, e.g., the foreign registration documentation if a priority is claimed based upon it.

Priority rights are granted as of the date of the filing of the application and *not* as of the date of the granting of the registration (the registration is deemed to retroactive as of the date of the filing).

As usual under Italian Intellectual Property Law, the check and controls by the IPTO are merely administrative and technical on the documentation filed and whether it complies with Article 31 of the IPC (whether the application is for a model or design), i.e., it does not check whether the application actually complies with Articles 32, 33, 34, 35 and 36 of the IPC. This check is de facto left to litigation that an interested party may file to have the application or the registration declared null and void. IPTO may request the applicant to supplement the application if the documentation filed is not clear or is incomplete.

In the event the application is denied, an appeal may be filed with an Appeal Commission.

Application for EU and international filings for the model or design may also be made with the IPTO.

[2] Utility Patents

Again the filing is quite simple.[83] The application for registration is filed with Italian Patent and Trademarks Office ("IPTO") or the locally competent Chamber of Commerce who will then forward the application to IPTO. Filing can be in the form of either paper or electronic filing. The application must be supported with an accurate description of the invention, including precise drawings, the claims, the name of the inventor, as well as any other useful and necessary documentation, e.g., the foreign registration documentation if a priority is claimed based upon it.

81. Articles 147-193 of the IPC.
82. Article 167 of the IPC.
83. Article 167 of the IPC.

Priority rights are granted as of the date of the application is made available to the public (through registration or otherwise) and *not* as of the date of the granting of the registration.

As usual under Italian Intellectual Property Law, the check and controls by the IPTO are merely administrative and technical on the documentation filed and whether it complies with Articles 45, 50 and 82 of the IPC (whether the application is for a model or design), i.e., it does not check whether the invention is null and void pursuant to Articles 45, 46, 48, 49, and 50 of the IPC. This check is de facto left to litigation that an interested party may file to have the application or the registration declared null and void. IPTO may request the applicant to supplement the application if the documentation filed is not clear or is incomplete.

In the event the application is denied, an appeal may be filed with an Appeal Commission.

Clearly, this mere formal review of application for models, designs, inventions and trademarks is the main reason why Italy is often the center of litigation involving intellectual property.

[3] Copyright

Italian Copyright Law protects any original work in the various fields of art, including literature, science, music, architecture, theater, cinema, dramatic and figurative arts, such as sculpture, painting, design, drawing, and engraving.[84]

According to the Copyright Law,[85] authors seeking protection for their work must deposit a copy of their work (or a picture or a phonographic record if the work cannot be deposited) with the Literary, Artistic, and Scientific Department of the *Presidenza del Consiglio dei Ministri* (the Italian Prime Minister's cabinet office).

The deposit is not mandatory, since it is required only for administrative purposes; thus, the omission of the same would not prejudice the rights of the author over his/her work.[86] However, it creates a rebuttable presumption in favor of the author, that the work exists and has been published, and secondly, that the person depositing the work is the author of the work.[87]

Under Italian Law, copyright consists of both moral and economic rights over the work. The moral rights over the work consist of the right to claim the authorship of the work, to withdraw the work from the market, prohibit and prevent any amendment to or distortion of the original work, as well as any other act that harms the reputation of the work or author.[88] Moral rights are of a personal nature, not assignable, perpetual

84. Articles 1 through of 5 of Law 633 of Apr. 23, 1941.
85. Articles 105 through 107 of Law No. 633 of Apr. 23, 1941.
86. Article 106 of Law No. 633 of Apr. 23, 1941.
87. Article 103, para. 4, of Law No. 633 of Apr. 23, 1941.
88. *See* Articles 20 through 24 of Law No. 633 of Apr. 23, 1941.

and upon the death of the author may be enforced by his/her heirs. Economic rights last until seventy years after the death of the author[89] and are freely assignable.

§10.08 EXPUNGING/CANCELLING OR VARYING REGISTRATION

According to Article 172 of the IPC before the final decision on the application for registration of a model or design, or the granting of a utility patent that the applicant may withdraw the application or amend the particulars (but not the main and substantial portions and claims) of the application.

IPTO may also request the applicant to integrate, complete, amend or rectify the application in order to provide a better understanding of the application. In respect of an application for a utility model the applicant may submit new examples, as well as reduce the description, the claims and the drawings originally filed.[90]

§10.09 OTHER

[A] Interface between Industrial Design and Trademarks

See above.

[B] Relationship between Industrial Design and Copyright/Design Copyright

See above.

[C] Industrial Design and Unfair Competition

In Italy, an attempt was made to protect non-patented or otherwise protected industrial design products from infringement by maintaining that such infringement violates the unfair competition rules and specifically, the prohibition of slavish imitation of products provided for by the Italian Civil Code.[91]

However, the Italian courts have consistently held that in the lack of a valid patent or other form of protection granted by other forms of intellectual property, the reproduction or imitation of the shape of industrial design products does not amount to an act of unfair competition or a violation of the prohibition against slavish imitation provided by Article 2598 No. 1) of the Italian Civil Code, since any shape not protected by a patent (or by any other exclusivity right) may be reproduced without committing

89. Article 25 of Law No. 633 of Apr. 23, 1941. The original (general) term was fifty years, but it has been extended to seventy years by Article 17 of Law No. 52 of Feb. 6, 1996. With respect to industrial design *see* article of the IPC as amended by Article 44 of Legislative Decree 131/2010.
90. Article 172, para. 2 of the IPC.
91. Articles 2598 through 2601 Civil Code ("c.c."). The prohibition against slavish imitation of the products of a competitor is provided by Article 2598 (1) c.c.

an act of unfair competition[92] provided, of course, that such reproduction does not amount to counterfeiting, it being sufficient that a different and not confusingly similar trademark be applied, as required by fair and generally accepted behavior in the market.[93]

In short, according to the Italian courts, in the lack of a registration (design or model) or a patent (or any other exclusivity right), an industrial design, is, in practice, of public domain and the shape of the same may be reproduced without committing an act of unfair competition.[94]

92. *See*, for example, the decision concerning the Wassily chair designed by Breuer issued by the Tribunal of Milan, May 24, 1976, *Giurisprudenza Annotata di Diritto Industriale*, Giuffrè, Milano, 1976, no. 836, 492; Appeal Court of Milan, February 9, 1979, *Giurisprudenza Annotata di Diritto Industriale*, Giuffrè, Milano, 1979, no. 1166, 337; and Corte di Cassazione, Jun. 26, 1980, *Giurisprudenza Annotata di Din'tto Industriale*, Giuffrè, Milano, 1980, no. 1258, 61; *contra* Tribunal of Siena, Oct. 30, 1985, reversed Appeal Court of. Florence, Jun. 1, 1990, *Fora Italiano*, Giuffrè, Milano, Rep. 1991, item "Diritto d'Autore" nos 56–58, appellate decision confirmed by Corte di Cassazione., Dec. 7, 1994, no. 10516, *Foro Italiano*, Giuffrè, Milano, 1995, I, 811 (case concerning the "Chaisse-long" chair designed by Le Corbousier).
93. G. Bergomi, *op. cit.*, end note no. 2; *see also* the cases cited in endnote no. 28.
94. *Ibid.*

CHAPTER 11

Industrial Design Rights: Japan

Hiromichi Aoki

§11.01 INTRODUCTION

[A] Highlights of New Developments in This Jurisdiction since Last Edition

Locarno Agreement establishing an International Classification for Industrial Designs came into effect on September 24, 2014.

Japan joined the Hague Agreement Concerning the International Registration of Industrial Designs Geneva Act (1999) on February 13, 2015 and it came into effect on May 13, 2015.

Secret Design System under Article 14 of Design Law is not applicable for International Design Application.[1]

§11.02 CONVENTIONS AND LEGISLATION

- Paris Convention for the Protection of Industrial Property, Stockholm Act.
- The Hague Convention Abolishing the Requirement of Legalization for Foreign Public Documents (Apostille), effective from July 27, 1970.
- Convention Establishing the World Intellectual Property Organization (WIPO), effective from April 20, 1975.
- WTO's TRIPS Agreement, effective from January 1, 1995.

1. Article *60novies* of Design Law.

- Locarno Agreement Establishing an International Classification for Industrial Designs, effective from September 24, 2014.
- Hague Agreement Concerning the International Registration of Industrial Designs Geneva Act (1999), effective from May 13, 2015.

§11.03 SUBJECT MATTER

[A] Definition/What Constitutes a Protectable Design?

[1] Shape, Configuration

"Design" under the Design Law[2] means the shape, pattern or color or any combination thereof in an article (including a portion of an article), which produces an aesthetic impression.

The shape, patterns or colors, or any combination thereof, of a part of an article as used in the preceding paragraph (Article 2.1) shall include those in a graphic image on a screen that is provided for use in the operation of the article (limited to the operations carried out to enable the article to perform its functions) and is displayed on the article itself or another article that is used with the article in an integrated manner.[3]

Variations of design are as follows:

- Shape (including a portion of article).
- Shape and color.
- Shape and pattern.
- Shape, color and pattern.

According to this definition, the following are not designs under the Design Law:

- Immovable objects such as buildings (however, a prefabricated house is a design because it is mass-produced and sold to be installed later as a fixture).
- Liquid (e.g., fountain), gas, electricity, light (fireworks).
- Non-visible things like a unit of powder.
- Things not causing an aesthetic impression.

[2] Ornamental or Functional

A design composed of only shapes indispensable to secure the functions of an article is not registrable.[4] Must Fit designs and functional designs will be rejected.

2. Article 2.1 of Design Law.
3. Article 2.2 of Design Law.
4. Article 5.3 of Design Law.

[B] Originality/Novelty

[1] Test for Originality Generally

See "Novelty" below.

[2] Novelty

If a design is publicly known anywhere in the world prior to the application date or priority date, it is not registrable.[5]

If a design is described in a publication distributed anywhere in the world prior to the application date (priority date), it is not registrable.[6]

If a design is similar to a design that is publicly known anywhere in the world prior to the application date (priority date) or is described in a publication distributed anywhere in the world prior to the application date (priority date), it is not registrable.[7]

[a] Exceptions to Lack of Novelty of Design

In the case of a design that: (1) lost novelty against the will of the person having the right to obtain a design registration or (2) lost novelty due to an act on the part of the person having the right to obtain a design application, such a design shall be deemed not to lose novelty provided that a design application with a written statement seeking exceptions was filed within six months from the date on which the design became publicly known or described in a publication. In the case of the above (2), it is necessary to submit a document proving that the design in the design application is a design that is publicly known or described in a publication, e.g., certification by a third party with photos, within thirty days from the application date.[8]

[b] Creativity

Where a design could easily have been created, prior to the filing of the design application (Priority Date), by a person having ordinary skill in the art (PHOSITA) to which the design pertains, on the basis of a shape, pattern or color or combination thereof publicly known anywhere in the world, a design application shall be rejected.[9] For example, if a design application for soap in the shape of the Eiffel Tower is filed, it will be rejected because of lack of creativity.

5. Article 3.1.1 of Design Law.
6. Article 3.1.2 of Design Law.
7. Article 3.1.3 of Design Law.
8. Article 4 of Design Law.
9. Article 3.2 of Design Law.

[C] Unobviousness (if Appropriate)

Not applicable

§11.04 DURATION OF PROTECTION

The duration of a design right shall be twenty years at a maximum from the date of registration of its establishment.[10] However, the duration of the Associated Design Right is the same as that of the Principle Design Right. Similar designs owned by same applicant have to be registered as Principle Design and Related Design.[11]

§11.05 INFRINGEMENT

[A] Scope of Infringement Action

A holder of a design right shall have the exclusive right to use the registered design and designs similar thereto as a business.[12]

The scope of a registered design shall be determined based upon the design stated in the application and depicted in the drawing or represented in the photograph, model or specimen attached to the application.[13]

Whether a registered design is identical with or similar to another design shall be determined based upon the aesthetic impression that the designs would create through the eyes of their consumers.[14]

[B] Test for Infringement/Criteria to Constitute Infringement

Pattern of Similarity of designs is as follows:

- Article is identical and Design is identical.
- Article is similar and Design is identical.
- Article is identical and Design is similar.
- Article is similar and Design is similar.

There is no Supreme Court decision with respect to how to determine the similarity of articles.

If there are common features with respect to usage and function of both articles, the two articles are similar (Sports shirts case, IP High Court, July 18, 2006).

"Similar articles" means that such articles that would cause, if the registered design or other similar designs are applied thereto, confusion with the article sold by

10. Article 20 of Design Law.
11. Article 10 of Design Law.
12. Article 23 of Design Law.
13. Article 24.1 of Design Law.
14. Article 24.2 of Design Law.

the design right holder among the consumers of such articles (Karabiner case, IP High Court, October 31, 2005, *see* below).

[1] Karabiner Case, IP High Court, October 31, 2005

The following designs are similar. However, articles "Karabiner" and "Accessory (Key Chain)" are not similar. Defendant design for "Accessory" would not cause, if the registered design for "Karabiner" or other similar designs are applied thereto, confusion with "Karabiner" sold by the design right holder among the consumers of such articles. Thus, defendant design does not constitute design infringement of plaintiff registered design.

(Plaintiff/Karabiner) (Defendant/Accessory)

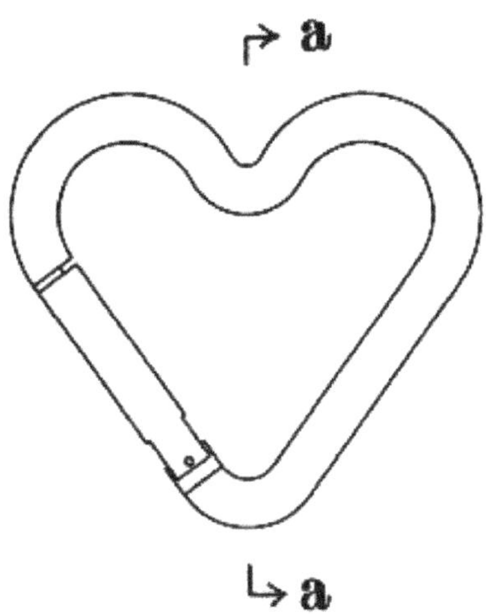

Whether a registered design is identical with or similar to another design shall be determined based upon the aesthetic impression that the designs would create through the eye of their consumers.[15]

A court identifies the characteristic (essential) features by considering which parts of registered design would attract a consumer's attention, considering: (1) nature, usage embodiment of articles under the registered design and (2) novel and creative parts not shown in prior art.

If the same characteristic features of registered design are located in the accused design and common parts of the two designs including characteristic features overcome different parts of the two designs, the accused design is similar to the registered design and hence infringes it.

15. Article 24.2 of Design Law.

[2] Rough Terrain Crane Case (Tokyo High Court, June 18, 2004)

The characteristic features of the registered design (Rough terrain crane) are: (1) hexagonal shape cabin, (2) long storage box, (3) position of boom between (1) and (2), and (4) arrangement thereof. The accused design has the said characteristic features. When observing each of the designs as a whole, the two designs each give consumers the same aesthetic impression. The different parts of the two designs do not overcome the common parts including characteristic features. Thus, the accused design is similar to the registered design and infringes it.

(Plaintiff) (Defendant)

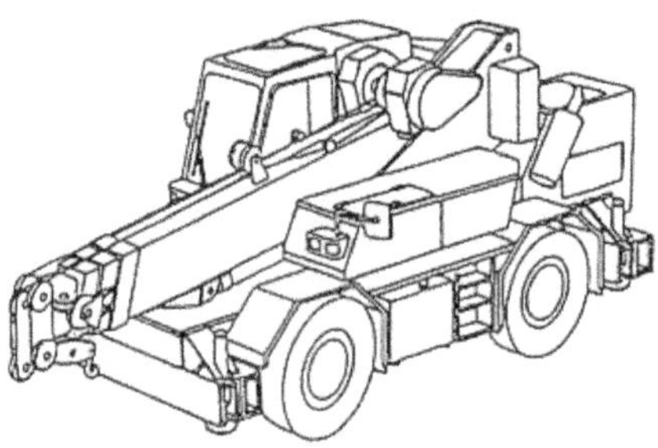

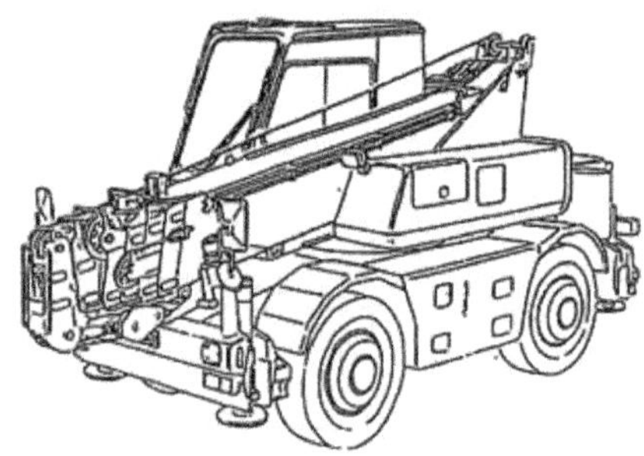

[C] Standing to Sue

Registered design holder or their registered exclusive licensee can sue.

[D] Remedies for Infringement

Remedies for infringement are as follows:

- Injunction.
- Damages.
- Unjust enrichment.
- Recovery of reputation (e.g., apology in daily newspaper).
- Criminal sanction.

There is no clause regarding Acquiescence under the Design Law and there is no court decision.

With respect to damages, it is necessary to take a court action within three years after becoming aware of design infringement for torts or ten years for unjust enrichment. Thus, compensation for damages can be retroactive for three years or ten years from the time compensation is demanded by a court action.

With respect to Secret Design, a design holder can claim an injunction after sending a warning letter because the secret design is not publicly known.[16] Secret Design will be kept secret without publication for maximum three years from the registration date.[17]

§11.06 DEFENCES TO INFRINGEMENT

[A] Validity

Where, in litigation concerning the infringement of a design right or an exclusive license, the said design registration is recognized as one that should be invalidated by a trial for design registration invalidation, the rights of the design right holder or exclusive licensee may not be exercised against the adverse party.[18]

Where a design registration falls under the requirements of any of the following items, a design invalidation trial may by demanded for the invalidation of the design. Any person may demand a design invalidation trial:

- A design is not a kind of design defined in the Design Law,[19] i.e., it must be an article with a shape and have visual and aesthetic qualities.[20]
- No industrial applicability.[21]
- A design has no novelty.[22]
- A design has no creativity.[23]
- A design is identical or similar to part of any earlier design published in the Design Bulletin.[24]
- A design is liable to damage public order or morality, or creates confusion with an article pertaining to another person's business, or a design only having a shape that is indispensable for ensuring functions of the article.[25]
- A design does not have the earliest application date if two or more applications are filed for the registration of identical or similar designs.[26]

[1] Not Novel /Original

As shown above

16. Article 37.3 of Design Law.
17. Article 14 of Design Law.
18. Article 41 of Design Law, Article 104-3 of Patent Law.
19. Main paragraph of Article 3.1 of Design Law.
20. Article 2.1 of Design Law.
21. Main paragraph of Article 3.1 of Design Law.
22. Article 3.1 of Design Law.
23. Article 3.2 of Design Law.
24. Article 3*bis* of Design Law.
25. Article 5 of Design Law.
26. Article 9 of Design Law.

[2] Functional/Not Ornamental

As shown above

[3] Other Possible Criteria, Such as Non-obviousness

As shown above

[4] Faults in Application

Not applicable

[B] Marking

Marking is "Toroku Ishou" in Japanese, and needs to include Registration Number.
Toroku Ishou means Registered Design.
Marking is not obligatory.

§11.07 PROCEDURES FOR FILING APPLICATION FOR REGISTRATION

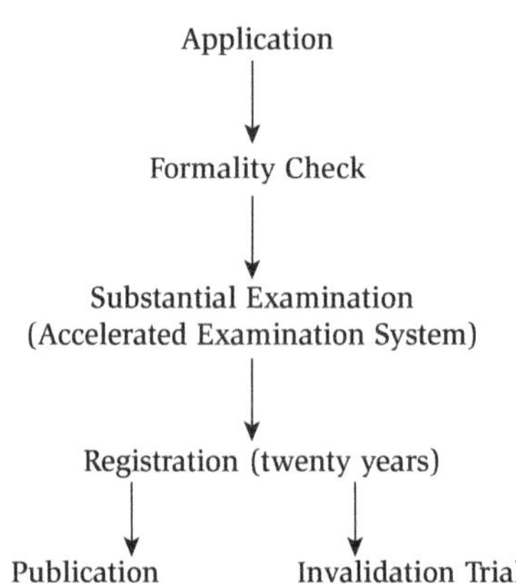

[A] Priority

When claiming a convention priority, it is necessary to claim it when filing an application.

It is possible to submit a priority document (original) within three months from the application date in Japan. This deadline is not extendible and a photocopy is not permitted.

[B] Drawings

It is necessary to submit one of drawings, photos, model or sample when filing an application. A model or sample is permitted under certain conditions (e.g., within 7 mm (9/32 inch) in thickness).

When preparing drawings or photos by Orthographic Projection, it is necessary to prepare drawings showing six views, namely, front view, rear view, left side view, right side view, top plan view and bottom view.

Shading on the drawings is possible provided the shaded parts are specified in writing in the Explanation of Design.

[C] Design Part of Another Article/Environment of Design

When filing a portion of an article, it is necessary to state in the application document that the subject design is a portion of an article. A portion of an article for registration is expressed by solid lines while the other portions are expressed by broken lines. It is necessary to state in the Description of Design that the portion expressed by solid lines is a portion of the article for registration.

[D] Description of Design (Where Required)

Description of design and description of article are optional.

Where the shape, patterns or colors of the article to the design are changeable based on the function possessed by the article, if the applicant intends to request a design registration of the shapes, patterns or colors, or a combination thereof as it appears before, during and after the said change, he/she shall state such an intention and include an explanation of said function of the article in the Description of Design.

If an article is new, it is necessary to submit a description of the article.

§11.08 EXPUNGING/CANCELLING OR VARYING REGISTRATION

To invalidate a design registration, it is necessary to file an Invalidation Trial against it before the Appeal Board of the Japan Patent Office.[27] A legitimate interest is not necessary. After a warning letter is received, an Invalidation Trial is often filed. With respect to the reason for invalidation trial, please see above §11.06[A].

27. Article 48 of Design Law.

§11.09 OTHER

[A] Interface between Industrial Design and Trademarks

It is possible to register a trademark consisting of the shape of goods or their packaging provided that they are distinguishable and are not indispensable for securing the function of the goods or their packaging (de jure functionality).

The following three-dimensional trademarks were registered by proving secondary meaning: (a) is for "electric touch" owned by Mag Instrument, (b) is for "cola drink" owned by Coca-Cola, (c) is for "fermented milk drink" owned by Yakult Honsha Co., Ltd., (d) is for "cosmetics" owned by Beaute Prestige International (Societe Anoynyme) and (e) is for "chair" owned by Carl Hansen & Son.

(a) (b) (c) (d) (e)

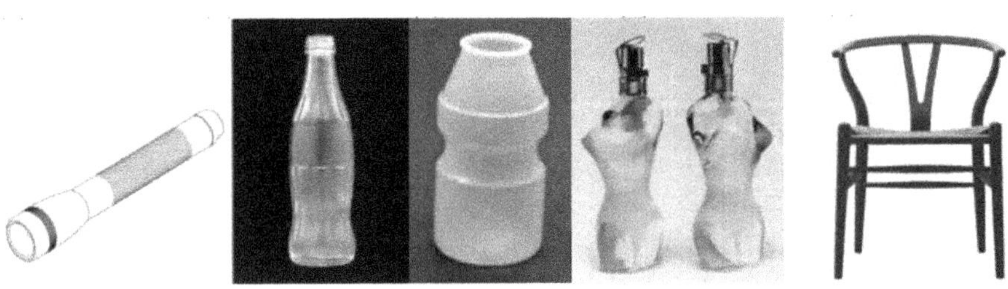

The following three-dimensional trademark owned by Chocolaterie Guylian N.V. were registered as being inherently distinctive for "chocolate, etc."

The following icons owned by Apple Inc. registered under the Trademark Law and Design Law:

(Registered design) (Registered trademark)

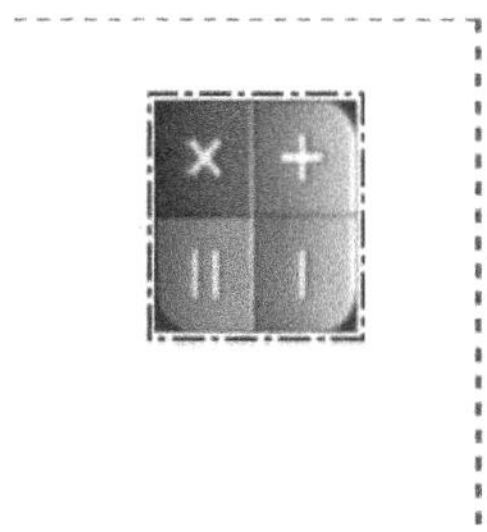

[B] Relationship between Industrial Design and Copyright/Design Copyright

Most of the related court precedents state implicitly or explicitly that copyrightability is not denied by the mere fact that a work is an industrial product. Cumulative protection may be allowed. However, for the recognition of copyrightability of an industrial product, such a product must be comparable in aesthetic qualities to fine arts, according to most court decisions. In many cases, copyrightability is denied on the grounds that, apart from its industrial applicability, a product in question cannot serve as an object of art. As a natural result, only a very limited number of industrial products can satisfy the requirements for cumulative protection by both rights.

Court decided that the following: (a) T-shirt can be copyrighted because it has aesthetic qualities to fine arts.[28] However, Court decided that the following (b) Chair cannot be copyrighted because it is not subject for aesthetic appreciation.[29]

However, recently the IP High Court adopted new criteria for copyrightability; namely, whether a design has individual characters or not, and held that the following (c) Chair has individual characteristics and therefore it can be copyrighted.

(a) (b) (c)

28. Tokyo District Court, Apr. 20, 1981.
29. Supreme Court, Mar. 28, 1991.

[C] Relationship between Industrial Design and Unfair Competition

The following are related to unfair competition under the Unfair Competition Prevention Act.

The act of using a good or other indication that is identical with, or similar to, another person's goods or other indication as would be widely recognized among consumers, or the act of selling, distributing, displaying for the purpose of selling or distributing, exporting or importing goods on which such goods or their indication is used, and thereby causing confusion with another person's goods or business (Article 2.1.1).

The act of using, as one's own, goods or other indication that is identical with, or similar to, another person's famous goods or other indication, or the act of selling, distributing, displaying for the purpose of selling or distributing, exporting or importing goods that use such goods or other indication (Article 2.1.2).

The act of assignment, lease, display for the purpose of assignment or lease, export or import of goods that imitate the form of another person's goods (excluding forms indispensable to ensuring the functioning of said goods) (Article 2.1.3).

The owners of the above-mentioned designs including unregistered designs may claim compensation of damages, cessation of infringement and criminal sanction against persons involved in the above-mentioned acts.

CHAPTER 12

Industrial Design Rights: Malaysia

Karen Abraham, Janet Toh & Tamara Lee

§12.01 INTRODUCTION

In Malaysia, protection for Industrial Design is governed under the Industrial Designs Act 1996 ('the Act') and Industrial Designs Regulations 1999 ('the Regulations') which came into force on 1 September 1999. Unlike the United Kingdom, Malaysia does not provide any specific protection for unregistered designs. Only industrial designs that are registered in Malaysia are protected in Malaysia. Thus, industrial design registration is an integral part of protecting design rights.

Prior to 1 September 1999, industrial designs were protected in Malaysia by way of registration in the United Kingdom under the UK Registered Designs Act 1949. The UK registration was then treated as extending to Malaysia under various enabling statutes. This system was repealed when the Act came into force, although it is still possible to maintain the protection conferred by UK registered designs granted prior to 1 September 1999.

[A] Highlights of New Developments in This Jurisdiction since Last Edition

In 2013, the Act was amended significantly with some of the main changes involving a change from national to worldwide novelty and an extension of the protection term from fifteen to twenty-five years. The extension period to twenty-five years is also applicable to industrial designs which were registered prior to the coming into force of the Industrial Designs (Amendment) Act 2013 on 1 July 2013. The recent amendments have also introduced provisions which recognize industrial designs as personal property which may be assigned, transmitted or dealt with in the same manner as other personal or movable property.

§12.02 CONVENTIONS AND LEGISLATION

Protection for industrial design in Malaysia is governed under the Industrial Design Act 1996 (IDA 1996) and Industrial Designs Regulations 1999 (IDR 1999). Malaysia is a signatory to the Paris Convention for the Protection of Industrial Property.

§12.03 SUBJECT MATTER

[A] Definition/What Constitutes a Protectable Design?

Requirements for a Registrable Design

It is an essential criterion for an industrial design registration that the said design must be new at the date an application for its registration is filed, or, if applicable, date of priority. The Malaysian Industrial Designs Act provides a provision for filing a multiple-design application wherein two or more industrial designs may be the subject of the same application, provided they relate to the same class of the International Classification (Locarno Classification) for Industrial Designs or to the same set or composition of articles. Each design in a multiple application is protected independently and have individual application numbers and consequently, individual registration certificates.

[1] Shape, Configuration

'Industrial Design' is defined as:

> features of shape, configuration, pattern or ornament applied to an article by any industrial process or means, being features which in the finished article appeal to and are judged by the eye, but does not include -
>
> (a) a method or principle of construction, or
> (b) features of shape or configuration of an article which -
> (i) are dictated solely by the function which the article has to perform;
> (ii) are dependent upon the appearance of another article of which the article is intended by the author of the design to form an integral part.

The position in Malaysia is that the ornamental or aesthetic aspects of an article are generally protected by way of the law of industrial designs. The protection of designs therefore does not extend to any method or principle of construction as it is confined to the aesthetic appearance of a specific article. The design may consist of three-dimensional features such as the shape and configuration of an article. These include things like industrial or handicraft items (other than a computer program).

[2] Ornamental or Functional

The design may also consist of two-dimensional features, such as pattern and ornamentation. These include a textile article, wallpaper or similar wall covering or lace, or to sets of such articles.

The definition of 'industrial design' in the Act specifically excludes a method or principle of construction, which recognizes that the protection conferred is confined to the aesthetic appearance rather than any technical aspects that are more suitable for patent protection. Also excluded from the legal definition of an industrial design are features of shape or configuration of an article which are: (1) dictated solely by the function which the article has to perform, or (2) dependent upon the appearance of another article of which the article is intended by the designer to form an integral part. These exclusions are often referred to as the 'must fit' and 'must match' exclusions, respectively.

Whilst spare parts are not explicitly excluded from registration, the definition of an industrial design under the Act seems to suggest that such parts are not registrable if they are either solely functional or if they fall within the must-match exclusion. Interestingly whilst a statutory interpretation of an industrial design suggests that spare parts may very often not be registrable, spare parts have, however, been allowed registration. This is probably due to the current practice of the Industrial Designs Registry as spelled out in the Act, which suggests that the Registrar's role is to determine whether the formality requirements of an industrial design application have been met, and if so, the industrial design may be recorded in the Register.

[3] Applied to an Article by Any Industrial Process

The design features must be applied to an article by any industrial process. Under the Act, 'article' means any article of manufacture or handicraft, and includes any part of such article or handicraft if that part is made and sold separately but does not include an integrated circuit or part of an integrated circuit within the meaning of the Layout-Designs of Integrated Circuits Act 2000, or a mask used to make such an integrated circuit.

A 'set of articles' means a number of articles that are of the same general character and ordinarily on sale together, or intended to be used together, being articles to each of which there is applied an industrial design that is the same as, or which differs only in immaterial details or in features commonly used in the relevant trade from, the industrial design applied to the other articles or to any of them. An example of a set of articles will be a set of cutlery, consisting of forks and spoons, bearing the same floral pattern.

[4] The Design Must Appeal to the Eye

The features of shape and configuration, pattern or ornament must have consumer or customer eye-appeal i.e., there must be aesthetic value in the design.

[5] Excluded Articles

Certain categories of product and types of articles are excluded from registration, for instance industrial designs that are contrary to public order or morality. When assessing morality, local sensitivities are taken into account. The layout design of an integrated circuit is not registrable under the Industrial Designs Act 1996. Instead, the protection of such a layout design is covered by the Layout-Designs of Integrated Circuits Act 2000.

[B] Originality/Novelty

[1] Test for Originality Generally

Not applicable

[2] Novelty

Prior to the amendment of the Act in 2013, novelty was only at the local level i.e., only local novelty needed to be satisfied. The current standard of novelty adopted as the registrability standard is worldwide novelty whereby a design shall not be considered as 'new' if it has, before the priority date or the application date, been disclosed to the public anywhere in Malaysia or elsewhere.

Under the new Act, the requirement for novelty extends internationally, as opposed to previously being local novelty. A design shall not be considered as 'new' if it has, before the priority date or the application date, been disclosed to the public anywhere in Malaysia or elsewhere. The only exception (as provided for in subsection 12(3) of the Act) is if it is disclosed to the public solely by reason of the fact that, within six months preceding the filing of the application for registration in either of the following circumstances:

(a) if the industrial design appeared in an official or officially recognized exhibition; or
(b) if it was disclosed by a person other than the applicant or his predecessor in title as a result of an unlawful act committed by that other person or another person.

Notwithstanding this, case laws have held that novelty is dependent on the nature of the article. In cases where the article is a generic product that is commonly used by members of the public, it is likely that the design features applied to an article are already individually in existence or in use prior to the formation of the said article. In this regard, in determining whether an article possesses novelty, the Courts have given paramount consideration to the totality of the design features taken as a whole and the overall appearance of the common article as compared to articles that were disclosed to public earlier.

[C] Unobviousness (if Appropriate)

Not applicable

§12.04 DURATION OF PROTECTION

Initially, the period of registration of an industrial design was fifteen years. The period of registration of an industrial design may now be extended for four further consecutive terms of five years each after the initial five-year term as opposed to only two further consecutive five-year terms under the Act prior to the amendments which came into force in July 2013. This shall also apply to any design that has been registered under the Industrial Designs Act 1996 before 1 July 2013 (except for designs registered in the United Kingdom before 1 September 1999 and which are subject to the repeal and saving provision of section 49 of the Industrial Designs Act 1996).

Under the new Act, the maximum period of protection is twenty-five years, subject to the payment of extension fees before the expiration of each five-year term. If a design application is filed claiming priority from an earlier application, the priority date is deemed to be the registration date of the design and for the purposes of determining the extension period, the expiry date is calculated from the priority date. The new term of protection of twenty-five years should make design registration a more attractive option.

Another relevant amendment provided under the amended regulations is that proprietors of registered designs may now submit applications for renewal of their designs *any time* before the expiration of the current term as opposed to the *six months* period before the expiration date. However, extension fees paid in advance are not refundable.

§12.05 INFRINGEMENT

[A] Scope of Infringement Action

The proprietor of a registered design enjoys the exclusive right to make or import for sale or hire, or for use for the purposes of any trade or business, or to sell, hire or to offer or expose for sale or hire, any article to which the registered industrial design has been applied. A design is infringed by a person who without the licence or consent of the owner of the industrial design commits the following things while the registration is still in force:

(a) applies the industrial design or any fraudulent or obvious imitation of it to any article in respect of which the industrial design is registered;
(b) imports into Malaysia for sale, or for use for the purposes of any trade or business, any article to which the industrial design or any fraudulent or obvious imitation of it has been applied outside Malaysia without the licence or consent of the owner; or

(c) sells, or offers or keeps for sale, or hires, or offers or keeps for hire, any of the articles described in paragraphs *(a)* and *(b)*.

[B] Test for Infringement/Criteria to Constitute Infringement

A certificate of registration will constitute prima facie evidence of the facts stated therein and of the validity of the registration.

In cases of infringement of a registered design, infringement extends beyond the use of an identical design to the application of a fraudulent or obvious imitation of the registered industrial design to any article in respect of which the design is registered. Based on settled cases, it is not necessary to adduce expert evidence during the trial to determine the issue of infringement under the Act. The court has to only make a visual comparison between the plaintiff's registered industrial design and the alleged infringing designs. In making the visual comparison between the registered industrial design and the alleged infringement by the defendant, the factor to be considered is the ordinary consumer and how his eye will, judge the design.

[C] Standing to Sue

The owner of a registered industrial design has the right to take legal action against an infringer within five years from the act of infringement.

Civil proceedings may be initiated by the owner of a registered industrial design against any person who has infringed or is infringing any of the rights conferred by the registration of the industrial design. For the purposes of this section, 'owner of a registered industrial design' means the registered owner and includes an assignee, a licensee or the beneficiary of a compulsory licence granted under the Act; but if any proceedings are instituted by a person other than the registered owner, it must be proved that that person had made a prior request to the registered owner to institute proceedings for the infringement complained of by him and that the registered owner had refused or failed to institute the proceedings within three months from the receipt of the request, without prejudice, however to the registered owner's right to join in such proceedings.

[D] Remedies for Infringement

Concurrent with the commencement of a civil suit, the plaintiff may wish to take pre-emptive measures to immediately stop the infringing action by sending a cease and desist warning notice, applying for an interim injunction, a *Mareva* injunction or an *Anton Piller* order. If the owner proves that an infringement has been committed, the court may award damages or an account of profits and may grant an injunction to prevent further infringement and any other legal remedies.

The following acts constitute a criminal offence:

(a) making or causing to be made a false entry in the Register;
(b) making or causing to be made a false entry in any certified copy of a document or publication in the Industrial Designs Registration Office;
(c) making or causing to be made a false document purporting to be a copy of an entry in the Register;
(d) producing or tendering or causing to be produced or tendered in evidence any document mentioned in paragraph (b) or (c) knowing such document or any entry therein to be false; or
(e) falsely making a claim that he is the owner of the registered industrial design or representing that he is acting in the capacity of, or on behalf of the owner of the registered industrial design.

[1] Interlocutory Injunctions

An interim (or interlocutory) injunction prevents the infringer from continuing infringing activities until the conclusion of the trial. Generally, the aggrieved person will seek an interlocutory injunction during the course of the pre-trial period, as opposed to a final or permanent injunction whereby the remedy is only granted at the conclusion of the trial. The principles on which interlocutory injunctions are to be granted are: (1) whether there is a bona fide serious issue to be tried; (2) having found that an issue has been disclosed that requires further investigation, he must consider where the justice of the case lies by taking into account all relevant matters, including the practical realities of the case before him and the harm the injunction would produce by its grant, against the harm that would result from its refusal; and (3) the judge must have in the forefront of his mind that the remedy he is asked to administer is discretionary, intended to produce a just result for the period between the date of the application and the trial proper and to maintain the status quo.

[2] Anton Piller Orders

In Malaysia, there are no specific criminal remedies available for infringement of a registered design. If there is strong reason to believe that the defendant may destroy the offending matter and other evidence of infringement, the plaintiff may make an ex parte application for an *Anton Piller* order to enter the defendant's premises to *inter alia*, search and seize any such evidence. The *Anton Piller* order, like the interlocutory injunction is a form of pre-trial relief.

The efficacy of an *Anton Piller* order therefore lies in its elements of surprise and speed because it is made ex parte. An *Anton Piller* order is a mandatory order permitting the applicant to enter specified premises to inspect and take into custody documents or articles relevant to the action which may be destroyed or otherwise concealed by the defendant. An *Anton Piller* order is not a search warrant but an order in personam to permit the plaintiff's entry or be in peril of proceedings for contempt of court. The criteria that must be satisfied before an *Anton Piller* order is granted is that the applicant must have an extremely strong prima facie case that the defendant is liable to the plaintiff for some wrongdoing; that the actual or potential damage to the plaintiff is very serious; that the defendant has in his possession incriminating documents or other materials; that there is a real possibility that such materials may be destroyed or concealed; that the order must not cause any harm to the defendant or his case; the applicant may be required to give a cross-undertaking in damages; the applicant should be accompanied by his solicitor and should act with 'due circumspection'; and the defendant should be allowed to consider the order and to consult his solicitor.

[3] Mareva Injunctions

Mareva injunctions are granted to the plaintiff to restrain infringers from dissipating their assets out of jurisdiction.

In order for a Mareva injunction to be granted, the plaintiff must establish that it has a:

- strong prima facie case;
- clear evidence that the defendant has assets within the jurisdiction;

there is a danger that the defendant's assets may be dissipated before the judgment.

§12.06 DEFENCES TO INFRINGEMENT

[A] Validity

Any ground on which the registration of an industrial design may be revoked is available as a defence in an action for infringement of a registered industrial design.

A registered design may be revoked on the following grounds.

[1] Not Novel

The registration of the industrial design was not new at its priority date.

[2] Solely Functional/Not Ornamental

The registered design consists of features of shape or configuration which are wholly dictated by the function for which it is to perform.

[B] Markings

In Malaysia, it is not mandatory to mark the goods prominently with the words 'Registered Design' upon registration. The owner may however still like to ensure that the articles which incorporate the design are properly marked. This is because the Act provides that, if a registered design is infringed, the owner of the design may be awarded damages. Nevertheless, if there is no design marking present on the allegedly infringing articles, and the infringer can show that he was not aware that the design was registered and that reasonable steps were undertaken to ascertain whether the industrial design had been registered, then the only relief available to the owner of the design may be an injunction to prevent further infringement of the registered design. In that case, the owner of the design would not be compensated for the monetary losses incurred by reason of the infringer's activities.

[C] Non-infringement

[1] Innocent Infringer

If the alleged infringer can prove that at the time of infringement, he was not aware that the industrial design was registered and that he had taken reasonable steps to ascertain whether the industrial design had been registered, the Court may refuse to award damages or to make an order for an account of profits in respect of an infringement. It would therefore be advisable that upon the registration of the industrial design, an advertisement be published in a newspaper.

[2] Prior User

Prior user rights are available if the industrial design was already disclosed to the public anywhere in Malaysia or elsewhere.

[3] Groundless Threats of Infringement

There are currently no provisions under the Act to protect victims of groundless threats of infringement.

[4] Compulsory Licence

There is no infringement if a person has applied to the court for the grant of a compulsory licence in respect of the industrial design on the ground that the industrial design is not applied in Malaysia by any industrial process or means to the article in respect of which it is registered to such an extent as is reasonable in the circumstances of the case.

[D] Other Defences

Not applicable

§12.07 PROCEDURES FOR FILING APPLICATION FOR REGISTRATION

[A] Priority

Malaysia is a signatory to the Paris Convention. The Act therefore provides for priority claims from one or more earlier national, regional or international applications filed pursuant to any international treaty or convention to which Malaysia is party. If an application for registration of an industrial design is first filed in another Paris Convention country within six months immediately preceding the filing date of the corresponding Malaysian application, the applicant may claim priority based on the earlier foreign application. If a design application is filed claiming priority from an earlier application, the priority date is deemed to be the registration date of the industrial design for the purposes of assessing the novelty of the design.

[B] Drawings

See §12.07[E] Requirements for Application.

[C] Design Part of Another Article/Environment of Design

See §12.07[E] Requirements for Application.

[D] Description of Design (Where Required)

See §12.07[E] Requirements for Application.

[E] Requirements for Application

Each application requires:

(i) Name and address of the author (the person who created the design) and the applicant. Kindly note that the author(s) must be one or more individuals, and not a corporation/company. However, the applicant(s) may be one or more individuals or corporations.

(ii) An indication of the kind of article of manufacture or handicraft to which the design shall be applied.

(iii) Agent's name and address (where an agent is appointed).

(iv) Locarno classification (class and sub-class number).

(v) A set of clear drawings or photographs of each of the views of the article which the applicant wishes to register.

(vi) Application or registration number of an earlier design which is the subject of the associated design (where the applicant seeks association).

(vii) The prescribed official fee(s). It is now the requirement of the Industrial Designs Registry for the Applicant to submit payment of the publication fee at the point of filing the application. The Designs Registry requires the applicant to state the number of views to be published in the Official Journal in the application form. In view of the recent amendment to the Malaysian Industrial Designs Regulations, the Industrial Designs Registry has taken the position that the applicant has a choice of whether to publish all views or only selected views in the journal.

(viii) Statement of novelty, setting out the features of the design for which the owner claims novelty.

Additionally applications will require one or more of the following, where appropriate:

(ix) If the applicant is not the author, a brief statement laying down how the applicant acquired the rights to the design for example, whether by virtue of a contract of employment or assignment.

(x) The filing number of the priority application of which the priority is claimed (if any) together with a certified copy of the relevant prior design application(s) together with the English translation, if relevant.

There is no minimum prescribed number of views and the representations show several views of the design applied to the article. Each view must be labelled with a brief description to indicate the type of view e.g., 'side view', 'back view', etc. The applicant should ensure that the views filed for registration are sufficient to confer novelty to the design. The drawings/photographs of each of the views must be of the required measurement, i.e., 12.5 cm x 9 cm.

A signed appointment of agent is required where the applicant's ordinary residence or principal place of business is outside Malaysia. A failure to appoint an agent in such circumstance may result in the refusal to proceed with the application.

Because there is no formal official search or substantive examination, once filed, the application undergoes only a formal examination. Objections of a substantive nature are however occasionally raised. If there are objections, an Office Action will be issued whereby the applicant will be given an opportunity to respond. In the absence of any objections, the application will proceed directly to registration if the application is in order. The registration details will be published in the Official Journal (previously the Government Gazette).

[F] Who May Apply for Registration

Under the Act, the original owner of an industrial design is entitled to make an application for the registration of the industrial design. In this context, the original owner, subject to contrary agreement, is the author i.e., creator of the industrial design. Thus, where:

(1) an industrial design is created in pursuance of a commission for money or money's worth, the person commissioning the industrial design shall, subject

to any contrary agreement between the parties, be treated as the original owner of the industrial design;

(2) an industrial design is created by an employee in the course of his employment, his employer shall, subject to any contrary agreement between the parties, be treated as the original owner of the industrial design;

(3) an industrial design, or the right to apply an industrial design to any article, becomes vested, whether by assignment, transmission or operation of law, in any person other than the original owner, either alone or jointly with the original owner, that other person or, as the case may be, the original owner and that other person shall be treated for the purposes of section 11 as the original owner of the industrial design or as the original owner of the industrial design in relation to that article.

In the case of an industrial design generated by computer in circumstances such that there is no human author, the person who made the necessary arrangements for the creation of the industrial design is considered the author. Where two or more persons own interests in an industrial design, all of the persons owning such interests, acting jointly, are, subject to any agreement among such persons to the contrary, entitled to make an application for the registration of the industrial design.

§12.08 EXPUNGING/CANCELLING OR VARYING REGISTRATION

[A] Rectification/Revocation/Grant of a Compulsory Licence

Any person aggrieved by or interested in the non-inclusion in or omission from the Register of any entry, or by or in any entry made in the Register without sufficient cause, or any entry wrongfully remaining in the Register, or any error or defect in any entry in the Register, may apply to the Court for an order to include, make, expunge or vary any such entry or for the correction of any such error or defect as the Court deems fit.

Any person may apply for the revocation of an industrial design at any time, on the ground that the industrial design has been disclosed to the public prior to the priority date of the application for registration of the industrial design, or apply for the cancellation of the registration of the industrial design on the ground that the registration of the industrial design has been procured by unlawful means, or apply for the grant of a compulsory licence in respect of the industrial design on the ground that the industrial design is not applied in Malaysia by any industrial process or means to the article in respect of which it is registered to such an extent as is reasonable in the circumstances of the case.

[B] Associated Designs

Due to the novelty requirement, a modified version of an industrial design cannot subsequently be registered for another article. However, in certain instances, the registered proprietor of a design is allowed to make a subsequent application to register the same design in respect of other articles, i.e., a subsequent application which attempts to associate a later registered design with an earlier one. In this respect, any prior registration or publication of a design will not be taken into account in relation to novelty. The period of registration of an industrial design registered as an associated design shall not extend beyond the period of registration of the original registered industrial design and any extended period thereof.

§12.09 OTHER

Registration of a design may not be the only form of relevant intellectual property protection available, or it might not be relevant to a particular design. Other potentially relevant intellectual property rights include trademarks, passing off and patents.

[A] Interface between Industrial Design and Trademarks

The existing *Malaysian* law does not specifically provide for the registration of three-dimensional marks. The existing definition of a trademark is a '*mark used or proposed to be used in relation to goods or services for the purpose of indicating or so as to indicate a connection in the course of trade between the goods or services and a person having the right either as proprietor or as registered user to use the mark....*'. It must therefore be distinctive and distinctiveness is judged by reference to its inherent ability to distinguish and acquired distinctiveness by reason of use.

A mark in this context is defined in *inclusive* terms, i.e., a mark includes a device, brand, heading, label, ticket, name, signature, word, letter, numeral, etc. Three-dimensional marks are currently not recognized as a 'mark' because the references to the use of a mark under the Trade Marks Act are use of a printed or other visual representation of the mark; and in relation to goods, it can include 'references to the use thereof upon, or in physical or other relation to, goods'.

In light of the above, two-dimensional line drawings of three-dimensional products are registrable. Despite the prevailing view that three-dimensional marks are not registrable, some have been filed taking advantage of the non-exhaustive definition of the term 'trade mark'. The Malaysian government itself is aware of the need to widen the scope of the Malaysian Trade Marks Act 1976 by introducing non-traditional trademarks to meet changing market demands. It is anticipated that when Malaysia accedes to the Madrid Protocol, the Trade Marks Act will be amended to expand the definition of a 'trade mark'.

[1] Passing Off–Unregistered Designs

The fundamental principle of the tort of passing off is that no trader should be allowed to compete unfairly with another trader by passing off the former's goods and/or services as and for those of the latter by way of misrepresentation and, thereby, damaging the goodwill and reputation of the latter. The elements of passing off are: (i) goodwill and reputation of the plaintiff, (ii) misrepresentation by the defendants in the course of trade, and (iii) damage or the likelihood of damage will arise as a result. A misrepresentation is actionable as a passing off only if it gives rise to confusion, which the court can infer from the surrounding facts and from whether the average reasonable person, with characteristics reflective of the relevant section of the public, is likely to be confused by the misrepresentation, rather than requiring evidence of actual confusion. If the owner of the unregistered design has commenced use of its design, and the owner can establish that the design is distinctive, any obvious imitation of the design is likely to amount to misrepresentation to the public at large in Malaysia.

[B] Relationship between Industrial Design and Copyright/Design Copyright

As of 1 September 1999, by virtue of section 7(5) of the Copyright Act 1987 ('Copyright Act'), 'copyright shall not subsist under the Copyright Act in any design which is registered under any written law relating to industrial design'. Based on this provision, this means that after 1 September 1999, the law has provided that copyright shall not subsist in any design which is registered under any written law relating to industrial design.

The Copyright Act contains provisions that deal with the exploitation of artistic works by an industrial process. Under such circumstances, section 13(B) of the Copyright Act provides as follows:

> (1) This section applies where an artistic work has been exploited, by or with the licence of the copyright owner, by-
> (a) making, by an industrial process or means, articles falling to be treated for the purposes of this Act as copies of the work; and
> (b) marketing such articles in Malaysia or elsewhere.
> (2) After the end of the period of twenty-five years from the end of the calendar year in which such articles are first marketed, the work may be copied by making articles of any description, or doing anything for the purpose of making articles of any description, and anything may be done in relation to articles so made, without infringing copyright in the work.

The effect of sections 7(5) and 13(B) of the Copyright Act when read together seems to suggest that an unregistered design can enjoy copyright protection for twenty-five years from the end of the calendar year in which such articles are first marketed (whether in Malaysia or elsewhere).

Pursuant to the Copyright (Amendment) Act 2012, section 7(6) which dealt with the subsistence of copyright in any design which is capable of being registered but which was not registered has been deleted and all such designs are now dealt with under section 13(B) of the Copyright Act.

[C] Relationship between Industrial Design and Unfair Competition

Not applicable

[D] Relationship between Industrial Design and Patent

Unlike industrial designs, patents protect inventions which may relate to a product or process - basically the functional aspects of an article. If a design has solely functional features, those may be excluded from the monopoly given by a design registration, because they fall foul of the definition of an 'industrial design', and it may be worth protecting them via a patent if the patent satisfies the criteria of patentability i.e., novelty, inventive step, capable of industrial application. Such can be filed directly in Malaysia or as a Patent Cooperation Treaty (PCT) application. Protection for some articles may be sought under both patent and industrial design law in order to effectively protect both the functional and the aesthetic aspects of their design.

[E] Relationship between Industrial Design and Layout-Design Protection

As is mentioned above (under 'excluded articles'), it is not possible to protect layout-designs as registered designs. Instead those are protected through the Layout-Designs of Integrated Circuits Act 2000 (the Layout-Designs Act).

The Layout-Design of Integrated Circuits Act came into force on 15 August 2000 and it creates a sui generis protection for layout designs. There is no requirement for formal approval or registration. Protection is accorded to a layout design which is original, i.e., the result of the creator's own intellectual effort, and which is not commonplace among creators of layout designs and manufacturers of integrated circuits.

The protection given to a layout-design under the Layout-Designs Act is available to a natural person who is a national of, or is domiciled or ordinarily resident in, Malaysia or in a qualifying country, or a legal entity that is incorporated or formed in Malaysia or in a qualifying country. A qualifying person also includes a person or legal entity with a real and effective industrial or commercial establishment for the creation of layout-designs or for the production of integrated circuits in Malaysia or in a qualifying country. A qualifying country is a country or territory that is a Member to the World Trade Organization Agreement. The Minister may, also by order published in the *Gazette*, designate a country or territory as a qualifying country if he considers that provision has or will be made under the law of that country or territory that will give a right holder adequate protection in that country or territory in respect of a protected layout-design in which he has rights.

The duration for protection of a layout design is ten years from the first commercial exploitation in Malaysia or elsewhere or fifteen years from the date of the creation of the design, whichever is shorter.

The owner of a layout design enjoys the right to reproduce and to authorize the reproduction of all or a substantial part of the layout design, whether by incorporation into the integrated circuit or otherwise and to commercially exploit and to authorize such commercial exploitation of the layout design, an integrated circuit in which the protected layout design is incorporated, or an article that contains an integrated circuit in which the layout design is incorporated. An infringement occurs if a person performs any of the aforesaid acts without the authorization of the owner.

It is no infringement to commercially exploit, or to authorize the commercial exploitation of:

(a) a copy of the protected layout-design;
(b) an unauthorized integrated circuit, being an integrated circuit in which the protected layout-design is incorporated; or
(c) an article that contains an unauthorized integrated circuit referred to in paragraph (b),

if, at the time when the person acquired the copy, integrated circuit or article, the person did not know, and could not reasonably be expected to have known, that the copy was unauthorized, that the integrated circuit was unauthorized or that the article contained an unauthorized integrated circuit, as the case may be.

[F] Nature of Registration

A registered industrial design may be the subject of a security interest in the same way as other personal or movable property. Any rights in the registered design may be enforced in like manner as in respect of any other personal or movable property.

An assignment or transmission or other operation of law in respect of a registered design or an industrial design application shall not have effect against third parties unless recorded in the Industrial Designs Register. Therefore, in an infringement proceeding, the Court shall refuse to award costs to the owner of a registered industrial design who becomes entitled to the registered industrial design by way of an assignment or transmission or by operation of law or by a security interest transaction unless:

(a) an application to record his title or interest in the Industrial Designs Register is made before the end of the period of six months beginning from the date of the relevant transaction; or
(b) the court is satisfied that an application to record his interest was made as soon as practicable.

[G] Licensing Agreements

The owner of a registered industrial design may grant a licence to any person to use that registered industrial design. In respect of joint ownership of designs, each owner may not grant a licence of the design right without the consent of each of the other owners. There are however no provisions for a recordal of licence with the Designs Registry.

[H] Protection of a Pending Design Registration

An action for infringement can only be initiated after the industrial design is registered. Accordingly, the creator of a design is very vulnerable to infringers during the pendency of his application, as there is a window of opportunity for infringers to apply the design prior to issuance of the registration certificate. It is therefore advisable for the details of the design to be kept as confidential as possible until the design is registered. Ideally, details of the design should therefore only be made available to those who have a definite need to know in order to further the commercialization of the designs. A 'confidentiality agreement' should be signed by the recipients of such details in all instances prior to their receiving the details.

[I] Transitional Provisions (under the 2013 Amendment Act)

Any legal proceedings which are pending before the date of coming into operation of the 2013 amendments shall be continued as if the amendments had not been enacted. However, it is crucial to note that section 12 of the Amendment Act states that any application for the registration of an industrial design pending on the date of coming into force of the amendments will be dealt with under the Act as amended by the Amendment Act. This being the case, it will be necessary to ensure henceforth that designs for which registration is sought are novel on a global basis.

§12.10 CONCLUSION

The extension of the total period of protection for an industrial design to twenty-five years and the clarification that the rights in respect of a registered industrial design may be assigned by way of security and be licensed to other users pursuant to the 2013 amendments is greatly welcomed as the system enhances the protection of the rights of owners of registered industrial designs further. The amendments to the Industrial Designs Act 1996 mark the first intellectual property (IP) legislation to recognize that the laws in relation to ownership and devolution of personal property are applicable in relation a registered industrial design marks. This recognition signifies the Government's commitment to encourage the use of IP as collateral in financial transactions. It is very likely that this commitment will be extended to other IP related acts in the near future, such as the Trade Marks Act which is currently being amended.

It is also hoped that with the switch of the registrability criteria from local novelty to worldwide novelty, this will reduce incidences of the designs of foreign owners being copied by third parties who may claim the design as their own in Malaysia and then attempting to enforce their design rights against the original design owner.

Owners of registered designs should also keep abreast with developments in trademark laws in Malaysia so that they can also benefit from the provisions of the Trade Marks Act once it is amended to expressly recognize the protection of three-dimensional marks.

CHAPTER 13

Industrial Design Rights: Mexico

Juan Carlos Suarez

§13.01 INTRODUCTION

[A] Highlights of New Developments in This Jurisdiction since Last Edition

The Mexican Industrial Property Law created the '*Instituto Mexicano de la Propiedad Industrial*' (IMPI), or Mexican Patent and Trademark Office (MPTO), as an independent government agency devoted to the registration and enforcing of intellectual property rights. Industrial Designs are encompassed therein.

In this connection, to improve its services, the IMPI offers now a new Filing System Online for Industrial Design applications at the IMPI's portal site 'http://eservicios.impi.gob.mx/seimpi/action/rdudi'.

§13.02 CONVENTIONS AND LEGISLATION

Article 28 of the Constitution provides that, in Mexico, monopolies are not permitted, with certain exceptions for strategic areas such as mail, telegraph, and etc.

Article 28 also provides that privileges may be granted to authors and artists for the production of their works and for the exclusive use of their inventions and improvements.

Based on this constitutional provision, the Industrial Property Law (the 'IPL') was enacted in 1994.

The IPL is the last in a series of laws which have been enacted to protect inventions, designs, models, trademarks, trade names, commercial slogans, appellations of origin, industrial secrets and to prevent unfair competition.

Mexico is a member of a number of international treaties, conventions and agreements which protect industrial and intellectual property. Mexico is a member of the Paris Convention, Patent Cooperation Treaty, TRIPS, Lisbon Convention on Appellations of Origin, The Hague Convention, Interamerican Copyright Convention, Universal Copyright Convention, Bern Convention, Geneva Convention, UPOV, Madrid Protocol and others.

The Federal Copyright Law (the 'FCL') came into force in December 1996 and has been amended several times to add certain provisions required to address recent developments, such as the protection of computer programs.

Other laws such as the Federal Law of Economic Competition may also affect the protection of intellectual property.

[A] The Mexican Industrial Property Institute

The IPL has, as its primary purpose, the protection of industrial property through the regulation and issuance of patents of invention, the registration of utility models, industrial designs, and etc.

The Mexican Industrial Property Institute (the 'IMPI') is the administrative authority in charge of the protection of industrial property.

Although it is an independent entity, the IMPI is coordinated with the Ministry of Economy (the 'ME'). Government fees are payable directly to the IMPI.

The IMPI is in charge of prosecuting and granting patents of invention, the registration of models of utility, industrial designs, trademarks, trade names, appellations of origin, renewals, assignments, licences, and all other matters dealing with the protection of industrial property.

The IMPI is also in charge of prosecuting cancellation actions, investigating infringements, ordering and carrying out inspections to verify compliance with the IPL and its regulations, gathering information, ordering and executing any preliminary measures required to stop infringement, imposing fines, closing down plants or places of business where infringing products are manufactured or distributed, and imposing any other necessary sanctions against infringement.

The IMPI may act as an arbitrator in industrial property cases at the request of the parties involved.

§13.03 SUBJECT MATTER

[A] Definition/What Constitutes a Protectable Design?

Industrial designs comprise industrial models and industrial drawings.

[1] Shape and Configuration

Industrial models, provided that they do not imply technical effects, are three-dimensional models that serve as molds to manufacture industrial products, and give a special appearance.

[2] Ornamental or Functional

Industrial drawings are any combination of figures, lines or colors, incorporated to an industrial product as an ornament, giving it a peculiar aspect of its own.

The protection conferred on an industrial design shall not cover those elements or features whose exact reproduction has been necessary to enable the product incorporating the design to be mechanically assembled or connected to another product of which it constitutes an integral part or component; however, this limitation shall not apply to goods in which the design element lies in a shape or form intended to permit the multiple assembly or connection of the goods or their interconnection within a modular system.

[B] Originality/Novelty

[1] Test for Originality Generally

The protection conferred on an industrial design shall not cover those elements or features whose exact reproduction has been necessary to enable the product incorporating the design to be mechanically assembled or connected to another product of which it constitutes an integral part or component; however, this limitation shall not apply to goods in which the design element lies in a shape or form intended to permit the multiple assembly or connection of the goods or their interconnection within a modular system.

[2] Novelty

The IP law requires absolute novelty of the design to be registrable in Mexico. However, the statute stipulates a grace period of twelve months counted from the date of public disclosure of the design (work, manufacture, publication, sale, exhibition) by the inventor or the assignee to file the design application in Mexico, provided that the applicant states the date of disclosure in the design application and provides the relevant documents to prove it.

In order to be registrable as industrial designs, designs must be new and capable of industrial application.

In order to determine whether an invention or design is new, the state of the art, as of the date of filing the respective application, shall be considered. In the case of a priority application, the state of art as of the priority date shall be considered.

In considering the state of the art, all applications filed prior to the application in question shall be considered.

Design applications are not published in Mexico and there is no interference procedure. Only design registrations are published.

Disclosure of the design by a foreign Patent Office or WIPO would not benefit from the grace period provided in the law and would constitute a bar for the registerability of the design.

Any person who has made an invention, utility model or design, or his or her assignee, has the right to its exclusive use upon obtaining a registration, in the case of utility models or industrial designs.

The right to obtain a registration lies with the inventor or designer, or his or her assignee.

If a design has been made by two or more persons independently of one another, a preferential right is given to the first one to file.

The right to apply for a registration certificate may be assigned to others by agreement or by inheritance.

The owner of a registration may be an individual person or a corporation.

Anything that is not within the state of the art is considered new. State of the art is the sum of technical knowledge made public by a written or verbal description, through working or through any means of publication, in Mexico or abroad.

Registrability of industrial application (or utility in the US) is deemed to be a possibility of a design being manufactured or used in any branch of economic activity.

It is presumed that the inventor or designer is the person who claims to be such in the first filed application. The inventor or designer, even if he or she is not the applicant, has the right to be named in the certificate.

[C] Unobviousness (if Appropriate)

Not applicable

§13.04 DURATION OF PROTECTION

Industrial designs may be registered with the IMPI and are granted protection for a non-renewable term of fifteen years if they are new, and are used as a type or mold, to make industrial products.

§13.05 INFRINGEMENT

[A] Scope of Infringement Action

The exclusive right of exploitation of the registered design shall confer the following prerogatives on its owner: the right to prevent others from manufacturing, using, selling, offering for sale or importing the patented product without his consent.

[B] Test for Infringement/Criteria to Constitute Infringement

There are no Infringement Criteria to constitute infringement specifically set forth under the Mexican IPL, but in general an industrial design may be considered to be infringed if an alleged infringing product comprise all features recited by a claim and depicted in the drawings to literally infringe the industrial design registration.

[C] Standing to Sue

Any patentee or licensee (unless expressly forbidden from doing so) has the right to bring forth a lawsuit against a third party infringing his or her rights. A distributor may not bring a lawsuit for infringement. An accused infringer may counterclaim patent invalidity under formal or technical considerations upon receiving the infringement suit before IMPI, but it is not possible to request an additional judicial ruling or declaration. The plaintiff must prove that there is an actual or potential infringement of the design registration. In this case, filing a sample of the infringing product would be sufficient.

[D] Remedies for Infringement

The IMPI has authority to decide on patent infringement cases.

An administrative action with the IMPI is the most usual way to oppose design infringement. If successful, the IMPI would impose the infringer a fine up to about USD 100,000.00 and would order it to stop the infringing activities.

The IMPI has no authority to award damages or legal expenses.

The administrative infringement action involves a full administrative trial, with a claim, an answer, reports from expert witnesses if necessary, final arguments and a decision.

An official from the IMPI serves the summons. The defendant has ten business days to answer the infringement action and oppose with all the available defences.

The IMPI usually takes an extremely long time to decide patent infringement cases (from three to five years).

§13.06 DEFENCES TO INFRINGEMENT

In design infringement cases, the defendant is entitled to challenge the validity of the plaintiff's design on any grounds, including lack of novelty, or due procedural flaws during the prosecution of the design application.

The nullity of a design registration has retroactive effects. If the design registration is declared invalid, the IMPI would dismiss the infringement action without reviewing the merits.

[A] Validity

[1] Not Novel/Original

In order to be registrable as industrial designs, an design must be new and capable of industrial application.

In order to determine whether a design is new, the state of the art, as of the date of filing the respective application, shall be considered. In the case of a priority application, the state of art as of the priority date shall be considered.

In considering the state of the art, all applications filed prior to the application in question shall be considered.

[2] Functional/Not Ornamental

The protection conferred on an industrial design shall not cover those elements or features whose exact reproduction has been necessary to enable the product incorporating the design to be mechanically assembled or connected to another product of which it constitutes an integral part or component; however, this limitation shall not apply to goods in which the design element lies in a shape or form intended to permit the multiple assembly or connection of the goods or their interconnection within a modular system.

[3] Other Possible Criteria, Such as Non-obviousness

Only novelty and industrial applicability criteria are sought in industrial designs by the IMPI.

[4] Faults in Application

Procedural flaws during the prosecution of the design application may be the basis to declare the nullity of a design registration in Mexico.

[B] Marking

According to the Mexican IPL, in order to apply a civil or criminal action, it is a requirement that the owner of the industrial property right had marked the product as 'registered' or announced it by any acceptable mean to the general public the existence of the design and that the products are protected with an industrial property right.

The law does not state the acceptable means to inform the public about the design in the absence of marking on the product. In some cases the lack of marking has been cured with notices published in some main newspapers and the IMPI has accepted such notices to issue preliminary measures.

It should be understood that according to the IPL, it constitutes an administrative infringement the offer of non-registered products as if they were registered, as well as to sell or put into circulation products indicating they are protected by a design registration, when the products are in fact not protected by any design registration.

[C] Non-infringement

The basis of this defence is that the proper interpretation of the design claim does not cover the alleged infringing product.

It also constitutes an administrative infringement, by a design registration owner, the offer of non-registered products as if they were registered, even if the design registration has lapsed or has been cancelled. However, infringement activities are permitted merely one year after the lapse date, or when applicable, as from the date in which the cancellation decision becomes final and conclusive.

§13.07 PROCEDURES FOR FILING APPLICATION FOR REGISTRATION

[A] Priority

Priority shall be recognized for any filing that is equivalent to a regular national filing under the domestic legislation of any country of the Union for the protection of industrial property under Paris Convention. The period of priority shall be six months for industrial designs. A certified copy of the priority application, if priority is claimed, must be lodged within ninety days from filing.

[B] Drawings

Every industrial design application must include either a drawing or a photograph of the claimed design. The design must be represented by a drawing which must contain a sufficient number of views to constitute a complete disclosure of the appearance of the design. Informal drawings may be submitted at the time of filing. The applicant has two months to file formal drawings counted from the date the IMPI issues the office action requesting them, or at any time before the issuance of said office action.

[C] Design Part of Another Article/Environment of Design

It is acceptable to include environment — subject matter that is not part of the article — in a single view, provided that the environment helps to make clear what the article is or what its design features are. It should be used broken lines to show parts of the article that are not part of the design, but the design must be shown in solid lines. No written matter is allowed on the drawing, such as the title, descriptive text or dimensions.

[D] Description of Design (Where Required)

The description submitted in the application shall refer briefly to the graphic or photographic reproduction of the design, with a clear indication of the view from which the illustration is seen.

The following information are needed to file any industrial design application in Mexico:

(i) Specification and claim.
(ii) Informal drawings may be submitted at the time of filing.
(iii) Name, nationality and address of the applicant(s).
(iv) Name, nationality and address of the designer(s).
(v) State or Country of incorporation of the applicant if it is a corporation, partnership, limited liability Company, etc.
(vi) Date, title, serial number and country of origin of the priority to be claimed, if any.
(vii) Date and circumstance of any prior disclosure of the design, if any.

The following documents may be submitted after the filing of the application:

(viii) Formal drawings.
(ix) Power of attorney.
(x) Assignment, if the applicant is not the designer.
(xi) Documentation showing prior disclosure of the design, if applicable.
(xii) Certified copy of the priority application, if priority is claimed (ninety days from filing).
(xiii) The applicant may file a copy of the issued design from any counterpart filed with any examination office as provided by the Patent Cooperation Treaty (PCT) Office to expedite the issuance of the design registration.

The applicant has two months to file the above-mentioned documents (viii–x) counted from the date the IMPI issues the office action requesting them, or at any time before the issuance of said office action.

All documents drafted in languages different from Spanish must be fully translated into Spanish.

A Formal Examination of the industrial design application is carried out within the next three months from filing. Two opportunities are given to the applicant to submit missing documents such as Power of Attorney, Assignment, etc.

After twelve months from completing formal examination approximately, the IMPI will carry out a Substantive Examination of the industrial design application to determine if novelty and industrial applicability requirements are complied with.

If a prior art document is considered to be a bar for registerability of the design, the applicant will be given two opportunities to overcome the objection raised by the IMPI. A two-month term is regularly set forth for filing a response to an Office Action during substantive examination procedure. In addition, an extension of time from thirty to sixty days is generally available.

The applicant may expedite the issuance of the design registration by filing evidence of a corresponding design registration or patent design issued by any examining office authorized by the PCT Office.

An industrial design application may not include multiple designs in Mexico. An industrial design must relate to a single design.

If more than one design is included in a single design application, it shall be required the filing of one or more divisional applications as many designs are present in the initial application.

When the industrial design application is allowed, a notice of allowance is issued by the IMPI in order for the applicant to submit proper copies for publication of the registration, issue fee, and the first five annuities for maintenance of the industrial design registration.

§13.08 EXPUNGING/CANCELLING OR VARYING REGISTRATION

Changes to a design registration are allowable to correct any obvious errors of form, or to limit the scope of the claim.

§13.09 OTHER

[A] Interface between Industrial Design and Trademarks

Industrial Design and Trademarks may coexist. However, a published trademark may be cited as a prior art reference against the industrial design if the latest is not filed on time.

[B] Relationship between Industrial Design and Copyright/Design Copyright

Industrial Design and Copyright may coexist. However, publication of the copyright may be cited as a prior art reference against the industrial design if the latest is not filed on time.

[C] Relationship between Industrial Design and Unfair Competition

According to Article 213 of the IPL, the following shall constitute patent/design infringements:

(i) Unfair competition.
(ii) Placing products on sale or offering products with the indication that they are protected by a patent/design when they are not.
(iii) Manufacturing or processing products covered by a patent without the consent of the patent owner or without the respective licence.

(iv) Offering for sale or put in circulation products covered by a patent/design knowing they were manufactured or processed without the consent of the patent owner.

(v) Using patented processes without the consent of the patent owner or without the corresponding licence.

CHAPTER 14
Industrial Design Rights: The Netherlands

Kurt Stöpetie & Rutger Stoop

§14.01 INTRODUCTION

Industrial designs were not protected as such in the Netherlands until 1975, when the Uniform Benelux Act on designs or models ('UBDA') came into force.[1] Until that time, design protection in the Netherlands could only be based on copyright law and unlawful act (slavish imitation). Even though these two legal grounds are still offering juxtaposed protection to designs in the Netherlands today, designs are now protected in more efficient and harmonized ways under dedicated Benelux and European law.

The UBDA introduced uniform legislation, harmonizing law on designs and models in the Benelux countries (Belgium, the Netherlands and Luxembourg). With the appointment of the Benelux Office for Designs or Models as its administrative body ('BOD', seated in The Hague), the harmonization of the trademark laws of these three countries has been ensured. The UBDA was changed by protocols three times, the last time on 20 June 2002. This change aligned the UBDA with the Community Design Directive ('CDD') of 1998.[2]

[A] Highlights of New Developments in This Jurisdiction since the Last Edition

As can likely be said for all Member States of the European Union, a lot has changed since the CDD and the Community Design Regulation ('CDR')[3] entered into force in

1. Uniform Benelux Act on designs or models, entry into force (as between Belgium, Luxembourg and the Netherlands) on 1 Jan. 1974; entry into force of the Uniform Law: 1 Jan. 1975.
2. Directive 98/71/EC of the European Parliament and of the Council of 13 Oct. 1998 on the legal protection of designs, as published in the Official Journal L 289, 28 Oct. 1998 pp. 28–35, entry into force on 17 Oct. 1998.
3. Council Regulation (EC) No. 6/2002 of 12 Dec. 2001 on Community designs, published in Official Journal L 003, 05 Jan. 2002 pp 1–24, entry into force on 6 Mar. 2002.

1998 and 2002 respectively. Compared to where our last edition left off, Benelux legislation has changed profoundly on the back of the CDR and CDD.

§14.02 CONVENTIONS AND LEGISLATION

Although the CDD was never meant to fully harmonize the legislation of the EU Member States (see also the European chapter), the implementation of the CDD into the national law of the Member States brought substantial changes to Benelux law. Rather than the Benelux Court of Justice, the Court of Justice of the European Union ('CoJ EU') is now the highest judicial body.

Primarily for cost and legislative efficiency reasons, the UBDA and the Uniform Benelux Trademark Act ('UBTA') were replaced and merged into the Benelux Convention on Intellectual Property ('BCIP') on 1 September 2006.[4] The BCIP also merged the BOD and the administrative body pursuant to the UBTA (the Benelux Trademark Office) into the Benelux Office for Intellectual Property ('BOIP') which is seated in The Hague. It did not, however, change any substantive law.

Because of their longtime common legislation on trademarks and designs, the Benelux countries are regarded as a single Member State in a European trademark and design context. The most telling manifestation of this in practice is the appointment in the CDR of the BOIP as the single central industrial property office for these three countries (Article 35(1)(c) CDR). Thus, while an application for a Community-wide design is made on the national level elsewhere in Europe (see also the European chapter), in the Benelux countries applicants can apply for such registered Community design ('RCD') with the BOIP (in addition to the OHIM).

Other international agreements to which the Netherlands is party and relevant specifically to industrial designs are the Hague Agreement concerning the international registration of industrial designs and the Locarno Agreement establishing an international classification for industrial designs. More general agreements on industrial property applicable to the Netherlands as well, are the Paris Convention on the protection of industrial property and the Agreement on Trade-Related Aspects of Intellectual Property Rights (TRIPs).

§14.03 SUBJECT MATTER

[A] Definition/What Constitutes a Protectable Design?

Under Article 3.1(1) BCIP, a design is protected to the extent that it is new and has individual character. Article 3.1(2) provides that the appearance of the whole or a part of a product is regarded as a design. The latter falls into two distinctive categories: 'designs' (two-dimensional) and 'models' (three-dimensional). The BCIP uses 'design' to cover both categories.

---

4. Benelux Convention on Intellectual Property (Trademarks and Designs) of 25 Feb. 2005, entry into force on 1 Sep. 2006.

Within the definition of 'design', two key elements are of interest; appearance and product. Article 3.1(3) provides that a product's appearance is determined, in particular, through the features of the lines, contours, colours, shape, texture and/or materials of the product itself and/or its ornamentation. This is not an exhaustive list. The appearance of a product has to be new and in possession of individual character (Articles 3.1(1) and 3.3 BCIP, see further §14.03[B]). According to Article 3.1(4) BCIP, a product means any industrial or handicraft item, including, for example, parts intended to be assembled into a complex product, packaging, get-up, graphic symbols and typographic typefaces (also see §14.03[B]).

[1] Shape and Configuration

The appearance of a product is pivotal for the protection of a design. Under current case law, it seems that only visual elements are protectable, and not elements like taste, sound, touch and smell.

No right to a design will be granted if the filing does not sufficiently reveal the characteristics of the design (Article 3.6(f) BCIP). The basis of protection is not the actual traded product but the appearance of the product *filed* as a design. The filing is decisive and has to demonstrate that the appearance of the product meets the requirements of protection. This does not necessarily mean that the product as marketed and sold is not significant at all. The traded product to which the design was applied can be taken into account in legal proceedings as far as this helps to clarify the filed design.

The design right can be based on the entire appearance of a product. However, the design right can also be related to a component part of a complex product (Article 3.4(1) BCIP). A complex product is defined by Article 3.4(2) BCIP as a product composed of multiple parts that may be replaced in such a way as to allow a product to be taken apart and reassembled. Designs consisting of such component parts of a complex product may be registered as a design, but they must satisfy the requirements of Article 3.4 BCIP. The first requirement is that the component part, once incorporated into the complex product, must remain visible during normal use (Article 3.4(1)(a) BCIP). Normal use means use by the end user with the exception of maintenance, servicing or repair (Article 3.4(3) BCIP). The second requirement is that the visible features of the component part must in themselves satisfy the requirements of novelty and individual character (Article 3.4(1)(b) BCIP). Both requirements are cumulative.

[2] Ornamental or Functional

Functionality is no longer required for a product since the last amendment of the UBDA entered into force in 2003. Products only suitable for ornamental use can now also be registered as a design (e.g., sculptures). However, the right to a design still has to relate to a product *in concreto*. A design that is merely filed for ornamental purposes, but which can be used for all kinds of applications, is void.

This raises the question of what impact all this has on the 'doctrine of specialty'. Before 2003 the three aspects of appearance, product and functional use could not be assessed separately. Under the doctrine of specialty, the known outer appearance of a product could be protected as an independent design right if it had a different functionality. The doctrine was developed on the basis of the Supreme Court's judgment in the *Kinderkapperstoel* case, in which a pedal car mounted on the bottom of a barber chair was considered a different product than a pedal car. The barber chair was therefore entitled to independent design protection.[5]

In their joint comments to this last amendment of the UBDA, the governments of the Benelux countries noted that, in their view, the doctrine of *Kinderkapperstoel* was still applicable, because design protection is linked to a product through the definitions of design and product. However, this view was criticized in literature.

In the Netherlands, the Hague Court of Appeal held in its *I-Drain* decision that the fact that the appearance of a product has been applied to some other product no longer implies that a design right can be obtained for that other product. The Court of Appeal acknowledged that this meant a deviation from the *Kinderkapperstoel*-doctrine, but held that the *Kinderkapperstoel* judgment was made before 2003 and that the doctrine was no longer supported by literature.[6]

Considering these divergent views, and the lack of guidance from the CoJ EU, it remains unclear what is left of the 'doctrine of specialty' under current (EU) design law.

[B] Originality/Novelty

[1] Test for Originality Generally

A design merits protection to the extent that it is new and has individual character. In assessing novelty and individual character, the filed design must be compared with earlier designs that were disclosed to the public before the date of filing.

Pursuant to Article 3.3(2) BCIP, a design is considered to have individual character if the overall impression it produces on 'the informed user' differs from the overall impression produced on such a user by any design which was disclosed to the public before the date of filing or the date of priority. In assessing individual character, the degree of freedom of the creator in developing the design is taken into consideration. Accordingly, three subtests have to be applied in determining the individual character, which will be discussed successively: the overall impression, the 'informed user' and the degree of freedom of the creator.

The 'overall impression' test requires a comparison between the design for which an application has been filed and prior designs, made from the viewpoint of the 'informed user'. If there is no similarity, the design is deemed to have individual character. Citing the CoJ EU: 'the overall impression which that design produces on the informed user must be different from that produced on such a user not by a

5. Dutch Supreme Court 10 Mar. 1995, NJ 1995, 670, (*Kinderkapperstoel*).
6. Court of Appeal The Hague 17 May 2011, IER 2011/59, p. 442 (*I-Drain*), para. 10.

combination of features taken in isolation and drawn from a number of earlier designs, but by one or more earlier designs, taken individually'. Thus, the assessment has to be based on the overall impression of the design considered as a whole.[7]

The (fictitious) 'informed user' is not the average consumer, but a particularly observant one, either because of his personal experience or his extensive knowledge of the sector in question. As regards the informed user's level of attention, the qualifier 'informed' suggests that, without being a designer or a technical expert, this user knows the various designs which exist in the sector concerned, possesses a certain degree of knowledge with regard to the features which those designs normally include, and, as a result of his interest in the products concerned, shows a relatively high degree of attention when he uses them.[8]

The third subtest is that of the degree of the freedom of the creator in preparing the design. If the creator's freedom is limited by functional requirements, even relatively small differences can contribute to the individual character of the design.

[2] Novelty

According to Article 3.3(1) BCIP, a design is considered new if no identical design has been disclosed to the public on the filing or priority date. 'Disclosure to the public' by the creator is also prejudicial to novelty. The novelty criterion must be considered objectively: the creator's knowledge is not relevant.

Designs must be deemed to be identical if their features differ only in immaterial details. The overall impression produced on the informed user is not relevant in assessing the novelty of the design. The only criterion is whether the designs are identical or not, without any reference to a specific public.

[a] Prior Use

Article 3.3(4) BCIP regulates circumstances under which 'disclosure to the public' is not prejudicial to novelty and individual character. In order to assess novelty and individual character, the disclosure of a design will not be taken into consideration if a design for which protection is claimed under a registered design right has been disclosed to the public during the twelve-month period preceding the date of filing of the application or the date of priority. This period of time is the so-called grace period, which can be characterized as an exception to the requirement of novelty.

Article 3.3(4)(a) BCIP deals with the disclosure to the public by the creator, his beneficiary, or a third person, based on information provided or acts performed by the creator or his beneficiary. This provision covers cases in which the creator markets his design without filing it first. Such marketing does not entail that the design will lose its novelty if he files the design within the twelve-month period. This period of time gives

7. European Court of Justice 19 Jun. 2014, C-345/13, ECLI:EU:C:2014:2013 (*Karen Millen v. Dunnes*), para. 26.
8. Court of Appeal The Hague 13 Aug. 2013, ECLI:NL:GHDHA:2013:3356, (*B & O/Loewe*) para. 8, with reference to ECJ 20 Oct. 2011, C-281/10 P, ECLI:EU:C:2011:679 (*PepsiCo*), paras 53 and 59.

the creator the opportunity to experiment and market his product without loss of novelty.

Whereas Article 3.3(4)(a) BCIP concerns disclosure by the creator himself, Article 3.3(4)(b) BCIP deals with designs that have been disclosed as a consequence of an abuse in relation to the creator or his successor in title. These disclosures of the design will not undermine the novelty and individual character of the design.

[b] *Prior Publication*

Pursuant to Article 3.3(3) BCIP, existing designs (also known as the 'existing design corpus'[9]) are prejudicial to novelty and individual character. In assessing novelty and individual character, a design is deemed to have been disclosed to the public if its application has been published or has been exhibited, used in trade or made public in any other way, unless such events could not reasonably be known in the normal course of business to specialists in the sector concerned, who operate within the European Community or the European Economic Area, before the filing or priority date. This does not necessarily mean that only designs that have been disclosed to the public within the Community or the European Economic Area belong to the existing design corpus. The design could also be disclosed to the relevant public outside of the Community. Further, the phrase 'reasonably be known' is aimed at taking out of the equation disclosures to an immaterial public that could not have reached the sector concerned, such as disclosures in some remote foreign area. In their joint comments to the last amendment of the UBDA, the governments of the Benelux countries also referred to the specialists in the sector concerned as the 'collective memory'.[10]

However, the design will not be deemed to have been disclosed to the public only because it was disclosed to a third person under explicit or implicit conditions of confidentiality.

[c] *Prior Application or Registration*

In accordance with the Paris Convention, the BCIP provides in Article 3.3(5) that it is possible to obtain a right of priority for six months for the registration of a design, as from the filing date. Consequently, any subsequent filing, in any of the other countries that signed the Paris Convention, before the expiration of the period is not invalidated by reason of any acts performed in the interval. Rights acquired by third parties before the date of the first application that serves as the basis for the right of priority are reserved in accordance with the domestic legislation of each country of the Union, Article 4B Paris Convention. The priority period starts only from the date of filing of the first application (Article 4C(2) Paris Convention).

9. As used in recital 14 in the preamble to the CDR.
10. Joint Comments of the Governments of the Benelux countries to the Protocol of 20 Jun. 2002 amending the Uniform Benelux Act on designs or models, p. 11.

Furthermore, a design can be invalidated if the design is in conflict with a prior design which has been the subject of disclosure to the public after the filing or priority date and which has been protected, since a prior date, by an exclusive right deriving from a Community design, the registration of a Benelux filing, or an international filing (Articles 3.6(a) and 3.23(1)(a) BCIP).

§14.04 DURATION OF PROTECTION

Whereas former Benelux legislation granted protection for a maximum duration of fifteen years, the CDD provides for a maximum duration of protection of twenty-five years, in line with the CDD and CDR. A registration does not immediately grant protection for the entire period of twenty-five years, but for an initial period of five years, which may be extended for four more consecutive periods of five years each (Article 3.14 BCIP).

§14.05 INFRINGEMENT

[A] Scope of Infringement Action

Article 3.16 BCIP is the key article regarding the infringement of a Benelux design. Under this article, the owner of the right in the design may challenge the use of a product in which the design is incorporated, to which the design is applied, or which has an identical appearance to the design as filed. In particular, this use covers the manufacture, offering, marketing, sale, delivery, hire, importing, exporting, exhibiting, use or holding for any of those purposes. Pursuant to the design right, the holder may challenge any third person using a product with an outer appearance that does not produce a different overall impression on informed users, including the third party holder of a design registered later.[11]

The exclusive right of the holder is subject to limitations and does not imply the right to contest acts performed privately and for non-commercial purposes, acts done for experimental purposes, or acts of reproduction for the purposes of illustration or teaching, provided that such acts are compatible with fair commercial practice and do not unduly prejudice normal use of the design, and that the source is mentioned (see Article 3.19(1) BCIP).

[B] Test for Infringement

Whether the holder of the design right can challenge the use of a product by a third party depends on the appearance of the product. If its appearance is identical to the registered design or does not produce a different overall impression on the informed user, the use can be challenged. In determining the overall impression on an informed

11. European Court of Justice 16 Feb. 2012, C-488/10, para. 52 (*Celaya v. Proyectos*).

user, the creator's degree of freedom in developing the design must be taken into account (Article 3.16(1) BCIP). It is not required that the product was intended to copy the design. An identical appearance or the same overall impression is enough to constitute an infringement.

[C] Standing to Sue

Article 3.16 BCIP allows the holder of a design right to enforce his design right if the requirements for infringement from the previous paragraphs are met. Licensees of the holder of the design may also enforce their licenses to some extent. Article 3.26(4) BCIP provides that the licensee has the right to intervene in an action brought by the holder of the design right as mentioned in Article 3.17(1)-(4) BCIP (on compensation for loss). The licensee may then seek compensation for a loss directly incurred by him or to be allocated a proportion of the profit made by the defendant. The licensee may bring independent action as mentioned in Article 3.17(1)-(4) only if it has obtained permission from the holder of the design right for that purpose. In addition, Article 3.26(5) BCIP provides that the licensee may exercise the powers mentioned in Article 3.18(1) (additional claims such as recalls or destruction of infringing goods), provided that these serve to protect the rights which it has been permitted to exercise and provided that it has obtained permission from the holder of the exclusive right in a design for that purpose.

[D] Remedies for Infringement

Besides an injunction covering the whole of the Benelux on pain of forfeiting a penalty for non-compliance, the holder of a design may claim full compensation for loss as a result of unauthorized use if the design was used after the publication of the design registration. Under Article 3.17(1) BCIP, this publication has to disclose the characteristics of the design adequately. In addition to the claim for compensation, the holder of the design may request the court to order a recall from the channels of commerce, definitive removal from the channels of commerce or destruction of the infringing goods. Where appropriate, the destruction of materials and implements principally used in the manufacture of those goods may be requested, as well as disgorgement of profits made by trading an infringing design (Article 3.18(1) BCIP).

§14.06 DEFENCE AGAINST INFRINGEMENT

As a defence against an infringement claim, the alleged infringer may challenge the validity of the registered Benelux design. Because any defence against infringement is aimed at the validity of the Benelux design, the validity checks of this section correspond with §14.03 of this chapter.

[A] Validity

At the application of a Benelux design, the Office obviously checks whether the application meets the formal requirements (Article 3.9 BCIP, see also §14.07). Besides that, it only checks whether this design is in compliance with public policy and accepted principles of morality in the Benelux countries (Articles 3.13(1) and 3.6(e) BCIP).

[1] Not Novel/Original

If a registered design does not meet the requirements for a valid design, it has been registered by the Office in breach of the requirements as set out by the BCIP. Article 3.23(1) BCIP provides that any interested party may then invoke the invalidity of that registered Benelux design. In case of an infringement action, the alleged infringer may be regarded as an interested party. One of the grounds for invalidity is the lack of novelty or individual character of the design (Article 3.23(1)(b) BCIP). For a Benelux design to be valid, it needs to be novel and in possession of individual character (Article 3.1(1) BCIP). If it lacks either of these requirements it cannot be a valid Benelux design (see also §14.03).

[2] Functional/Not Ornamental

If the features of appearance of a product are solely dictated by its technical function, these features are excluded from protection. So, even if registered nonetheless, any interested party may invoke invalidity on this ground (Article 3.23(1)(c) BCIP). In principle, the same goes for the features of appearance of a product which must necessarily be reproduced in their exact form and dimensions in order to permit the product in which the design is incorporated or to which it is applied to be mechanically connected to or placed in, around or against another product so that each product can perform its function (Article 3.2(1)(b) BCIP).

[3] Other Possible Criteria

In addition to the CDR and CDD, the BCIP provides another validity defence. Article 3.6(f) BCIP explicitly states that the filing of the design has to sufficiently reveal the characteristics of the design. If the design registration does not meet this requirement, it lacks validity. This means that, for protection to be granted, all the characteristics of the design must follow from the document filed with the BOIP.

Furthermore, if the filing of a design is invalidated because a prior trademark or copyright is used in the design without consent of the owner of that right (Article 3.6(b) and (c) BCIP), or because the design improperly uses flags or emblems of a state party to the Paris Convention (Article 3.6(d) BCIP), or because a design is contrary to morality or public policy in one of the Benelux countries (Article 3.6(e) BCIP), or, finally, if a design is invalidated based on the grounds mentioned hereinbefore under

§14.03 [A] and [B], then the filing can be maintained in an amended form. According to Article 3.24(2) BCIP, this is only possible if the amended form of the design still meets the requirements for protection and the identity of the design is retained. All the other invalidation defences apply to the design in its entirety (Article 3.24(1) BCIP). This results in the entire invalidity of the Benelux design.

[4] Faults in Application

In case of lack of clarity, invalidity may be invoked if the filing does not sufficiently reveal the characteristics of the design (Articles 3.23(1)(d) and 3.6(f) BCIP).

[B] Marking

Marking a design means that the design is marked with a design right mark to demonstrate its design right. The BCIP does not require the marking of a design to grant protection, nor does it mention the option of doing so. However, although it is not a requirement, the marking of a Benelux design can be helpful when seeking compensation in an infringement action, because the marking makes it easier to prove that the infringing party should have been aware of the fact that a product was protected by a design right and that it therefore was not allowed to copy it.

[C] Non-infringement

The exclusive right granted by a Benelux design is subject to limitation. Consequently, the right may not be exercised in respects of acts performed privately and for non-commercial purposes, acts for experimental purposes and acts of reproduction for the purposes of illustration or teaching (Article 3.19(1) BCIP). Furthermore, the rights may not be exercised in respect of equipment on ships and aircraft in a third country when these temporarily enter the territory of the European Community, the importation in the Community of spare parts and accessories for the purposes of repairing such craft and performing repairs on such craft (Article 3.19(1) BCIP). Because these types of use do not fall within the protective scope of the Benelux design right, such use is not considered to be an infringement.

[D] Other Defences against Infringement

The CDD provides for three optional defences to infringement, which are all implemented in the BCIP. The first of these defences focuses on the use of a trademark in the design. If this prior trademark is used in the design as filed without the consent of the holder of that trademark, the design registration is liable to be invalidated (Article 3.6(b) BCIP). The same goes for a design registration incorporating a work protected by copyright. If this work is incorporated in the design without the consent of the copyright owner, the registration is liable to be invalidated (Article 3.6(c) BCIP). Note that all these defences can only be invoked by the holder of the respective prior rights (Article 3.23(3) BCIP).

Article 3.19(4) BCIP provides that the exclusive right does not entitle the holder to contest its use (as set out in Article 3.16 BCIP) if the products in which the design is incorporated or on which it is applied has been put on the market in the Community by the holder or with his consent. To this extent, the design right is 'exhausted' within the Community. This means that the holder can still challenge the use of products which are placed on the market in a country outside the Community. Unlike a Community design holder, however, the holder of a Benelux design can only enforce his right in the Benelux, but the right is exhausted when the product is placed on the market elsewhere within the Community by the holder or with his consent.

§14.07 PROCEDURES FOR FILING APPLICATION FOR REGISTRATION

The procedure for filing applications for registration is provided by the BCIP, and further specified in the practical Implementing Regulations under the BCIP. Part II of these Implementing Regulations is dedicated to the application process for design rights.

A Benelux design right in the Netherlands is obtained by application with the BOIP (Article 3.9 BCIP) and payment of the fees due. The BOIP only examines the formal requirements for application (Article 3.9(5) BCIP), along with contravention of public policy and morality (Article 3.13 BCIP), and has to register the filing without delay. If the BOIP considers - which it rarely does -, that the design contravenes public policy and/or morality, it will notify the applicant and grant a period of two months to withdraw the application. If the applicant does not withdraw its application in that time, the BOIP will notify the Public Prosecutor and ask it to commence an action to invalidate the application.

While a dedicated Benelux registration can be obtained in this way, an international registration can also be filed with designation of the Benelux at the WIPO office in Geneva (Article 3.5(1) BCIP). Such an international application results in a bundle of national registrations for the countries designated in this registration. As a result, an international registration with designation of the Benelux is governed, as regards the Benelux part, by the BCIP.

[A] Priority

Article 3.3(5) BCIP provides for a right of priority for applications for designs. The BCIP mirrors the definition of priority right as given in Article 4(C) of the Paris Convention, which provides for a maximum priority period of six months. This period starts the day after the date of the initial application.

The applicant must expressly claim its right to priority at the time of filing or in the month following filing by means of a special declaration. If the applicant does not do so, the right of priority will lapse (Article 3.10(2) BCIP). The Implementing

Regulations specify this special declaration further in Rule 2.4, requiring, *inter alia*, a copy of the documents supporting the right of priority.

[B] Drawings

A Benelux application may comprise either a single design (single filing), or several designs (multiple filing, Article 3.9(1) BCIP). The Implementing Regulations (Rule 2.1) further specify the requirements of the application: an image or images showing the appearance of the product, an indication of the product embodying the design and a description of the colour or colours of the design. Furthermore, a detailed description of the product embodying the design should be provided, preferably using the wording in the Alphabetical List of International Classification referred to in the Locarno Agreement Establishing an International Classification for Industrial Designs (the 'Locarno Agreement'). In a multiple filing, a maximum of fifty designs is allowed: the requirements applicable to a single filing are equally applicable to each individual design of a multiple filing.

It is important to file an application which clearly shows the protected elements of a design. Failing to do so sufficiently will, in principle, be disadvantageous to the holder of the right. It can limit the scope of the protection (e.g., design features which are not clearly shown in the drawings fall outside the scope of protection) but - and most notably - nullity might be invoked if the filing does not sufficiently reveal the characteristics of the design (Articles 3.23(1)(d) and 3.6(f) BCIP).

[C] Design Part of Another Article/Environment of Design

Neither the Implementing Regulations nor the BCIP mentions any particulars regarding the filing of a design for part of another article. Because the right to a design can easily be nullified if the filing does not sufficiently reveal the characteristics of the design (Article 3.6(f) BCIP), the importance of filing a clear application is self-evident.

Considering the harmonization of European design law, it might be advisable to take into account the Examination Guidelines of the OHIM, which in fact do state specific guidelines for the registration of a design for part of another article (see the chapter on Europe, §14.03[C]).

[D] Description of the Design (Where Required)

The Implementing Regulations state that the application should contain an indication of the product embodying the design, and a description of the colour or colours of the design. In relevant cases, the corresponding colour code can be provided as well (Rule 2.1(1) Implementing Regulations). Optionally, the application may contain a description (of at most 150 words) explaining the characteristic features of the new appearance of the product.

Rule 2.1(4) of the Implementing Regulations furthermore requires that a detailed description should be given of the product embodying the design, preferably using the wording in the List of International Classifications referred to in the Locarno Agreement.

§14.08 EXPUNGING/CANCELLING OR VARYING REGISTRATION

Unlike the CDR, the BCIP only requires payment of the registration fees for a registration to be renewed (Article 3.14 BCIP). These fees must be paid within the six months preceding expiry of the registration. If payment is not made within these six months, the fee may still be paid within six months after the expiry date of the registration, but only if an additional fee is paid simultaneously (Article 3.14(3) BCIP).

The holder of the Benelux registration may request cancellation of his registration at any time, unless third parties have legal contractual rights which have been notified to the BOIP (Article 3.21(1) BCIP). The cancellation will have effect for the entire Benelux territory (Article 3.21(4) BCIP). A holder of the Benelux registration may cancel all or part of the designs included in a multiple filing (Article 3.21(2) BCIP).

If the Benelux design registration may be cancelled, it may yet be maintained in an amended form. The amended form of the design must then still comply with the requirements for protection and the identity of the design must be retained (Article 3.24(2) BCIP).

§14.09 OTHER

[A] Interface between Industrial Design and Trademarks

A design may also be protected as a trademark, even if it has a three-dimensional form. In the Netherlands, as well as in Belgium and Luxembourg, trademarks are also covered by the BCIP. A trademark need not be new, although it may not conflict with older trademarks designated for similar goods or services. Trademark protection may in some cases provide a suitable means of escape from the requirement of novelty of the right to a design under the BCIP. This means that, although perhaps difficult to obtain and uphold, it may be worthwhile to file a design as a trademark if it has been proven to be a commercial success.

According to Article 2.1(1) BCIP, any sign can be a valid trademark if it is capable of being represented graphically and of distinguishing goods or services of a company from those of other companies. This would include, for example, names, letter or number combinations, logos, colours, and even sounds and the shapes of products. A non-distinguishing sign is capable of being a trademark if it has acquired distinctive character (i.e., by prolonged, active and intensive use as a trademark).

However, signs consisting solely of a shape that results from the very nature of the product, which gives the product substantial value or which is necessary to obtain a technical result cannot be trademarks (Article 2.1(2) BCIP).

[B] Relationship between Industrial Design and Copyright/Design Copyright

The registration of a design does not preclude copyright protection. Accordingly, registered designs can be protected by both design law and copyright law.

Any work of literature, art or science, including works of applied art, may be subject to copyright. Abstract ideas are not protected as such; they must take some sort of form. The only test for copyright protection is that the work, as a result of creative labour, has its own original character and bears the personal stamp of the author. Copyright is obtained automatically without any formalities. Designs that are not, or no longer, protected under design law (with a maximum term of twenty-five years), but qualify as a 'work' under copyright law, may thus enjoy prolonged protection (in principle seventy years after the author's year of death). This is particularly important for a creator who has not registered his design and for a creator whose registration has expired.

In Articles 3.28 and 3.29 BCIP, the legislator has provided for convergence with copyrights to ensure that third parties will not have to deal with multiple right holders. Under Article 3.28(1) BCIP, if the creator of a work protected by copyright authorizes a third party to register a design in which that work is incorporated, this implies assignment of the copyrights in that work.

The party filing a design is presumed to also be the owner of the related copyrights but this presumption does not apply in respect of the true creator or his beneficiary (Article 3.28(2) BCIP). Furthermore, Article 3.28(3) BCIP provides that the assignment of the copyrights relating to a design results in assignment of the right in the design and vice versa, without prejudice to the application of Article 3.25 BCIP.

Article 3.29 BCIP provides that where a design is created by a worker or an employee in the course of his employment or created on commission (referred to in Article 3.8 BCIP), the copyrights relating to the design belong to the party deemed to be the creator, in accordance with the provisions of that article. This is quite remarkable, because Dutch copyright law does not entitle a party commissioning the creation of a design to the copyright in that design. The Benelux Court of Justice decided in the *Electrolux* case that this provision also relates to designs which are not registered. Otherwise, it would only be possible for a party commissioning the creation of the design to use the design after it has been registered.[12] In a more recent case, the Dutch Supreme Court found that the application of Articles 3.29 in conjunction with 3.8 BCIP does not require a 'design' which is eligible for design protection as referred to in Article 3.1(1) BCIP, but that a 'design' as provided for in Article 3.1(2) BCIP is sufficient, consisting of 'the appearance of a product or a part of a product'. This means that novelty and individual character are not required for this particular purpose, although the design does need to meet copyright requirements.[13]

12. Benelux Court of Justice, 22 Jun. 2007, IER 2007, 309 (*Electrolux/Sofam*).
13. Dutch Supreme Court 25 Oct. 2013, IER 2014/6 (*S&S/Esschert*).

[C] Relationship between Industrial Design and Unfair Competition

If a design is not, or no longer, protected by a design right, the creator may be able to rely on protection of his design against slavish imitation, which is based specifically on the principles of Dutch unlawful act (Article 6:162 Dutch Civil Code). The difference between these two regimes is that whereas design rights offer protection against the use of designs which produce a similar impression on the informed user, slavish imitation protects against imitations that cause unnecessary confusion amongst the relevant public.

Slavish imitation of another product is only prohibited if the imitated product has a distinguished position on the Dutch market and the resemblance is such that it causes a likelihood of confusion or association, while the imitation could have been shaped differently without affecting its soundness and usability.

The most important remedies available in the event of slavish imitation are injunctions and compensation of damages.

In conclusion, design is soundly protected in the Netherlands, in some ways more so than in many other European countries. Design rights are protected via the Community Design Regulation as well as the Benelux Convention on Intellectual Property, complemented by robust additional ways of protection like broad copyright protection and protection against slavish imitation. If the holder's rights are infringed, ample remedies are available, such as injunctions and the compensation of damages.

CHAPTER 15

Industrial Design Rights: Denmark and Other Nordic Countries

Klaus Ewald Madsen & Arly Carlquist

§15.01 INTRODUCTION

Denmark and the other Nordic countries have long been well known for manufacturing high quality and user-friendly products with an outstanding and often evolutionary design. Nordic design is typically characterised as being pure and minimalistic and as such an attractive object for imitations.

Hence, it can hardly be surprising that Nordic law has long acknowledged the need for protection of industrial design and for nearly a century, statutory rules have provided a broad basis for protection of industrial design.

Traditionally the Nordic intellectual property laws have been based on joint Nordic preparatory works and considerable legal unity among the Nordic countries exists. However, there are differences in the laws and administrative practices and as well as case law are not always entirely consistent It is not possible, within the scope of the text in this chapter, to highlight all characteristics pertaining to the laws in each of the five countries. Therefore, while this chapter seeks to provide an overall description of the protection of industrial designs in the Nordic countries it is based on Danish law and may not contain an accurate presentation in all respects of the legal position in the other Nordic countries, and legal advice should always be obtained in the relevant country in each specific case.

[A] New Developments

Since publication of the first edition of *Industrial Design Rights, An International Perspective* in 2001, there have been some very important developments relevant for the protection of industrial designs.

First of all, a new Danish Design Act came into effect in October 2001. This act implemented EU Directive 98/71 on the legal protection of designs. The act has been amended several times since then, among other things to implement the Directive on the enforcement of intellectual property rights (see right below)

Second, Council Regulation No. 6/2002 on the Community Design has come into effect. Today, this regulation provides an important vehicle for the protection of industrial designs.

Finally, Directive No. 2004/98 on the enforcement of intellectual property rights has been adopted. The directive, which was to be implemented by April 29, 2006, requires all EU Member States to apply effective, dissuasive and proportionate remedies and penalties against those engaged in counterfeiting and piracy. It means that all Member States will have a relatively similar set of measures, procedures and remedies available for right holders to enforce their intellectual property rights.

§15.02 CONVENTIONS AND LEGISLATION

In Denmark, industrial designs are protected through comprehensive national intellectual property law. Most relevant are the Design Regulation, the Design Act and the Copyright Act. They will be discussed further in the following. However, the Trade Mark Act and the Marketing Practices Act may also prove useful in the protection of industrial designs, *see* §15.09[A] and §15.09[C] below.

While purely functional products are generally exempted from design protection (see §15.03[A][5] below) such designs may have inventive qualities which qualify for patent protection. The two tubes inside the famous Lego brick is an excellent example of an invention which has actually qualified for grant of patent protection and which at the same time adds to the harmonic appearance of the brick. However, since purely functional designs are relatively uncommon, patent protection of industrial designs is equally rare. Consequently, the Patent Act will not be further dealt with below.

Furthermore, EU Directives and Regulations are obviously a very important part of the intellectual property laws of Denmark, Sweden and Finland. The most important are Directive No. 98/71 on the legal protection of designs, Council Regulation No. 6/2002 on the Community Design, and Directive No. 2004/98 on the enforcement of intellectual property rights.

The above-mentioned EU Directives have all been implemented in Danish, Swedish and Finish Law.

Norway and Iceland are not members of the European Union, but pursuant to the 1992 EEA Agreement, Norway and Iceland are obliged to harmonise their legislation with EU Directives and Regulations and within the intellectual property area the harmonisation seems to be up to date.

Finally, industrial designs are globally governed by a number of international conventions which in general has been signed and ratified by the Nordic countries. The relevant conventions are listed in the Table 15.1 below.

Table 15.1 International Conventions Signed and Ratified by the Nordic Countries

	Denmark	*Sweden*	*Norway*	*Finland*	*Iceland*
Paris Convention for the Protection of Industrial Property (Paris convention)	+	+	+	+	+
World Intellectual Property Organization (WIPO convention)	+	+	+	+	+
Convention Concerning the Classification of Goods and Services (Nice Agreement)	+	+ excluding 1979 amendments	+	+	+
Agreement Concerning International Classification of Figurative Elements of Marks (Vienna arrangement)	Signed but not ratified	+	Signed but not ratified	-	-
Arrangement Concerning the International Registration of Marks (1891) and the Protocol (1989) (Madrid arrangement)	+ Protocol ratified	+ Protocol ratified	+ Protocol ratified	+ Protocol ratified	+ Protocol ratified
Convention for the Protection of Literary and Artistic Works (Bern Convention)	+	+	+	+	+
Agreement Concerning the International Registration of Industrial Designs (Hague Agreement)	+	-	+	+	+
Trade related aspects of intellectual property rights (TRIPS)	+	+	+	+	+
Locarno Agreement Establishing an International Classification for Industrial Designs	+	+	+	+	+

§15.03 SUBJECT MATTER

[A] What Constitutes a Design?

In its broadest sense, the term *"design"* may refer to all aspects of the arrangement and appearance (internal or external) of the whole or part of a product including any preliminary drawings or models etc.

However, this chapter will focus primarily on the protection of *"industrial design"*, meaning the design of industrial products and parts thereof, i.e., those features of shape and/or pattern which give the products a particular appearance.

In practice, most industrial design will start life in two-dimensional form as drawings and then proceed into three-dimensional form, first as a production model and then in the form of the commercial product itself. Other designs will exist in three-dimensional form from the outset and some designs will by virtue of their nature always be two-dimensional. It makes no difference for the protection whether a design is in the two-dimensional or three-dimensional form in the Nordic countries.

Industrial designs comprise a broad variety of products such as furniture and other applied art, decorated fabrics, clothes, jewelry, toys, machines and their components etc. Even clothing may be protected as a design. The significance of relevant design features may vary substantially from product to product and often the various features even serve different primary purposes depending on the product in question. For instance, when creating a new decorated fabric, the artistic and aesthetic intent will normally dominate, whereas the development and design of a new gear may primarily serve a technical and functional purpose. These differences are best illustrated by dividing industrial designs into the three categories noted below.

[1] Purely Artistic/Aesthetic Designs

Purely artistic/aesthetic designs comprise products designed exclusively or primarily with artistic/aesthetic intent, e.g., carefully chosen designed shape and/or configuration. Examples would be paintings, decorated fabrics, sculptures, engravings, all traditional works of art, constituting the subject matter of copyright. However, this category also includes the work of craft persons such as silversmiths, glassblowers and potters, as well as various accessories, all of which may well be created solely for artistic and/or aesthetic reasons. Such products are often subject to industrial exploitation and where that is the case, they would also be categorised as an industrial design.

[2] Purely Functional Designs

Entirely functional products usually produce little or no aesthetic reaction (eye appeal), and although such products may still look attractive to potential buyers, their appearance, whether attractive or not, plays no major part in their purchase. They are bought merely because of their functional or technical qualities or features and in this context price and performance are the decisive factors, not the attractiveness of the product. Examples may include tools, packaging, spare parts, etc. where the design is solely dictated by its function. In a specific case the court ruled that the basic form of a T-shaped diaper was not eligible for design protection.

[3] Designs Combining Functional and Artistic Features

Most commercial products comprise both functional and artistic/aesthetic features, and in fact it is not always easy to distinguish between different aspects of an industrial design (e.g., artistic, aesthetical, shaping, configuration and/or functional aspects). The vast majority of today's industrial designs are specifically designed to meet certain functional purposes, but in order to make them look more attractive and thereby appealing to the potential customer, design features have been added to the products.

The significance of the functional elements and the design features, respectively, may well vary significantly from product to product and also from customer to customer. Often the design of a product is less important than the price, technical performance and general quality of the product, but in other cases the design may be the determining factor when a customer has to choose between two or more competing products, all of which may perform equally well and be similarly priced.

Obvious examples of products falling within this category are furniture, lamps, household articles, tools, electronics and other kinds of consumer products, packaging, spare parts etc. Even clothes may on rare occasions enjoy protection.

[4] Shape and Configuration

The protectable subject matter under the Design Regulation and the Design Act are designs, meaning the appearance of a product or part of a product resulting from the features of lines, contours, colors, shapes, texture or materials. Shape and configuration are thus essential features in a protectable design.

The general idea of a design is not protectable. It is only the specific appearance of a design that may be protected.

[5] Ornamental or Functional

It does not matter whether the design is dictated by aesthetic or functional considerations. It is the appearance of the design that matters.

However, design protection is not available with respect to features of the appearance of a product, which are solely dictated by the technical function of the product.

Where a product has been designed and/or configured solely for its functional qualities or to accommodate certain technical features and the ultimate functional or technical solution at a given time is applied, other manufacturers wishing to obtain the same (ultimate) qualities or features in a competing product would have to produce a similar design. Such products will usually not be subject to protection (besides patent protection), but, they may however, in certain circumstances, enjoy protection against slavish plagiarism under the Marketing Practices Act, *see* §15.09[C] below.

Thus, entirely functional products are generally excluded from protection as an industrial design under the Design Regulation as well as the Design Act. Also under the Copyrights Act entirely functional products would normally be excluded from protection because they do not fulfill the originality requirement.

Functional products may have inventive qualities and if that is the case and they meet the appropriate requirements for patent protection or protection as a utility model, protection under these acts may be available.

However, even products without inventive qualities may be designed to serve a technical function, leaving little or no room for aesthetic appeal. Such products will usually not be granted any protection as an industrial design, but as it will be explained below, they may, in certain circumstances, be entitled to protection against slavish plagiarism under the Marketing Practices Act.

[B] Originality/Novelty

[1] Test for Originality Generally

Originality (in Danish "værkshøjde") is a requirement in order for an industrial design to be protected by the Danish Copyright Act. In order to reach the required level of originality, the industrial design must be the result of the creator's own personal intellectual creation. This simply means that some, even limited, effort has gone into the creation of an individual work.

The purpose and the artistic quality of the work do not influence the originality assessment. The originality assessment is made purely on the design and shape of the work, while the underlying ideas and technical functions are irrelevant.

Generally, Danish courts have imposed a fairly low originality requirement resulting in protection of industrial designs, the creation of which has required only limited intellectual effort, and of designs where functional aspects are prominent.

Protection under the design regulation and the Design Act may only be obtained if the design has individual character. A design shall be considered to have individual character if the overall impression produced by the design on the informed user differs from the overall impression produced by any design which has been available to the public before the date of filing of the application. In assessing the individual character of a design, the degree of freedom of the designer in developing the design shall be taken into consideration. This is of particular relevance in areas such as industrial design.

The Trade Marks Act contains a "*distinctive character*" requirement, which is quite different from the novelty and originality requirements of the Copyright and Design Acts. A trademark with distinctive character has the ability to create a connection between product and company. A trademark which consists exclusively of either a shape which results from the nature of the goods themselves, a shape of the goods which is necessary to obtain a technical result or a shape which gives substantial value to the goods, does not have distinctive character.

[2] Novelty

Protection under the Design Regulation and the Design Act may only be obtained if the design is new. A design shall be considered to be new if no identical design has been made available to the public prior to the filing date of the application or, if priority is claimed, the date of priority. Designs shall be deemed to be identical if their features differ only in immaterial details. This applies to both registered and unregistered designs.

A design shall be deemed to have been made available to the public if it has been published following registration or otherwise, or if it has been exhibited, used in trade or otherwise disclosed. In theory, this is both an objective and global novelty requirement. The design shall, however, not be deemed to have been made available to the public if publication of the design could not reasonably have become known in the normal course of business to the circles specialised in the sector concerned, operating within the Community.

[a] Prior Use

Contrary to most other intellectual property rights, copyright is not a right of priority but a protection against someone else copying the work (subjective novelty requirement). Thus, if an identical work is produced independently by two separate persons they will both have copyright in their respective works.

To obtain a design right the design must be novel. Novelty can be defeated by the applicant's own prior use/publication.

However, in order to protect potential design rights, the Design Act and the Design Regulation contain a twelve months grace period according to which a design shall not be deemed to have been made available to the public if it has been published during the twelve-month period preceding the date of filing of the application by the designer as a result of information provided or action taken by the designer or his successor in title.

Thus, the grace period provides the designer with an opportunity to use or publish the design and thereby assess its potential before the designer decides whether to register the design or not.

Furthermore, under the Design Regulation (but not under the Design Act) any new design with sufficient individual character will be protected as an unregistered design, if made available to the public.

Copyright is automatically established at the creation of the work. Accordingly, prior use/publication is not an issue.

[b] *Prior Publication*

See above regarding prior use.

[c] *Prior Application or Registration*

A design shall be refused from registration if the design is in conflict with a prior design application or registration. Priority may be claimed up to six months, *see* §15.07[A] below.

§15.04 DURATION OF PROTECTION

The duration of protection varies under each of the intellectual property laws, which provide protection of industrial designs.

The Copyright Act provides protection for industrial designs for a period of seventy years after the death of the creator of the industrial design. If the industrial design has two or more creators, without the individual contributions being separable as independent works, the protection lasts for seventy years after the year of death of the last surviving creator.

The Design Regulation and the Design Act provide protection for industrial designs for a period of up to twenty-five years from the date of application for registration. If the design is unregistered, the protection period is three years from the date the design is published in the EU.

§15.05 INFRINGEMENT

[A] Scope of Infringement Action

The scope of protection for industrial designs is based on the appearance of the product. When assessing a possible infringement of an industrial design, the courts are likely to make an overall evaluation of the appearances of the original product and alleged infringing product. Minor deviations from or changes of details compared to the original product will normally not avoid infringement, if the overall appearance of the alleged infringing product is similar.

As far as works of applied art and other industrial designs are concerned, it should be emphasised that Nordic practice is fairly liberal in that such works in addition to being protected pursuant to the Design Regulation and the Design Act may often be granted protection also pursuant to the Copyright Act even when they have a considerable functional purpose of application, e.g., furniture, and the like.

However, it is also generally acknowledged that a too wide protection of applied art or other industrial designs whether based on copyrights, design rights and/or pursuant to unfair competition principles is not necessarily desirable, the reason being that a too wide protection restrains innovation and competition. Hence, while a relatively great number of phenomena are protected, case law have established that the scope of the protection, in terms of what is deemed to be an infringement, shall be limited to products which are (nearly) identical to the original work. This is particularly true with regard to functional designs, for example furniture whereas more artistic designs, such as a decorated fabric, would probably enjoy a slightly wider protection.

This enables the courts to consider all relevant facts on a case-by-case basis and thereby to make an individual assessment so as to reach a reasonable scope of protection in each case. However, it is often difficult to predict the outcome of a given dispute and as a consequence, case law is quite extensive in this area.

[B] Test for Infringement/Criteria to Constitute Infringement

Infringement of industrial designs will occur only when all or a substantial part of the industrial design has been copied. Substantiality will be determined by reference to qualitative rather than purely quantitative criteria. Thus, taking only small parts of an industrial design may still amount to infringement provided the part taken constitutes an essential element of the industrial design.

The courts often use the criteria "risk of confusion between products" in their assessment of infringement. This is a market related criteria acknowledging that industrial designs are commercial rights. When using the criteria, the courts compare the original product with the alleged infringing product and assess whether there is a risk that the average consumer will mistake one product for the other.

According to case law, a relatively large degree of similarity between the products is necessary in order to establish infringement. The degree of similarity is assessed in each individual case, however, generally it can be stated that products with great originality will be given a broader scope of protection than products with limited originality and/or products where the functional aspects are prominent.

[C] Standing to Sue

Infringement of intellectual property laws is subject to both civil and criminal sanctions. Depending on the circumstances e.g., the severity of the infringements such may be subject to private and/or public prosecution.

For the majority of intellectual property infringement matters, legal proceedings shall be instituted by the right holder and in practice primarily civil sanctions are applied. However, in more severe cases criminal sanctions may be considered and applied and to a limited extent infringements of intellectual property laws are subject to public prosecution.

None of the intellectual property laws govern if a licensee may commence legal actions without the consent of the right holder. Thus, it depends on the agreement between the right holder and the licensee, whether the licensee may take legal action.

If the license agreement is exclusive it is generally assumed that the licensee may take legal action without consent of the right holder, provided that the right holder has not reserved this right.

On the contrary, if the license agreement is non-exclusive it is generally assumed that the licensee may not take legal action without consent of the right holder, unless the license agreement explicitly grants a right (or obligation) for the licensee to enforce the intellectual property rights.

[D] Remedies for Infringement

The remedies available in case of infringement are generally the same under all of the intellectual property laws.

Save for obvious counterfeit cases, an attempt to stop or prevent infringements will usually be commenced by a cease and desist letter requesting the actual or potential infringer to immediately cease the infringing activities, e.g., marketing and sale of the infringing products, which in the right holder's opinion, constitute an infringement of the right holder's rights or if the infringement has not yet taken place, to abstain from doing so.

Having said that, obviously there may be situations where legal steps have to be taken with or without any previous notification of the infringer, be it by way of application for a preliminary injunction, securing of evidence or otherwise.

The most commonly applied remedies for infringement of intellectual property rights are briefly described in the following.

[1] Preliminary Injunction

Pursuant to the Danish Administration of Justice Act, a preliminary injunction is available in cases concerning infringement of intellectual property rights to prevent future infringements. The right holder must prove on a balance of probabilities that an infringement is ongoing or will take place in the near future.

Preliminary injunction proceedings may take place without prior notice to the defendant if there is a risk that infringing copies will be destroyed or removed. Save for cases concerning counterfeit products, usually the infringer is notified before the proceedings take place in court. In addition to granting a preliminary injunction, the court may upon request of the right holder seize infringing products to ensure compliance with the preliminary injunction.

Usually, it will take one to three months following filing of the application for grant of preliminary injunction until a preliminary injunction is actually granted, but in certain cases e.g., counterfeit cases performed without prior notice to the infringer, a preliminary injunction may be obtained within a few weeks. Upon request the courts may assist the right holder in upholding the injunction e.g., by seizure of infringing products.

In general, grant of a preliminary injunction (and eventual seizure of infringing products if so requested by the right holder) will be conditional upon provision of security by the right holder. The purpose is to secure a potential claim for damages by the infringer if it turns out during a subsequent lawsuit to be filed by the right holder that the conditions for granting the preliminary injunction were not satisfied.

Provided that a preliminary injunction is granted, the right holder must file a lawsuit within two weeks following grant of the preliminary injunction. The purpose of such lawsuit is to establish that the conditions for grant of a permanent injunction are satisfied. In connection with the lawsuit, the right holder may include a claim for damages and other relevant claims e.g., for criminal sanctions and seizure and destruction of infringing products.

If the lawsuit is not filed by the right holder in due time, the preliminary injunction will not be upheld and the infringer will be awarded costs. The infringer may waive the right of a lawsuit and if so the injunction will be final and remain in force without the need for a lawsuit.

Preliminary injunction proceedings may be settled out of court at any stage of the proceedings.

[2] Civil **Ex Parte** *Search (Securing of Evidence)*

Pursuant to the Danish Administration of Justice Act, evidence may be secured without prior notice to the alleged infringer in cases regarding infringement of intellectual property laws. Likewise, the rules will apply to infringing product imitations comprised by the Marketing Practices Act.

The purpose of such securing of evidence proceedings is to secure evidence of the infringements including the scope and nature hereof. Any relevant material e.g., invoices, purchase orders, products, correspondence etc. may be seized and any equipment of the infringer e.g., servers and PCs as well as physical locations may be examined.

The right holder must render probable that an infringement has taken or will take place and that evidence of such infringement is in the alleged infringer's possession. As with respect to a request for grant of a preliminary injunction, an application specifying the claim and the reasons for the claim must be filed with the enforcement court.

In general, securing of evidence will be conditional upon provision of a security by the right holder. The purpose is to secure a potential claim for damages by the alleged infringer should it turn out that the conditions for securing the evidence were not satisfied. The amount would usually be limited.

Provided that a search order is granted, the search will usually take place within two to four weeks from the date of filing the request for securing of evidence. The enforcement court will appoint an impartial expert to conduct the search. The impartial expert will draft a report following the securing of evidence proceedings and the right holder and alleged infringer will each receive a copy.

Following the right holder's receipt of the report prepared by the impartial expert, the right holder must within four weeks file a lawsuit on the merits unless the alleged infringer waives filing of such lawsuit. In connection with the lawsuit, the right holder may include a claim for damages and other claims e.g., seizure and destruction of infringing products and criminal sanctions.

If the right holder fails to file such lawsuit, the evidence seized cannot be used in a subsequent lawsuit for damages and the alleged infringer will be awarded costs and may in certain circumstances have sufficient grounds for filing a lawsuit for damages.

The costs incidental to the securing of evidence proceedings i.e., fees to the impartial expert and eventual storage costs while legal proceedings are pending must initially be paid by the right holder. However, provided that the right holder prevails, the infringer will be liable to reimburse such costs.

Securing of evidence proceedings may be settled out of court at any stage of the proceedings.

[3] Right of Information

Pursuant to the Danish Administration of Justice Act subject to filing a lawsuit concerning infringement of intellectual property rights, the right holder may request the court to order the alleged infringer or a third party to provide information which is in the possession of the alleged infringer or the third party and which is relevant to clarify the scope and nature of the alleged infringements alleged in the lawsuit.

The said rules do not apply in connection with filing for grant of a preliminary injunction or securing of evidence.

[4] Compensation

According to the intellectual property laws, any natural or legal person, who intentionally or negligently infringes intellectual property rights is liable to pay compensation to the infringed right holder. The provisions of the intellectual property laws were harmonised in connection with the implementation of the Enforcement Directive and are thus based on identical general principles.

An infringed right holder is entitled to claim a reasonable remuneration and damages for any additional economic loss suffered. Moreover, the infringed right holder may claim compensation for non-financial loss i.e., loss of image or the like related to the infringement.

The right holder's right to a "*reasonable remuneration*" entails that the infringer must always pay a reasonable fee for the infringing exploitation, irrespective of whether or not the right holder may establish that the right holder suffered any economic loss. Such fee is generally to be calculated based on the license fee that should have been paid to the right holder had the right holder granted a license to the infringer.

A claim for damages based on additional economic loss being suffered in excess of the reasonable remuneration is subject to the right holder being able to substantiate that such economic loss has actually been suffered by the right holder. In principle, the right holder is entitled to claim damages in terms of any economic loss, including any consequential loss suffered by the right holder provided that such economic loss is causal and generally foreseeable given the nature of the infringement. When establishing an eventual claim for damages for additional economic loss, the infringed right holder's loss of profit and the infringer's unjust profits should be taken into consideration.

In practice, it is often difficult to establish what a reasonable remuneration (i.e., license fee) and additional damages amounts too. The courts tend to be conservative in their assessment and generally compensation awarded is somewhat lower than the economic loss suffered by the right holder. This fact should always be considered before initiating legal proceedings instead of entering into an out-of-court settlement negotiations.

[5] Border Detention

The border detention regime concerning customs enforcement of intellectual property rights as set out in Regulation No. 608/2013 of June 12, 2013 (repealing Council Regulation No. 1383/2003 and previous versions hereof) apply. Accordingly, a right holder may request suspension of release of infringing industrial design products pursuant to the said Regulation as applicable throughout the EU.

The right holder must prove on a balance of probabilities that the products infringe the industrial design rights of the right holder. There will be no oral pleadings and the application will be handled by the customs authorities based on the documentary evidence provided by the right holder.

Provided that the customs authorities accept the request for suspension of release of the alleged infringing products and detain them, the right holder must file a lawsuit against the infringer within ten working days for the purpose of obtaining a final court decision confirming that suspension of release was justified i.e., that the products were in fact infringing. If a lawsuit is not filed within the said deadline, the products will be released.

In connection with the lawsuit, the right holder may include a claim for damages and other relevant claims e.g., for criminal sanctions and seizure and destruction of infringing products.

The costs incidental to storage of suspended products while legal proceedings are pending must initially be paid by the right holder. However, provided that the right holder prevails, the infringer will be liable to reimburse such costs.

Border detention proceedings may be settled out of court at any stage of the proceedings.

[6] Criminal Sanctions

Criminal sanctions are available under all of the intellectual property laws and were harmonised in connection with the Danish implementation of the Enforcement Directive.

Infringements of intellectual property laws caused by gross negligence may be punished with a fine only.

Fines are usually of relatively low amount. Intentional infringements committed under aggravating circumstances may be punished with a maximum of up to one and a half years of imprisonment. "*Aggravating circumstances*" are typically deemed to be present if the infringement is of a commercial scale committed to obtain profits and/or concerns a considerable number of infringing products.

If infringements are of a particularly aggravating and serious nature and provided that such infringements have been committed intentionally such are punishable with a maximum penalty of up to six years of imprisonment pursuant to the Danish Criminal Code.

The least serious infringements are subject to private prosecution by the infringed right holder. Criminal proceedings pursuant to the Criminal Code are subject to public prosecution.

Save for criminal proceedings being subject to private prosecution, the right holder cannot in its own discretion recall a request filed by the right holder for criminal prosecution i.e., if the public prosecutor has decided to commence criminal proceedings, the public prosecutor will control these criminal proceedings. Hence in this situation the right holder and the infringer cannot settle the criminal sanctions and typically any civil action will be stayed pending a final decision with respect to the criminal proceedings.

§15.06 DEFENSES TO INFRINGEMENT

[A] Validity

All intellectual property rights must fulfill certain requirements regardless of whether such rights come into existence automatically or through registration.

If a registered intellectual property right is subsequently deemed not to fulfill the requirements for registration, or if an unregistered intellectual property right (such as an unregistered Community design or a copyright) does not meet the criteria for such right to be established (novelty, individual character or originality) then no valid right will exist (the registered design right may be revoked) and no infringement can occur. Therefore, invalidity is commonly used as a defense in infringement cases regarding intellectual property rights.

[1] Not Novel/Original

Lack of novelty (for design rights), lack of individual character (for design rights) and lack of originality (for copyrights) are all commonly used defenses to infringement.

While the assessment of novelty is essentially an objective one (i.e., that no identical design has previously been made available to the public) and only rarely gives rise to dispute, the assessment of individual character and originality is more subjective in the sense that it is not confined to one or a few objectively verifiable criteria and therefore often more complex.

[2] Functional/Not Ornamental

A design being entirely functional is also one of the common arguments used as a defense against infringement, as entirely functional products may not be registered as a design under the Design Act. This also applies to products protected under the Danish Copyrights Act.

[3] Other Possible Criteria Such as Non-obviousness

The "obviousness" criterion is taken from patent law, where it is often used as an invalidity defense, however, it is not relevant for copyright, design and trademark rights.

[4] Faults in Application

If an application for a registered design does not comply with the requirements laid down in the Design Act, the applicant is notified accordingly and invited to file any observations or to take steps to correct the application within a specified time limit. Unless any deficiencies are remedied in due course the application will be refused.

An application may on some occasions be resumed after expiry of the prescribed time limit if the applicant requests resumption and files observations or otherwise corrects the application.

[B] Marking

There are no requirements in Denmark as to the use of symbols such as "©", "®" or "TM".

[C] Non-infringement

The most common defense is probably non-infringement i.e., that the design is not sufficiently similar to the protected design to constitute an infringement. The test for infringement is briefly described in §15.05[B].

Furthermore, there are a number of specific exceptions where certain acts are excluded from intellectual property law protection and therefore not infringing. These include e.g., reproduction for private use and other non-commercial use.

[D] Other Defenses

In general, intellectual property rights will not be infringed by acts concerning original products put on the market within the European Union and with the consent of the proprietor.

According to the Design Regulation, the Design Act and the Copyright Act exhaustion of rights takes place where a product has been put on the market within the European Economic Area (EEA) by or with the consent of the proprietor. With respect to countries outside EEA, exhaustion does not occur.

§15.07 PROCEDURES FOR FILING APPLICATION FOR REGISTRATION

As already mentioned, copyrights are not subject to registration under the Danish Copyright Act.

With regard to designs standard application forms are issued by the Danish Patent and Trademark Office but they need not necessarily be used. The application should be accompanied by an illustration picture or a model of the design. The official fee (2014: DKK 1,200) has to be paid in conjunction with the submission of the application.

Generally, the design application must fulfill the formality requirement set forth in the Design Act. The Danish Patent and Trademark Office does not examine whether the design application fulfills the material requirements, such as novelty. Instead, the applicant must declare under penalty of law that the design, to the best of his or her knowledge, is "new" in terms of the Design Act. Even so, it is undoubtedly the situation that many registered designs do not in fact meet the high threshold for design registration.

Further, as a consequence of Denmark ratifying the Hague Agreement Concerning the International Deposit of Industrial Designs in 2008, a design with effect in Denmark can be registered through World Intellectual Property Organization (WIPO).

[A] Priority

An application for a design right (under the Design regulation or under the Design Act) may upon request enjoy a right priority of six months from the date of filing of the first application for the same design in or for any state that is a party to the Paris Convention or to the WTO Agreement.

Similarly, six months priority may be claimed from an application made in of for any state which is not a party to the Paris Convention or to the WTO Agreement provided that such state grants – on the basis of Community design or national Danish filings a right of priority having equivalent effect.

Furthermore, if an application for the registration of a design is filed in this country no later than six months after the design has been displayed for the first time at an official, or officially recognised international exhibition, the application shall at the request of the applicant enjoy priority from that date.

[B] Drawings

A prerequisite for protection under the Design Regulation and the Design Act is that the design is visible. Therefore, it is important that the design application is accompanied by a representation of the design (i.e., a graphic or photographic reproduction of the design illustrating the design from all relevant angles). The protection provided by the design registration is limited to what is depicted on the graphic or photographic representation of the design.

According to the Design Act the application may also be accompanied by a specimen of the design. If so, the specimen shall constitute the basis for the registration design.

[C] Design Part of Another Article/Environment of Design

The Design Regulation and the Design Act incorporates the "*must fit*" rule stating that a design right may not be obtained with respect to features of the appearance of a product, which must be reproduced in their exact form and dimensions in order to permit the product to which the design is incorporated or to which it is applied to be mechanically connected to or placed in, on, around or against another product so that either product may perform its function.

As an exception to the "*must fit*" rule, however, design rights may be obtained to a design which is new and has individual character, provided that the design allows the multiple assembly or connection of products within a modular system consisting of mutually interchangeable parts (e.g., Lego bricks).

According to the Design Regulation (Article 110) protection as a Community Design shall not subsist in designs which constitute a component part of a complex product in which such design is incorporated (such as body parts for cars). Contrary to the EU regulation, no "*must match*" exception exists in the Danish Design Act. Thus, under the Danish Design Act, protection of designs which constitute a component part (e.g., a body part of a car) of a complex product (e.g., a car) may be obtained even though such design must be reproduced 1:1 for the purpose of repair of that complex product so as to restore its original appearance. However, very importantly, the duration of the protection has been specifically limited to fifteen years.

[D] Description of Design (Where Required)

When filing the standard design application form to the Danish Patent and Trademark Office, the applicant must include a product specification. Beyond this, no further description requirements applies.

§15.08 EXPUNGING/CANCELLING OR VARYING REGISTRATION

Any person may at any time file a request with the Patent and Trademark Office for an entire or partial cancellation of a national design registration which has not been effected in accordance with the Design Act. Decisions of the Patent and Trademark Office may be appealed to the Patent Board of Appeal no later than two months following notification of the decision.

Alternatively, any person may institute proceedings before the courts claiming that a design registration be revoked entirely or partially by a court decision.

If a case before the courts concerning a design remains to be finally decided upon, a request for cancellation may not be filed with the Patent and Trademark Office concerning the design in question. If proceedings concerning a design are instituted before the courts prior to a final decision having been made by the Patent and Trademark Office relating to the same design, the Patent and Trademark Office shall suspend the examination of the request until the court case has been finally decided upon, unless the request has been filed by the holder of the design.

Registered Community designs may be cancelled (declared invalid) by OHIM upon application by any third party. OHIM's decisions may be appealed within two months after the date of notification of the decision.

Alternatively, registered Community design may be declared invalid by decision of a Community design court (in Denmark the Maritime and Commercial Court) in connection with an infringement action where a counterclaim for a declaration of invalidity has been raised.

Unregistered Community designs may be declared invalid by decision of a Community court in an action for a declaration of invalidity of an unregistered design or in connection with an infringement action where a counterclaim for a declaration of invalidity has been raised.

§15.09 OTHER

As noted above, industrial designs cover a broad range of products, each of which will be subject to different forms of protection. This is due to the particular nature of the product, the qualities embodied in and the features added to the product. Consequently, the scope of protection of industrial design, as well as the duration thereof, depends on the product in question. Other ways in which industrial designs may be protected are discussed below.

[A] Interface between Industrial Design and Trademarks

According to the Trade Mark Act, *"a trade mark may consist of any sign capable of distinguishing the goods or services of one undertaking from those of other undertakings and capable of being represented graphically...."* Among the examples of marks listed are *"the shape, equipment or packaging of goods."*

In general, the get up of goods is registrable. The standard example of get up protected as a trademark is the Coca Cola bottle, but many other examples exist.

Also, the product itself or for example the decor of restaurants and retail outlets of franchising operations may be subject to trademark registration provided it satisfies the requirements mentioned above.

A trademark right may be established either by registration of a trademark in accordance with the Trade Mark Act or by commencement of use of a trademark in the respective country. Unfortunately, the latter is quite a common occurrence, despite the fact that lack of proof of commencement of and/or use often creates substantial problems when an infringement situation eventually occurs. Thus, registration is always strongly recommended.

The proprietor of a trademark shall be entitled to prohibit all persons who do not have his or her consent from using, in the course of trade, any sign which is identical with or similar to the trademark if the use relates to goods or services which are identical or similar to those comprised by the trademark right, and there exists a likelihood of confusion. This includes the likelihood of an assumed connection between the trademarks.

The grant of a patent, for any technical feature embodied in a product, or the registration of a "new" design under the Design Act or the protection of any artistic element under the Copyright Act, do not hinder the registration of a trademark. Hence, a trademark right may subsist concurrently with one or more other exclusive rights.

A trademark is valid for an initial term of ten years from the date of registration. However, the registration may be renewed for ten-year periods from the expiry of the previous registration period. Thus, trademark protection may be extended indefinitely and for obvious reasons this feature makes it a very attractive way of protecting industrial designs as compared with the other intellectual property laws which are all limited in time. There are no requirements as to the use of trademark symbols such as "®" or "TM."

However, since a wide protection of products and get ups under the Trade Mark Act would inevitably restrict innovation to a non-acceptable degree, it is generally acknowledged that the protection must not become too wide. Consequently, section 2.2 of the Danish Trade Mark Act states that "a trade mark right shall not be obtained for signs which consist exclusively of a shape which results from the nature of the goods themselves, a shape of goods which is necessary to obtain a technical result or a shape which gives substantial value to the goods." This wording is based on the EU Directive and the exact contents must be determined in accordance with EU case law.

In a ruling by the Danish Maritime & Commercial Court, it was stated that the famous Lego brick could not be subject to trademark protection. Although not expressly mentioned in the judgment, there is not much doubt that the decision was based on section 2.2 of the Danish Trade Mark Act. The Swedish Supreme Court reached the same result in a decision from 1987.

Moreover, it shall not be deemed infringement to use, in the course of trade and in accordance with honest practices in industrial or commercial matters: (i) one's own name and address, (ii) indications concerning the kind, quality, quantity, intended purpose, value or geographical origin of the goods or services, the time of production of the goods or of rendering the services or other characteristics of the goods or services, (iii) the trademark where it is necessary for the indication of the intended purpose of the goods or services, in particular as accessories or spare parts, or (iv) for private use.

The rules about exhaustion of rights described in §15.06[D] above also apply to trademarks. Exhaustion shall not occur, however, if there are legitimate reasons for the proprietor to oppose further commercialisation of the goods, especially where the condition of the goods has been changed or impaired after the goods have been put on the market.

An application for a trademark shall be filed with the Danish Patent Office. Standard application forms are issued by the Patent Office but need not necessarily be used. The application should state the class(es) (1-45) in which the mark should be registered and must be accompanied by the official fee. The registered trademark is published in the official Gazette.

A trademark may at any time be administratively revoked upon request if the registration is not in accordance with the Trade Mark Act or if within five years following the date of completion of the registration procedure, the trademark has not been put to a genuine use in the respective country, either by the proprietor or with his or her consent, in relation to goods or services covered by the registration, or if such use has been suspended for an uninterrupted period of five years, unless there are proper reasons for non-use.

Trademarks may also be revoked by court decision.

[B] Relationship between Industrial Design and Copyright/Design

Initially, the copyright system was not aimed at protecting industrial design. In practice, however, copyright has turned out to be an important method of protecting industrial design alongside with the Design Regulation and the Design Act.

The Copyright Act, the Design Regulation and the Design Act are overlapping, but according to an explicit provision in the Copyright Act, protection under the Design Regulation and/or the Design Act does not exclude or limit copyright protection.

[C] Relationship between Industrial Design and Unfair Competition

The purpose of the Marketing Practices Act is first and foremost to ensure that business be carried out in a way acceptable to society and thus with due regard to the interests of consumers and businesses. Thus, the objective of the Marketing Practices Act differs substantially from that of the different intellectual property laws e.g., the Design Regulation and the Design Act.

The main purpose of the Marketing Practices Act is not to provide a designer with additional protection where the intellectual property laws do not provide for protection, but rather to support that marketing and sales of products takes place on principles acceptable to society including to protect such businesses from unfair acts of their competitors. Such unfair acts may include e.g., systematical slavish plagiarism of a company's industrial designs. Accordingly, the Marketing Practices Act does not provide for protection against acts which are explicitly allowed pursuant to the intellectual property laws e.g., private copying or non-commercial use as such exemptions to the exclusive right of the right holder are intended by the legislators.

According to the so-called *general clause*, compare section 1 of the Marketing Practices Act, "*no acts contrary to good marketing practice may be undertaken.*"

The exact scope of this basic normative standard is obviously not easy to assess but instead of introducing a more detailed regulation, the preferred route has been to regulate by means of this general clause with a view to obtaining flexibility and a continuous evolution of practice in accordance with the emergence of new marketing methods. In assessing whether a given marketing or other sales activity comply with good marketing practice, the business interests, as well as the interests of consumers and society in general, have to be balanced. However, at the end of the day, it is for the courts to decide whether an action violates the Marketing Practices Act. A substantial amount of case law within this area has been of great assistance by laying down various general guidelines and basic principles. The term "good marketing practice" has been applied in a broad sense and includes not only marketing in the technical sense but any activity performed for commercial purposes.

In terms of industrial design and its protection against infringing behavior, the Marketing Practices Act may be a useful instrument, provided that it is possible to establish that the act in question (e.g., copying) is not in accordance with good marketing practice (which would usually be the same as an act of unfair competition). It should be emphasised, however, that copying of a product that for example, does not meet the requirements for the grant of a design right or for which such protection has expired, is not in itself a violation of the general clause in the Marketing Practices Act. In addition, it is a requirement that the act is (subjective) blameworthy. Sponging or other improper use or exploitation of the achievements of other business enterprises is not regarded as acceptable. According to case law this would typically be the case in the situations described below:

(1) When all products or a complete range of products marketed and sold are systematically copied (slavish plagiarism) by a competitor predominantly in order to benefit from the developing efforts carried out and to save startup costs suffered by the owner of the industrial design.
(2) When an infringer deliberately takes advantage of the considerable marketing efforts of another by putting an identical product on the market knowing that it is likely to cause confusion or uncertainty among consumers as to the origin of the product and realising that the infringer is likely to benefit from such confusion due to the fact that the original product is well-known.
(3) Where there has been a contractual relation e.g., development, production and/or distribution agreement, it will often constitute a violation of the Marketing Practices Act if upon or shortly after the termination of such contractual relation or cooperation, one party starts to copy the products originally developed and sold by the other party (using know how or other relevant information originating from the other party which has been disclosed under the contract or during the cooperation).

The scope of the Marketing Practices Act is well illustrated by the LEGO case (U 1995.92 SH), where the Maritime and Commercial Court reached the result that the Lego brick did not in itself enjoy protection as a trademark. However, the Court found that the LEGO brick should enjoy protection against an almost identical building brick named Byggis. The Maritime and Commercial Court reasoned as follows:

> The Lego-brick's position today in the Danish market is the achievement of many years of extensive product innovation and marketing. The result of this effort ought to be protected according to the Marketing Act in such a way that very close product imitations should be opposed, unless the similarities between the products are due to technical/functional requirements.
> The Byggis-brick appears in its original form identical to the Lego-brick.
> It must be assumed that this is a conscious act and that Byggis has manufactured their brick without noticeable independent efforts. The technical differences are few and in view of appearances the differences are insignificant. The technical idea upon which the Lego-brick is based is described in the patent document, and this idea may be copied. However, this does not imply that Byggis as a matter of course is entitled to manufacture bricks which are so identical to Lego's bricks that they can substitute Lego's bricks and be assembled with them. It must be assumed that Byggis could have implemented the described idea in many other ways without causing malfunction to the brick and still resulting in an appearance different from the Lego-brick. Additionally, in the marketing of the Byggis-brick it has been indicated in various ways directly or indirectly that the Byggis-brick is adaptable to the Lego-brick, and moreover Byggis has pictured a building brick on several packages which calls attention to Lego's trademarks. Accordingly, the court finds that the Byggis-bricks are marketed in such a way that Byggis improperly exploits Lego's position in the Danish market.

Due to the nature of the protection there are no formal requirements, such as registration or the like, which have to be fulfilled in order for a product to enjoy protection pursuant to the Marketing Practices Act. Likewise it is not possible to say anything precise about the duration of the protection. Contrary to protection under the intellectual property laws which is limited in time (save for trademark rights), the protection provided by the Marketing Practices Act does not expire. In theory, the protection may last for as long as the product is on the market.

The Marketing Practices Act is commonly referred to as an alternative source of protection. If it is doubtful whether an industrial design meets the requirements for protection under any of the intellectual property laws, the Marketing Practices Act is certainly worth considering when seeking protection of an industrial design.

CHAPTER 16

Industrial Design Rights: Singapore

Gladys Mirandah & Juvelyn Cubilla

§16.01 INTRODUCTION

Registered Designs in Singapore are governed by the Singapore Registered Designs Act 2000 (the 'Act') that came into force on 13 November 2000, revised on 31 July 2005 and recently amended on 13 November 2014. The Act is implemented using the Registered Designs Rules 2000 (the 'Rules'), revised on 31 January 2002, and recently amended on 13 November 2014.

§16.02 CONVENTIONS AND LEGISLATION

Singapore is a party to the Paris Convention for the Protection of Industrial Property (1883), which governs priority claims from foreign design applications.

Singapore is also a party to the Agreement on Trade-Related Aspects of Intellectual Property Rights (the 'TRIPS Agreement'), which contains specific obligations regarding the substantive protection of industrial designs.

Furthermore, Singapore is a party to the 1999 Geneva Act of the Hague Agreement Concerning the International Registration of Industrial Designs (the 'Hague Agreement').

As mentioned in the Introduction, Registered Designs in Singapore are governed by the Singapore Registered Designs Act 2000, which is implemented using the Registered Designs Rules 2000.

§16.03 SUBJECT MATTER

Unlike copyright protection in Singapore, protection for industrial designs may only be acquired through registration.

[A] Definition: What Constitutes a Protectable Design

'Design' is defined as:

> features of shape, configuration, pattern or ornament applied to an article by any industrial process, but does not include -
>
> (a) a method or principle of construction, or
> (b) features of shape or configuration of an article which -
>   (i) are dictated solely by the function which the article has to perform;
>   (ii) are dependent upon the appearance of another article of which the article is intended by the designer to form an integral part; or
>   (iii) enable the article to be connected to, or placed in, around or against, another article so that either article may perform its function.

The basic characteristic of design law is that it is concerned with the visible characteristics which an article has: what it looks like, not how it works. If an article has features of shape, configuration, pattern or ornamentation, other than those necessary for it to match other articles or fit with them, then it has the basic requirements of a design.

The Singapore system assumes that an article can exist (at least theoretically) with no design applied to it. A designer then forms it with shape, configuration, pattern and/or ornament to produce a design. As the definition of 'design' indicates, a design only exists once it is applied to an article. The manner in which it is applied does not affect the issue. Therefore it cannot be registered by itself as distinct from an article. What is depicted in the registration is the article with the design incorporated. The same design may be applied to different articles but must be registered separately for each article to which it is to be applied. Others may still manufacture and sell an article for which a design has been registered, but they will not be allowed to imitate the features or visible effects for which registration has been granted.

The Act apparently does away with the requirement for the features of the design to be judged by or appeal to the eye. There is no need for distinctiveness or individuality of appearance (as long as the design is still novel). Even features with no eye-appeal at all fall into the definition. Perhaps the definition also brings into the protected design, features of the design which are not even seen in the finished article.

'Article' is defined as:

> any article of manufacture and includes
>
> (a) any part of an article if that part is made and sold separately; and
> (b) any set of articles.

The definition is supposed to cover items which have their own commercial life. For 'parts', it is not supposed to cover those which are made and sold by a component manufacturer to someone putting several components together. Thus representations for design applications should show complete articles, as they are sold, and not just portions of them.

A set of articles is two or more articles of the same general character ordinarily on sale or intended to be used together, to each of which the same design has been applied. The example normally given for this is the pattern on a set of crockery. It might also apply to the same shape applied to salt and pepper grinders. However, applying the same design to articles which are not to be sold or used together, e.g., a series of cars or telephones, would not make them a set.

The Act omits the requirement for the article to be of a type whose appearance has a material effect on the decision to buy. Thus, a design can still be registered even if no one cares what the article to which it is applied looks like. For example, in the past an electrical terminal for washing machines was held to be unregistrable because a potential customer would decide to buy it solely for its utility and not because of any attraction in its shape. Similarly, it might have been argued that a toner cartridge for a printer was not registrable, because people only bought such cartridges to fit specific machines with no regard to the appearance. Presumably, however, that is no longer the case.

Certain categories of product and types of articles are excluded from registration, for instance those, the publication or use of which would be contrary to public order or morality. No computer program or layout-design may be registered. Layout-designs are, instead, protected through the Layout-Designs of Integrated Circuits Act 1999. Additional excluded articles are those which are primarily literary or artistic in character and therefore usually the subject of copyright protection. The Rules list what cannot be registered as a design. These are: works of sculpture (other than casts or models used or intended to be used as models or patterns to be multiplied by any industrial process); wall plaques, medals and medallions; and printed matter primarily of a literary or artistic character including (but presumably not limited to) book jackets, calendars, certificates, coupons, dressmaking patterns, greeting cards, labels, leaflets, maps, plans, playing cards, postcards, stamps, trade advertisements, trade forms and cards, transfers, and similar articles.

[1] Shape, Configuration, Pattern and/or Ornament

Shape and configuration are the three-dimensional aspects of the design. Almost any aspect that can be felt falls within this term. These include any visible feature caused by grooves, contours, etc.

Pattern and ornament are the two-dimensional aspects of the design. These include tonal contrasts caused by changes in colour or hue, even including a change from an opaque portion to a transparent portion.

[2] Ornamental or Functional

The registration of a design is intended to give a monopoly in a particular, but arbitrary appearance given to an object. It is not intended to give a monopoly in what the object does, how it does it and how it was made. Those rights are properly in the purview of patent law. Others must be free to make the same article, which has the same functions, operates in the same way or is made in the same way. Thus the design of the lower part of the exit of a tea pot spout, which never drips, may not be a feature for which a monopoly will be given, if that design is chosen solely to avoid dripping.

[B] Originality/Novelty

To be registrable, a design must be new. Worldwide novelty is a requirement of the Act. Novelty in only the UK and Singapore is no longer sufficient.

[1] Test for Originality Generally

Not applicable

[2] Novelty

Novelty is judged at the date of application or at the priority date if the application is filed within six months of and claiming priority from an earlier first application in a Convention country or in a country having a reciprocal priority right agreement.

There are savings for designs disclosed in breach of confidence or good faith or disclosed to government departments and for textile designs when a first and confidential order for such goods is accepted. There are also savings for designs disclosed at a 'certified' exhibition, if the application is filed within six months of the opening of the exhibition. Additionally there are savings for artistic works, with copyright, that have been sold, let for hire or offered for sale or hire, provided the design has not been applied industrially (fifty copies or more made or made in lengths or pieces, not being handmade). However, where a design is only registrable thanks to this last saving, the term of the registration cannot extend beyond the expiry of the copyright.

All applications should be accompanied by a statement of the features of the design for which novelty is claimed (the 'statement of novelty') whether for shape, configuration, pattern or ornament, or a combination of all or any of these four features. The statement of novelty should also indicate if novelty is only claimed for part of an article (as opposed to all of it). If the statement of novelty only refers to the features of shape and configuration, then all features of pattern and ornament will be ignored when assessing novelty, even if the only novelty resides in the pattern and ornament, as was found for a specific rear light cluster for a car (*Valeo Vision S.A. v. Flexible Lamps Ltd.* [1995] RPC 205). If novelty is only claimed in the pattern and ornament, then the features of shape and configuration will be ignored. Likewise if novelty is only claimed in a portion of an article, for example the underside, then the rest of the article will be ignored.

A design would be clearly 'new' if it is totally unlike anything that has gone before, whether applied to the same article or to different ones. To determine if a design is new, one has to make reference to designs which have been published, registered or were otherwise available to the public at the date of the application and make reference to the novelty claimed in the application. When the design is a new shape, it is usual only to look at previous designs for the same shape, since the article has an effect on the shape. However, when the design is only a new pattern, designs for almost any other article may be relevant.

In judging novelty, the eye is the judge of the similarities and differences. However, it is the general impression of the object as a whole that may be more important. One would have to imagine a typical customer with the kind of object in question. If, for example, costume jewellery were in question, one would expect the customer to have an eye for a fair amount of detail in the object when he inspects it. He would probably look at the object both at close range and from a distance. If in contrast, the object in question were wallpaper, the examination of the design would probably not be at such close range.

In the process of registering designs, the Registrar does not conduct a novelty search. Therefore, there should be no assumption of novelty in a design simply because it has been registered.

[a] Prior Use

Under the Act, a design is not new if it differs only in immaterial details, or features, from other designs that are commonly found in trade.

[b] Prior Publication

Lack of novelty can be due to prior publication or other disclosure anywhere in the world.

[c] Prior Application or Registration

Lack of novelty can also be due to a prior application, which proceeds to registration, to register the same or a sufficiently similar design in Singapore, whether for the same or a different article.

In general, once a design has been registered for one article, it cannot subsequently be validly registered for a different article (nor can a trivially modified version of the registered design be registered for the same or any other article), because it is then no longer new. An exception to this is where the same proprietor 'associates' a subsequent registration with a first registration. In this case, the subsequent registration does not have a full life. It must expire no later than the prior registration. The proprietor for the later application does not need to be the proprietor of the first registration at the date he makes the later application, as long as he becomes the registered owner of the first registration whilst the latter is still pending. If an applicant wants to take advantage of this provision, he should mention so on filing.

§16.04 DURATION OF PROTECTION

Registration takes effect from the date of filing. This is termed the 'Date of Registration'. The initial term of a registered design is five years from the Date of Registration, renewable for further periods of five years, up to a maximum period of fifteen years. An application to renew cannot be made earlier than six months before the date of expiry of the current period of registration. It can be made up to six months late, upon payment of a penalty fee.

§16.05 INFRINGEMENT

Under section 36(3) of the Act, no proceedings can be taken in respect of infringement committed before the actual date on which the registration certificate is granted.

[A] Scope of Infringement Action

Registration gives exclusive rights in Singapore, to make or import for sale or hire, or for use for the purposes of a trade or business, or to sell, hire or expose for sale or hire articles in respect of which a design is registered. It also makes it an infringement to make anything for enabling such articles to be made in Singapore or elsewhere. It additionally gives similar rights over kits for the article and anything for enabling such kits to be made or assembled.

There are new savings against infringement for those who do any act for a private non-commercial purpose or for the purpose of evaluation, analysis, research or training. Additionally, there is no infringement by most actions in relation to articles placed on the market in Singapore or elsewhere by or with the consent of the registered owner.

[B] Test for Infringement/Criteria to Constitute Infringement

In cases of infringement of a registered design, the question is not one of copying but of resemblance. The registered owner does not have to prove copying but merely that features of the defendant's product and the 'claimed' novel features of his product look the same. This is a question of fact to be decided by the court.

[C] Standing to Sue

An infringement of a registered design is actionable by the registered owner. However, the holder of an exclusive licence shall also have the same right as the registered owner to bring proceedings for an infringement of the design committed after the date of the grant of the licence.

[D] Remedies for Infringement

Prior to trial, a registered proprietor or exclusive licensee may apply for an interlocutory injunction, Anton Pillar Order or Mareva Injunction. At the end of a trial a successful proprietor or exclusive licensee may obtain an injunction against the infringer as well as damages or an account of profits and an order for delivery up, possibly for disposal.

The normal procedure in civil proceedings involves the issue of a writ of summons. This is often coupled with a request for an injunction and a request for disclosure of the source of supply in cases where the evidence of infringement has been obtained from a retailer and not the manufacturers themselves.

In any proceedings where the validity of a registration has been contested and the Court finds the registration valid, the Court may certify the finding. If the validity is then ever unsuccessfully contested again, the proprietor may then be entitled to improved costs.

§16.06 DEFENCES TO INFRIGEMENT

[A] Validity

Any interested person may apply at any time to have a registration revoked on the grounds that at the Date of Registration it was not new or on any of the grounds on which the Registrar could have refused registration. Any interested person may apply at any time to have a registration revoked on the grounds that at the time it was registered, there was a corresponding artistic work in which copyright subsisted and the copyright has expired.

[B] Marking

It is advisable to mark the goods prominently with the words 'Registered Design' followed by the registration number. Not only will this help in claiming damages, but also may cause many third parties to refrain from copying the design.

[C] Non-infringement

There is no infringement if the Government wishes to use the design for its own purposes or to supply it abroad for defence purposes, although it is required to come to terms over such use.

[D] Other Defences

No damages or account of profits can be recovered in respect of any infringement of a design registration by an innocent infringer, that is a defendant who proves that, at the date of infringement, he was not aware and had no reasonable grounds for believing that the design was registered. Marking a product as 'registered' is not sufficient for someone to be taken as having been aware, or as having had reasonable grounds for believing that the design was registered, unless the marking is accompanied by the registration number of the design.

In the past, people have published cautionary notices in the local newspapers to rebut the defence of innocence and advise the public of the registered design. In so far as such an approach was of use in the past, it is presumably also of use under the Act.

Furthermore, third parties who, before the Date of Registration:

(a) do in good faith in Singapore an act which would have constituted an infringement if the registration had been in force at the time; or
(b) make in good faith effective and serious preparations to do such an act in Singapore, have the right to do the act or continue to do it. Such a right can be assigned or transmitted on death, but is not licensable.

Moreover, a person aggrieved by a threat that he will be sued for infringement of a registered design may claim relief from the person making the threats (whether or not that person has any interest in the registered design), in civil proceedings in the form of a declaration that the threats are unjustifiable, an injunction against their continuance, and damages for any loss the threats cause. The person aggrieved is entitled to such relief unless the person making the threats can justify them and the person aggrieved does not show that the registration is invalid. An advocate or solicitor is not liable to an action for groundless threats in respect of an act done on behalf of a client. Presumably he can, however, be sued by his client for negligent advice or for misdrafting the letter.

Such a declaration cannot be sought if the threat is to bring proceedings for infringement by making or importing anything. Mere notification that a design is registered does not constitute an actionable threat.

§16.07 PROCEDURES FOR FILING APPLICATION FOR REGISTRATION

Each application requires:

(i) name and address of the applicant;
(ii) design representations;
(iii) name of the article or set of articles;
(iv) Locarno classification (class and sub-class) of the article or set of articles;
(v) statement of novelty;

additionally, applications may require one or more of the following, where appropriate:

(vi) where the applicant is not the designer, a statement explaining the applicant's rights in relation to the design;
(vii) any priority details;

(viii) details of any allowable prior disclosures, e.g., by way of international exhibition, disclosure in breach of confidence, etc;
(ix) details of any prior applications or registrations (effective in Singapore) for the same or similar designs;
(x) if the present applicant was not the same for the priority application, details of any instruments whereby the right to priority was transferred to the present applicant.

A signed appointment of agent is not needed unless specifically requested.

The Registry may also request a statement of consent from a person entitled to give consent, where a design includes a reproduction of armorial bearings, insignia, orders of chivalry, decorations, emblems or flags etc. Likewise it may also request a statement of consent from a living person, where the name or portrait of that living person appears in a design or a statement of consent from a relevant legal representative, where the name or portrait of a person who has recently died appears in a design.

Once filed, an application undergoes a formalities examination. If there are any errors or omissions, the applicant will usually be given an opportunity to correct them. Whilst the Registry does not conduct a novelty examination, it may refuse registration if it deems the design not to be new or not to be registrable for any other reason. Once all the formal requirements have been satisfied, the design will be entered in the Register, a certificate of registration will issue and a notice of the registration and a representation of the design will be published in the Gazette. Publication may be deferred up to eighteen months from the filing date, but such request must be submitted at the time of filing.

[A] Priority

It is possible to file registered design applications claiming priority from one or more applications previously filed in a Convention country (a country or territory, other than Singapore, which is a party to the International Convention for the Protection of Industrial Property (the 'Paris Convention' or a member of the World Trade Organization). Any person (or his successor in title) who has made a first application for protection of a design in a Convention country can claim priority from the date of that application in a Singapore application, for the same design, filed within six months.

The six-month term allowed for filing a convention application should usually be calculated from the first application filed in any Convention country for the design in question. A second application for a design in the same Convention country as a first application can only serve as a basis for a priority claim if, by the time of the second application: (i) the first application had been abandoned, withdrawn or refused, without having been laid open to public inspection and without leaving any rights outstanding and (ii) the first application had not already served as a basis for claiming priority.

When claiming priority, the date of the earlier priority application becomes, in effect, the application date of the Singapore convention application. Upon registration, the priority date becomes the Date of Registration. There are the usual advantages of claiming priority, in that any disclosures or Singapore registered design applications (not claiming an earlier priority date) in the interval between the priority date and the actual date of filing are ignored.

[B] Drawings

The representations do not have to be views from any particular angles, nor is there a minimum number of views. However, they should show all sides of the article (or it should be clear that missing views would be the same as views that are present). All representations should fit within a space of 15 cm by 13 cm.

[C] Design Part of Another Article/Environment of Design

Designs cannot be registered if they are dependent upon the appearance of another article, of which it is intended by the designer to form an integral part of another article, so that either article may perform its function.

[D] Description of Design

Not applicable

§16.08 EXPUNGING/CANCELLING OR VARYING REGISTRATION

As mentioned above, any interested person may apply at any time to have a registration revoked on the grounds that at the Date of Registration it was not new or on any of the grounds on which the Registrar could have refused registration. Any interested person may apply at any time to have a registration revoked on the grounds that at the time it was registered, there was a corresponding artistic work in which copyright subsisted and the copyright has expired.

§16.09 OTHER

Registration of a design may not be the only form of relevant intellectual property protection available, or it might not be relevant to a particular design. Other potentially relevant intellectual property rights include trademarks, copypassing off and patents.

[A] Interface between Industrial Design and Trademarks

In some cases the pattern that is applied to an article can be registered simply as a two-dimensional trademark. Additionally, the Trade Marks Act 1998 now allows the registration of three-dimensional designs as trademarks. Thus, it is possible to register patterns as they appear on articles as trademarks and to register the shapes of the articles as trademarks. However, this does require that the designs are distinctive, although they do not need to be new. Indeed distinctiveness in many designs is usually only acquired after considerable use. Trademark registration has the benefit that there is no maximum term, provided renewals are paid. Additionally, there is the benefit of provisions which make it an offence to trade in counterfeits of a registered trademark. As such, a proprietor of a registered trademark is entitled to pursue actions both under criminal and civil laws. Additionally, trademark law allows proprietors to obtain search warrants to conduct raids on premises suspected of dealing in counterfeit products. Such searches take place with the aid of the police.

[B] Interface between Industrial Design and Copyright

Previously, industrial designs were also protected in Singapore through copyright. However, the Singapore legal system, on designs, no longer favours cumulative or dual protection. The author of a design cannot invoke both the law of protection of design registration and that of copyright. The Copyright Act has been amended with the Act. If a design is of a type that is registrable in Singapore (whether it has been registered in Singapore or not), it is not protected by copyright. As the Act extends what is registrable to almost anything new, other than spare parts, there is also no infringement of copyright in even non-registrable articles once the design has been industrially applied (anywhere). Thus, copyright law is not useful in protecting any designs once registered, or industrially applied.

CHAPTER 17

Industrial Design Rights: South Africa

Mike von Seidel

§17.01 INTRODUCTION

The Designs Act of 1993 provides separate protection for aesthetic and functional designs through registration at the Designs Office. Aesthetic designs relate to designs applied to articles of manufacture and are intended to afford protection for the pattern, shape, configuration or ornamentation of the article as is judged solely by the eye. Functional designs, however, have features which are dictated by the function which the article is to perform to which the design is applied and includes integrated circuit topography and mask works related thereto. Aesthetic designs must be new and original while functional designs must be new and not commonplace.

Whilst South African and British laws diverge substantially in many respects, South African law closely follows British law in other respects and for that reason, and the fact that South African law is woefully short of precedents in some areas, British cases are referred to in what follows as those are rather likely to be followed in South Africa.

[A] Highlights of New Developments in This Jurisdiction since Last Edition

A third category of designs, namely that of traditional designs, was introduced by way of legislative amendment at the end of 2013, but, as far as can be established, it has not yet been implemented at the Designs Office on a practical level.

§17.02 CONVENTIONS AND LEGISLATION

[A] International Conventions

(1) The Paris Convention for the Protection of Industrial Property.
(2) Agreement on Trade-Related Aspects of Intellectual Property Rights (TRIPS).

[B] South African Legislation

The registration of industrial design rights in South Africa is governed by the Designs Act No. 195 of 1993 as amended by the Intellectual Property Laws Amendment Act, No. 38 of 1997 and the Intellectual Property Laws Amendment Act No. 28 of 2013. The implementation of the Designs Act is regulated by the Design Regulations of 1999 that regulate proceedings in relation to the registration of designs in terms of the Act and define the activities that are to be recorded in terms of the Act in the Designs Register that is set up by the Act.

The Designs Register is supposed to be divided into three parts; namely an Aesthetic part 'A' of the register in respect of which design application and registration numbers are preceded by the letter 'A'; a Functional part 'F' of the register in respect of which design the application and registration numbers are preceded by the letter 'F'; and a yet to be implemented Traditional part 'T' of the register that is to be reserved for traditional designs that can be either aesthetic 'TA' or functional 'TF' in nature. A traditional design, which was introduced by the Intellectual Property Laws Amendment Act 28 of 2013 that became law in December 2013, is one that is recognized by an indigenous community as having an indigenous origin and a traditional character and includes 'derivative indigenous designs'. Traditional designs will not be considered at any length in what follows as they are unique to South Africa and unlikely to be relevant from an international perspective and the administrative infrastructure for traditional designs has, as far as can presently be established, not yet been set up.

§17.03 SUBJECT MATTER

[A] Definition/What Constitutes a Protectable Design?

A protectable design can be either an aesthetic design, which is new and original, or a functional design, which is new and not commonplace. The design must be for an article which is intended to be multiplied by an industrial process (section 14(1) and 14(4) of the Designs Act No. 195 of 1993, hereinafter referred to as the Act). Protection may be obtained for both the aesthetic and the functional qualities of a design (i.e., applied to a single article in combination) by filing two design applications, one of which will designate Part 'A' of the register and the other of which will designate part 'F' of the register.

[1] *Articles and Sets of Articles*

An article means any article of manufacture and also includes a part of such article if manufactured separately.

For example, a shoe and a shoe sole, if manufactured separately, constitute two different articles which when placed together make a third composite article. Where there is more than one article a separate design application may be made for each article and for the composite article itself. If a single article of a composite article is going to be design protected, the irrelevant portion of that composite article can be disclaimed in order to focus protection on the relevant portion of the article, e.g., '*A shoe irrespective of the shape and configuration of the shoe sole...*'. Disclaimers cannot be removed by amendment as this will have the effect of widening the scope of the protection.

A set of articles means a number of articles of the same general character which are ordinarily on sale together or intended to be used together, and in respect of which the same design, or the same design with modifications or variations not sufficient to alter the character of the articles or substantially affect the identity thereof, is applied to each separate article; provided that a series of mask works shall not be a set of articles.

Section 32 the Act, which states that registration of a design shall be granted for one design only, applies where a set of articles has been registered but it does not truly constitute a set and in this way more than one design has been registered. However, the section also states that no person may in any proceedings apply for the revocation of such a design registration on the ground that it comprises more than one design. In such a case, the only possible recourse is to take the Registrar of Designs on common law review given that he should not have granted a design where there was more than one design evident in the application.

[2] *Multiplied by an Industrial Process*

Designs for articles which are not intended to be multiplied by an industrial process shall not be registrable under the Act (section 14(4)). In this respect industrial does not mean that the process must be automated and hand-made articles fulfil this requirement. However, there must be an intention for the manufacturing process to be repeated many times so that many articles will be made. Where a single article is made, for example a sculpture, this will not be industrially applicable and hence such an item shall not be the subject of design protection.

[a] *Aesthetic Designs*

An aesthetic design is a design applied to an article, whether for the pattern, the shape, the configuration, the ornamentation thereof, or for any two or more of such purposes, and by whatever means it is applied, having features which appeal to and are judged solely by the eye, irrespective of the aesthetic quality thereof (section 1(i) of the Act).

[i] Pattern and Ornamentation

Pattern refers to a decorative or artistic design applied to the manufactured article and is repetitive in nature. It is usually two-dimensional but can be three-dimensional.

Ornamentation refers to the idea of adornment and there is no repetitiveness as in the case of a pattern but a single design applied to the article.

The natural meaning of these characteristics does not exclude colour (Smith, Kline and French Laboratories Ltd's Design Application 1974 RPC 253 at 260). However, colour per se that is not associated with the article to which the colour is applied is not a design by definition and cannot be registered as such. The weighted importance of colour in an aesthetic design was dealt with in *Marley v. Piping Innovations* 1989(2) SA 510 (T), which looked at whether the notion of a copper-coloured plastic pipe meets with the requirements of the definition to qualify as a design and whether such a design, is new or original. The court followed the dictum in the Smith Kline judgment and held that colour will be of importance in a design if it is of primary importance in determining what the design is. In this case the copper colour of a pipe was held not to be of primary importance to the design, as copper pipes have existed for centuries and the fact that this particular pipe is made of plastic rather than metal just changes the method of construction and not the design itself.

[ii] Shape and Configuration

Shape refers to the external form or contour and configuration is the arrangement of discrete parts to form a whole. These terms thus signify the three-dimensional contours of an article.

In judging the features of an aesthetic design which appeals to and are judged solely by the eye, the court in *Swisstool v. Omega Plastics* 1975(4) SA 379 (W) held that '*the eye*' is to be the eye of the Court and it should view the design through the spectacles of the customer. There is reliance on expert evidence but not complete reliance. It should also be borne in mind that the probing, tutored eye of the court may discern features which could well escape many customers.

The phrase 'which appeals to' has seemingly been introduced by the legislature to distinguish an aesthetic design from a functional design (Burrell's South African Patent Law and Design, 3rd edition, Butterworths, Durban, 1999).

Section 14(5) of the Act provides that an aesthetic design will not be granted protection if the feature for which protection is sought is necessitated solely by the function that the article is meant to perform or by the method of construction applied to make the article; in other words there is no component of the design that appeals to the eye. This clearly delineates the difference between aesthetic and functional design protection.

Method of construction refers to when the matter claimed is essentially a wish list of parts which can be used to make any number of different articles i.e., when the definitive statement tries to claim a method of invention or design (much like what is claimed in a patent). This occurs when no specific embodiment of the article is claimed;

a design should always claim a specific embodiment by the very nature of design protection.

[b] *Functional Designs*

A functional design is a design applied to an article whether for the pattern, the shape, the configuration, or for any two or more of such purposes and by whatever means it is applied, having features which are necessitated by the function to which the article to which the design is applied is to perform and includes an integrated circuit topography, a mask work and a series of mask works (section 1(xi) of the Act).

The idea behind a functional design is that the article is designed in a specific manner to enable it to perform a specific function. The protection by way of a functional design is unique to South Africa. In many other countries items that would constitute functional designs are usually protected as petty patents or innovation patents, which have a lower patentability requirement than patents.

[i] Integrated Circuit Topography and Mask Works

Protection of integrated circuits is required by TRIPS and South Africa does this by way of protection in the form of a functional design.

An 'integrated circuit' is an article, in final or intermediate form, containing electrical, electromagnetic or optical elements and circuitry, which render it capable of performing an electrical or an optical function, and in which at least a part of the electrical, electromagnetic or optical elements and circuitry are integrally formed, in accordance with a predetermined topography, in a semiconductor material. 'Integrated circuit topography' means a functional design which consists of the pattern, shape or configuration of the three-dimensional disposition of the electrical, electromagnetic or optical elements and circuitry of an integrated circuit.

'Mask work' means a functional design which consists of a pattern of an image however fixed or encoded, having or representing at least a part of an integrated circuit and a 'series of mask works' means a related group of mask works which together represent the three-dimensional disposition of the electrical, electromagnetic or optical elements and circuitry of an integrated circuit.

[ii] Spare Parts

Section 14(6) of the Act provides that in the case of an article which is in the nature of a spare part for a machine, vehicle or equipment, no feature of pattern, shape or configuration of such article shall afford the registered proprietor of a functional design applied to any one of the articles in question, any rights in terms of this Act in respect of such features.

The intention of this section is to assist the importers of spare parts such that their business is not curtailed by a registered design that will create a monopoly for a specific manufacturer (in items that are readily available and consumable in nature).

As this limitation to protection of spare parts applies to functional designs only, there was previously still room to file an aesthetic design for the part.

However, in a recent decision dealing with such parts (*BMW AG v. Grandmark International (Pty) Ltd and Another* 2014(1) SA 323) the court found that, while the whole of a vehicle can be protected by way of an aesthetic design, it is not permissible to obtain valid protection in this manner for body panels or parts. The court found that the shape of body panels or parts can only be functional as they are not selected by customers for their visual appeal but rather to replace a component so as to restore the vehicle to its original form. Thus they could only be registered as functional designs and this in turn is prohibited by section 14(6) of the Act.

[B] Originality/Novelty

[1] Test for Originality Generally

In terms of section 14(1) of the Act, the proprietor of a design may apply for the registration of such design if, *in the case of an aesthetic design, the design is new and original; or, in the case of a functional design, the design is new and not commonplace in the art in question.*

For the purpose of establishing novelty under the Act, cognizance must be taken of the definitions of 'date of application' and 'release date' as defined in the Act.

'Date of Application' means in respect of a convention application, the date on which the application in respect of the relevant design was lodged in the convention country concerned; and in relation to any other application, the date on which the application was lodged at the 'South African Designs Office'.

'Release Date' means 'the date on which the design was first made available to the public (whether in the Republic or elsewhere) with the consent of the proprietor or any predecessor in title'.

Consequent on the different requirements of 'original' in respect of aesthetic designs and 'not commonplace in the art' in respect of functional designs, different levels of originality are required for each different type of design.

As a first step, it must be determined whether a specific disclosure or other occurrence qualifies it to be a part of the prior art for the purposes of the particular design being tested.

If the answer is positive, then the second step is to assess whether that item of prior art is sufficiently similar to the design being tested such that it destroys the novelty of the design as qualified by the definitive statement. The definitive statement is a statement that is formulated by the applicant and forms part of the application for a registered design, the purpose of which is to set out the features of the design for which protection is claimed. The definitive statement is used to interpret the scope of the protection afforded by the design registration and also to determine if any parts of the relevant article should be disregarded from being considered to be part of the design. An example is a design of a contoured cut end of the top end of a one piece pale of a palisade fence is quite independent of the rest of the length of the pale, in which instance, the definitive statement would focus on the top end of the pale and disclaim the balance of the length of the pale.

[2] Novelty

Section 14(2) of the Act reads as follows:

> 14(2) A design shall be deemed to be new if it is different from or if it does not form part of the state of the art immediately before the date of application for registration thereof or the release date thereof, whichever is the earlier: Provided that in the case of the release date thereof being the earlier, the design shall not be deemed to be new if an application for the registration of such design has not been lodged-
>
> (a) in the case of an integrated circuit topography, a mask work or a series of mask works, within two years; or
> (b) in the case of any other design, within 6 months, of such release date.

In the light of section 14(2), a design will be novel prior to the application date or the release date if the release date is earlier, if it does not form part of the state of the art immediately before that date.

[a] Prior Use and Grace Period

From section 14(2) it will be quite apparent that the proprietor or predecessor in title can use the design commercially for a period of up to two years in the case of integrated circuit topography and mask works before lodging an application for registration of the design, and in all other cases a period of six months from the release date.

In practice, this means that person A may manufacture a container and put it on show at a fair on 1 February 2014 without filing an application for a registered design. If person A then wants to obtain a registered design in respect of the container, an application for such a registered design must be filed no later than 1 August 2014. If the application is not filed on or before 1 August 2014, person A would have destroyed the novelty of their own design, because the design would have been made available to the public with consent and then falls within the ambit of the state of the art (see below).

The release date and the application date are treated equally when it comes to validity of a design. Thus, where a person A has filed a design application on an earlier date on which person B releases his design at an expo, provided that person B subsequently files his design application within the grace period, both person A and person B would be proprietors of a registered design in which neither will have legal recourse against the other in the absence of any fraud or dishonesty being uncovered.

Unauthorized use of the design proprietor's own design is excused in certain cases as indicated below with regard to the state of the art.

It should be noted that there is no provision for any extension of the six-month or two-year periods for the filing of a design application after the release date.

[b] *Prior Publication: The State of the Art*

Section 14(3) of the Act provides that:

> The state of the art shall comprise-
>
> (a) All matter which has been made available to the public (whether in the Republic (of South Africa) or elsewhere) by written description, by use or in any other way; and
>
> (b) All matter contained in an application-
>
> (i) for the registration of a design in the Republic; or
>
> (ii) in a convention country for the registration of a design which has subsequently been registered in the Republic in accordance with the provisions of section 44,
>
> (iii) of which the date of application in the Republic or convention country, as the case may be, is earlier than the date of application or the release date contemplated in subsection (2).

The state of the art therefore includes anything that has been made available anywhere in the world. It is also irrelevant whether the disclosure was made orally (evidentially very challenging) or in writing. In fact, the Act goes even further, stating that disclosures made by '*use in any other way*' is also considered to form part of the state in the art.

The state of the art also includes the contents of certain 'closed' design applications (i.e., applications that are not yet open to the public in any way or manner).

If, for instance, person A files a South African design application on 1 July 2014 (without relying on a release date), and person B then files an application on 10 July 2014, the contents of person A's design application would count as prior art against the design application of person B even though person B would have had no knowledge of person A's application.

Another example of a 'closed' application that is considered to be state of the art, is an application filed in a convention country, but only if an application is subsequently made for a South African registered design on the back of such a convention application.

It is important to mention that the contents of 'closed' applications will only be considered to be state of the art if the application date thereof (whether in South Africa or in a convention country) is earlier than the relevant application date or the release date of the design application in question.

[i] Circumstances in Which Prior Disclosure, Use or Knowledge Will Not Destroy Novelty

In terms of section 17 of the Act:

> the registration of a design shall not be invalid by reason only of the fact that the design was disclosed, used or known prior to the release date if the proprietor proves that such knowledge was acquired, or such disclosure was made, without his knowledge or consent, and that the knowledge acquired or the matter disclosed

> or used was derived or obtained from him, and that he applied for and obtained protection for his design with all reasonable diligence after learning of the disclosure, use or knowledge.

Imagine a scenario where person A manufactures a container and keeps it in a cupboard in his room without showing it to anyone. His flatmate, person B, opens the cupboard, looking for a cardigan person A said he could borrow. While looking for the cardigan, person B discovers the container. Person B knows about an upcoming expo at which inventors can obtain funding for their inventions and he decides that displaying person A's container at the expo in an attempt to obtain funding would be a great way of doing person A a favour. Person B attends the expo without person A's consent or knowledge, displaying the container for everyone to see. No one is, however, willing to invest in the container and person B neglects to inform person A about displaying the container at the expo. If, after eight months, person A decides to file an application for the design, he would be within his rights to do so as the disclosure was made without his knowledge or consent. That being said, in order to avoid invalidity of the design, person A must: (1) show that the knowledge acquired or matter disclosed or used in respect of the container was derived or obtained from him, and (2) that he applied for and obtained protection for his registered design with all reasonable diligence after learning of the disclosure, use or knowledge.

[ii] The Anticipation Test

The Anticipation Test is a test that needs to be applied to the facts to establish whether an item of prior art destroys the novelty of a registered design. In terms of section 14(2) of the Act, the test is if the registered design in issue '*is different from or if it does not form part of the state of the art*'.

The Supreme Court of Appeal established the test for novelty in *Clipsal Australia (Pty) Ltd v. Trust Electrical Wholesalers ([2007] SCA 24 (RSA))*. For the sake of explaining the test the facts of this case were that Clipsal Australia (Pty) Ltd was the proprietor of the registered design and the South African exclusive licensee, Clipsal South Africa (Pty) Ltd (hereinafter referred to as 'the appellants') sought relief against Trust Electrical Wholesalers and Gap Distributor (hereinafter referred to as 'the respondents') on the ground that the respondents were infringing their registered design.

The court *a quo* dismissed an earlier application with costs on the ground that the design in question had not been validly registered because it was not new and original. Further, the court *a quo* held that the design was not infringed in any event.

The design in question was registered as an aesthetic design in class 13, which covers equipment for the production, distribution or transformation of electricity. The articles to which the design registration was applied were a set of electrical accessory plates with surrounds.

The respondents filed a counterclaim for revocation of the registered design, stating that the design was not new or original.

Harms ADP addressed the issue of 'a trade variant'. Although earlier case law established that absolute identity with prior art was not required to prove lack of novelty and that substantial identity would suffice, the concept of 'a trade variant' was introduced into South African law in the case of *Shultz v. Butt [1986] 2 All SA 403 (A)*. In the *Shultz* case, a design was held to be invalid because it was considered to be an ordinary trade variant.

Making reference to the *Shultz* case, Harms ADP stated the following with regards to a trade variant:

> This does not mean that absolute identity (with a piece of prior art) has to be shown; only substantial identity is required. Immaterial additions or omissions are to be disregarded, so, too, functional additions or omissions. That is why it is usually said that an ordinary trade variant is not sufficient to impart novelty. This principle is well illustrated by the facts in *Schultz v Butt*. The design in issue related to a boat and differed from a previous design by the addition of what was assumed to be a novel and original window structure. This addition did not make the claimed design new. Basically its function was to protect the occupants against spray and wind and since it was an ordinary trade variant and since the design as a whole was not substantially novel, the design was held to be invalid.

Harms ADP further addressed the issue of a trade variant when he dealt with infringement. Harms ADP, in finding that the differences between the registered design in question and the infringing product were insubstantial, posed the following question: If the respondents' product had been part of the prior art, would the registered design have been new in light of this?

The judge then answered this question as follows:

> The answer must be no because the move of the position of the switches and the removal of the steps on the narrow sides of the surrounds would have been regarded as trade variants. What anticipates if earlier, in general terms, infringes if later, the converse of the general rule mentioned earlier. It follows that the differences which are per se insubstantial, do not save the respondents from infringing.

Based on the reasoning of Harms ADP, one can conclude that a trade variant is a variant which differs in insubstantial respects from the state of the art.

The judge also addressed another important issue in stating the following:

> The argument for the respondents is that this document discloses the design in issue because it permits one to place any socket design within the surround. Although attractive at first blush, the argument has to fail because it means that the more general a prior disclosure is, the easier it anticipates, whereas the opposite is true: the more general the disclosure the less likely it renders the particular design identifiable. There is another aspect and that is that the inner border of this surround has a clearly defined frame, something lacking in the registered design which leads to the consideration of another test: that which infringes if later, anticipates if earlier. I find it difficult to envisage that this design could be said to be to be an infringement of the registered design in issue.

In essence, and taking all the above into consideration, the South African law is that a design must have more than insubstantial differences over the prior art in order to pass the test of novelty.

[iii] Originality

Section 14(1)(a) of the Act requires an aesthetic design to be new and original. The meaning of 'original' is, however, not defined in the Act.

Harms ADP, did, however deal with the issue of originality in the Clipsal case by stating the following:

> It is furthermore incorrect to equate (as the high court did) originality with not being commonplace in the art although that is how the concept is defined in the UK Copyright, Designs and Patents Act 1988. The reason is obvious. The 1993 Act requires that aesthetic designs must be new and original and that functional designs must be new and not commonplace. Originality and being 'not commonplace', consequently, cannot mean the same. The only other meaning 'original' can bear is one that is the same or akin to the meaning in copyright law, something that is not farfetched if regard is had to the fact that the 1916 Act spoke of design copyright. As was said by Mummery LJ in Farmers Build v Carier [1999] RPC 461 at 482:
>
>> The court must be satisfied that the design for which protection is claimed has not simply been copied (e.g. like a photocopy) from the design of an earlier article. It must not forget that, in the field of design of functional articles, one design may be very similar to or even identical with another design and yet not be a copy: it may be an original and independent shape and configuration coincidentally the same or similar. If, however, the court is satisfied that it has been slavishly copied from an earlier design, it is not an 'original' design in the 'copyright sense'.

In light of the comments of the judge, it is concluded that the concept of originality is similar to that as given in copyright law. This means that it must be the product of original thought and labour in execution. Thus, it can be copied, but the copy must show some form of original thought. It should be noted that mosaicing of prior art may be possible when determining originality. This has, however, never been tested in our courts.

[iv] Not Common Place in the Art in Question

Section 14(1)(b) of the Act requires a functional design to be new and not commonplace in the art in question, a concept that was first introduced into South African design law under the present Act together with the concept of functional design.

This expression should not be confused with its use in section 213(4) of the British Copyright, Designs and Patents Act 1988 where it serves to define the concept of 'originality'. As explained above, 'originality' has no relevance to functional designs in South Africa. Nevertheless, in the context of the British legislation, the idea of commonplaceness is defined in the British case of *Ocular Service Ltd & Another v. Aspect Vision Care Ltd and Others*.

In the *Ocular* case, Laddie J stated that *'any design which is trite, trivial, common-or-garden, hackneyed or of the type which would excite no particular attention to those in the relevant art is likely to be commonplace...'*

[c] *Prior Application or Registration*

Section 15(5) of the Act provides for another exception to the rule. This section states that, if an application for a registered design was filed in class X on a first date (or even if a design has already been registered in class X), the same proprietor may file a subsequent registration of the same design in class Y on a second later date, without invalidating the second application by virtue of the earlier application registration in class A. Section 15(5) goes even further in stating that public disclosure of the design applied to goods in class A, would not invalidate the second application for the same design in class B.

[C] Unobviousness (if Appropriate)

Obviousness is not a part of South African design law. This was confirmed in the *Clipsal* case in which the SCA dealt with the attempted use of 'obviousness' in determining novelty of a registered design. Mosaicing is a concept common in determining obviousness of patents in South African patent law. It entails combining two or more pieces of prior art in an attempt to show that an invention was obvious prior to the earliest priority date of a patent. In essence, the SCA held that obviousness, as well as mosaicing, *is not* part of design law in South Africa and that it is not permissible in determining novelty of a registered design. In the SCA, Harms ADP stated the following:

> Over the objection of the appellants the high court held that it was entitled to mosaic different pieces or prior art. This is a surprising conclusion. It is old law that one is not entitled to mosaic for purposes of novelty. This principle is also well established in patent law and as Pollock J had said more than a century ago, the Designs Act was intended to add to the Patent Act by making that which was not patentable the subject of a design. There is nothing in the Act to justify a departure from this principle especially since obviousness is not a ground of invalidity of a design. A design is not novel if it forms part of the prior art – meaning that it is to be found in the prior art – and not if it can be patched together out of the prior art.

§17.04 DURATION OF PROTECTION

The duration for protection of an aesthetic design is fifteen years, and in the case of a functional design is ten years, from the date on which the application was lodged or from the release date, whichever is earlier(section 22(1)). If priority is claimed, then the date on which the application was lodged will be the convention priority date. The term of protection cannot be extended.

The period of protection is subject to the payment of renewal fees, which are due three years from either the priority date, release date or filing date, whichever is earliest. Registration of the design will lapse at the end of the period prescribed for the payment of the renewal fees, if it is not paid within that period or with the payment of a fine, within an extension to that period of a maximum of six months (section 22(2)).

[A] Restoration

If a design has lapsed due to non-payment of renewal fees it can only be revived through an application for restoration to court together with evidence to the effect that there was not any intention on the part of the design proprietor that the design should be allowed to lapse and that there has been no undue delay in making the application for restoration. Unavailability of funds for paying the renewal fee(s) that were not paid is not a valid excuse and the reason why the design lapsed would usually hinge on some miscommunication or failure on the part of a responsible person to execute instructions to make the payment(s).

§17.05 INFRINGEMENT

A registered design is a legal monopoly granted by the Designs Office on behalf of the State which allows the proprietor of a registered design to prevent others from using the specific appearance of an article for a limited period of time. A design registration protects the physical appearance of an article, typically the shape and/or configuration and/or pattern and/or ornamentation (aesthetic design) thereof, and, unlike a patent, does not encompass the underlying principle of how the article functions.

In accordance with section 35 of the Act, proceedings for infringement of a design may be instituted by the registered proprietor immediately upon registration thereof. The registered proprietor must give notice to every licensee in respect of the registered design whose name is recorded in the register so as to enable a licensee to intervene as a co-plaintiff. A plaintiff in proceedings is entitled to relief by way of an interdict, surrender of any infringing product or a product of which the infringing part is inseparable, or damages. At the option of the plaintiff, the amount of damages may be calculated on the basis of a reasonable royalty which would have been payable by a licensee or sub-licensee in respect of the registered design.

[A] Scope of Infringement Action

In order to determine if infringement of a design registration has occurred, the following factors need to be considered:

(i) whether the alleged act is of such a nature that it constitutes infringement;
(ii) whether the article that allegedly infringes the registered design is included in the class in which the design was registered; and

(iii) the representations of the design and the definitive statement need to be construed to determine whether the infringing design is within the scope of the registration.

[1] Acts that Constitute Infringement

Section 20(1) of the Act states that the effect of a design registration is to grant to the registered proprietor in the Republic, for the duration of the registration the right to '*exclude other persons from the making, importing, using or disposing of any article included in the class in which the design is registered and embodying the registered design or a design not substantially different from the registered design*'. Section 35(9) provides that if any of the acts listed in section 20(1) above were performed *before* the date of registration of the design, the registered proprietor may not interfere or restrain the person from continuing with the acts. Consequently, an infringement action may not be instituted if a design is registered on the same day or any day after another party commenced with the making, importing, using or disposing of any article with the same or a similar design.

The registration of a design affords the proprietor thereof a statutory incorporeal real right which is negative in its application. There are certain limitations to the rights acquired by a registered proprietor upon registration of a design (Burrell TD, Burrell's South African Patent and Design Law, 3rd Edition 1999 Butterworths Durban pp 481–489):

(i) No feature of an article in so far as it is necessitated solely by the function which the article is intended to perform affords the registered proprietor of an aesthetic design any rights in terms of the Act in respect of that feature (section 14(5)(a)).
Purely functional features are in general not entitled to protection as aesthetic designs. In *Robinson v. D Cooper Cooperation,* it was held that the shape of a metal frame for a trap door was a purely functional feature and therefore the proprietor of the aesthetic design was not afforded any rights in respect of the shape of the registered design. (*Robinson v. D Cooper Cooperation of SA Ltd* 1984 (3) SA 699 (A)).

(ii) No method or principle of construction affords the registered proprietor of an aesthetic design any rights in terms of the Act in respect of that feature (section 14(5)(b)).
No design registration ought to be construed so widely as to give its proprietor a monopoly in a mode or principle of construction. The rights conferred by a design registration are limited to one particular and specific appearance of an article.

(iii) In the case of an article which is in the nature of a spare part for a machine, vehicle or equipment, no feature of pattern, shape or configuration of such article affords the registered proprietor of a functional design applied to any of the articles in questions, any rights in terms of the Act in respect of such features (section 14(6)). (See 'spare parts' above).

(iv) The rights of the registered proprietor of a registered design in the form of an integrated circuit topography will not be infringed by a person who:

   (a) Makes an article embodying the registered design or a design not substantially different from the registered design for the sole purpose of evaluation, analysis, research or teaching (section 20(3)(a).

   (b) Imports or disposes of an integrated circuit embodying the registered design which has been unlawfully produced or an article incorporating such an integrated circuit in the circumstances set out in section 20(3)(b) of the Act.

   Sections 20(3)(a) and (b) were introduced into the Act to comply with section 6, Articles 35–37 of the TRIPS Agreement. During an infringement action, section 23(b) may provide a defence that the infringer of a registered design was unaware of the registered design in relation to an integrated circuit. However if there is continued infringement the registered proprietor may claim payment of a reasonable royalty from the date the infringer was notified of the existence of the registered design.

(v) The rights of a registered proprietor are limited by the use of the design in relation to marine vessels, aircraft and land vehicles of convention countries that travel temporarily or accidently within the borders of South Africa. (Section 38).

(vi) The disposing of an article embodying a registered design by or on behalf of a registered proprietor or his licensee gives the purchaser the right to use and dispose of that article and so results in the exhaustion of rights.

In terms of section 26 of the Act, any of the limitations (i) to (vi) could form the basis of an application for a declaration of non-infringement.

Once it has been determined that an act constitutes infringement, the next step is to determine whether the article that allegedly infringes the registered design is included in the class in which the design was registered.

[2] Classification of Registered Designs

A proprietor of a registered design may prevent others from exploiting the design only if an alleged infringing design is applied to an article included in the class in which the design is registered. Correct classification is thus vital for the enforcement of a design registration against third parties.

Once it has been determined that the article in question falls within the class in which the design was registered, the next step is to determine the scope of design protection afforded to the proprietor thereof.

[3] The Scope of the Registration

The Designs Regulations provide that *'the definitive statement shall be used to interpret the scope of the protection afforded by the design registration'*. The definitive statement sets out the nature of the protection sought and may relate to an article as a whole or to particular aspects of an article.

A definitive statement may be in the following form: *'the features of the design for which protection is claimed reside in the shape and/or configuration of an [article] substantially as illustrated in the accompanying representations'*. In this example, the definitive statement is of the omnibus type and it only claims the shape and/or configuration of the article with reference to the representations. The scope of protection of the registration will therefore be assessed by a visual inspection of the representations of the design whilst having regard to the features of shape and/or configuration only, and disregarding any features of ornamentation or pattern that were not claimed.

If the definitive statement relates to the article as a whole, then the scope of protection is decided by considering the article as a whole, whereas if the definitive statement relates to a portion only of the article, then the scope of protection is decided by considering only that portion of the article. Having determined that the design relates to an article as a whole or to a portion only of the article, it is necessary to compare the alleged infringing design with the design registration according to the test for infringement.

[B] Test for Infringement

The test for infringement is whether the alleged infringing article has substantially the same appearance as the registered design. The court will make use of the definitive statement and the representations in support of the application for the registered design to determine the scope of the registered design and compare it with the appearance of the allegedly infringing design.

In determining the scope of an aesthetic design protection important aspects follow from the definition of an 'aesthetic design', namely, that design features have to appeal to and be judged solely by the eye (*Clipsal Australia (Pty) Ltd and another v. Trust Electrical Wholesalers and another* 2007 BIP 432 (SCA)). The court must have regard to the *'visual aspects'* of the registered design that will appeal to the likely customers of the goods in question. The visual aspects that *'may influence choice or selection'* by having an *'individual characteristic'* that *'attracts the attention of the beholder'*. Thus infringement is determined by the eye of the court through the spectacles of the likely customer. (Burrell TD, Burrell's South African Patent and Design Law, 3rd Edition 1999 Butterworths Durban p. 504). The two designs must be viewed side-by-side and also separately. In addition the doctrine of imperfect recollection may be taken into account when comparing two designs. *(Swisstool Manufacturing Co v. Omega Africa Plastics 1975 (4) SA 379 (W)*

In the case of *Homecraft Steel Industries v. S M Hare & Son,* in 1984 the court employed a logical test for determining the degree of similarity between two designs. *(Homecraft Steel Industries (Pty) Ltd v. S M Hare & Son (Pty) Ltd and another 1984 (3) SA 681 (A)).* The test for infringement is whether the article alleged to infringe embodies the registered design or a design not substantially different from the registered design. In determining whether or not a difference is substantial, other designs available to the public prior to the date of application or the release date of the registered design, i.e., the prior art, must be considered. If only small differences separate the registered design from the prior art, then relatively small differences between the alleged infringement and the registered design will be sufficient to avoid infringement. In other words, a registered design which possesses substantial novelty and originality will have a broader reading given to the monopoly which it affords than a design which is barely novel or original.

It is important to note that in design law, as in patent law, it is not permissible to mosaic or piece together prior art designs in order to destroy or reduce novelty. With regard to the originality of a design, a similar interpretation of the word 'originality' as that used in copyright law is implied, i.e., that the design was a product of creative labour (Burrell TD, Burrell's South African Patent and Design Law, 3rd Edition 1999 Butterworths Durban p. 500). Therefore, in assessing the differences between the registered design and the prior art to determine the scope of protection, only the differences between the single most similar prior art design and the registered design are considered and the differences will implicitly be original. It is simply a matter of assessing the similarity between a single prior art design and the registered design, which will be used, in turn, to assess the degree of similarity that is required to constitute infringement.

[C] Standing to Sue

According to section 35(1) of the Act, only the registered proprietor may institute proceedings for the infringement of a design. The proprietor, however, is required to give notice of such proceedings to every licensee in respect of the registered design, who are in turn entitled to intervene as co-plaintiffs in the matter.

In the case of joint ownership of a registered design, the joint registered proprietor may institute proceedings for infringement, and shall give notice to every other joint registered proprietor who may intervene as co-plaintiff.

The general rule, therefore, is that only the registered proprietor is entitled to institute proceedings for infringement in his own name and that a licensee has no independent right to institute such proceedings as sole plaintiff.

There is, however, the single exception that a licensee who has been granted an exclusive compulsory licence may institute any proceedings to prevent infringement, and be allowed to recover damages as a result of the infringement, provided the following provisions are adhered to:

(a) the registered proprietor concerned must be joined as a party to the proceedings;
(b) the registered proprietor may join the proceedings either as co-plaintiff or co-defendant; and
(c) the registered proprietor shall not be liable for any costs in connection with such proceedings unless he enters an appearance and takes part in those proceedings.

In the situation where a registered proprietor whose design is subject to an exclusive compulsory licence institutes proceedings for infringement, points (a) to (c) will apply but with the exclusive licensee in the place of the registered proprietor.

[D] Remedies for Infringement

A plaintiff in proceedings for infringement is entitled to various remedies, which include:

- an interdict;
- surrender of any infringing product or any article or product of which the infringing product forms an inseparable part;
- damages; or
- in lieu of damages, at the option of the plaintiff, an amount calculated on the basis of a reasonable royalty which would have been payable by a licensee or sub-licensee in respect of the registered design concerned.

[1] Limitation on Remedies

In terms of section 35(6) a proprietor/licensee shall not be able to claim damages or obtain an interdict from a person in the situation where:

(1) the proprietor/licensee made a representation that a design is registered; and
(2) the proprietor/licensee failed to disclose the number of such registered design; and
(3) the person, being unaware of the number, requested the number in writing by registered post.

This limitation applies for a period commencing with the representation and terminating two months after the person who requested the number receives the number in writing from the proprietor/licensee. The person who requested the number may also claim from the proprietor, with the court's discretion, any money, time or labour with the view of making, using or disposing of any article embodying the registered design.

§17.06 DEFENCES TO INFRINGEMENT

In proceedings for infringement, the defendant may deny infringement outright and leave it to the plaintiff to prove otherwise. Furthermore, section 35(5) of the Act provides that, in any proceedings for infringement, the defendant may counterclaim for the revocation of the registration of the design. By way of defence, the defendant may rely on any ground on which the registration may have been refused.

The grounds for revocation (according to section 31(1)) are:

(a) that the application for the registration of the design was not made by a person entitled thereto under section 14;
(b) that the registration of the design is in fraud of the rights of the applicant or of any person under or through whom he claims;
(c) that the design in question is not registrable under section 14;
(d) that the application for the registration of the design contains a false statement or representation which is material and which the registered proprietor knew was false at the time when the statement or representation was made;
(e) that the application for the registration of the design should have been refused in terms of section 16.

[A] Validity

When counterclaiming for the revocation of a registered design, the defendant may argue that the registered design is invalid. The defendant bears the onus to prove (*Swisstool Manufacturing Co v. Omega Africa Plastics* 1975 (4) SA 379 (W)), on a balance of probabilities, that a registered design is invalid in terms of section 31(1). A registered design may be invalid for a number of reasons, as is briefly set out below.

[1] Not Novel/Original

Section 14(1) requires a design to be new and original, in the case of an aesthetic design, and new and not commonplace, in the case of a functional design, in order to be registrable. Thus, a person being sued for design infringement may counterclaim for the revocation of the registered design on the ground that it is not new, or that it is not original (in the case of an aesthetic design) or commonplace (in the case of a functional design). The requirements for a design registration to be considered new and original or not commonplace are discussed above in section 2 and the defendant would be entitled to attack the design registration for not meeting these requirements.

Further to this, section 35(9) provides what may be considered a 'statutory' Gillette defence to an infringement action (Burrells T *Burrell's South African Patent and Design Law* Butterworths, Durban 1999). According to section 35(9), nothing in the Act shall entitle the registered proprietor to interfere with or restrain any person from performing any of the acts complained of in section 20(1) (making, using, importing or disposing of an article, or an article embodying a similar design, in the class in which the design was registered) where such person commenced performing any such act before the date of registration of the design. A defence being used in patent matters is the so-called Gillette defence whereby if a defendant can show that the relevant product or process is wholly covered by a piece of prior art then it cannot infringe the patent for the alleged invention. The reasoning here is that one is using a product or process that was known at the date of the patent and if one's product or process is an infringement, the patent is invalid on the grounds of lack of novelty (what infringes if later, anticipates if earlier). It is submitted that an equivalent defence should be available to a defendant in a design matter who is able to establish that, at the application date of the registered design, the allegedly infringing design was not novel, or was not original or was commonplace, as the case may be.

Section 35(9) certainly applies to those cases where (in terms of the six-month window period provided for in section 14(2)) a design is released prior to an application for a design registration being lodged. For example, where a person commences making or using an article to which the design in question has been applied before the proprietor of the design applies for registration of the design, upon registration of the design the proprietor will not be able to institute infringement proceedings against such a person.

However, section 35(9) is not limited only to cases such as the above and in fact, as indicated by the phrase 'nothing in this Act', applies broadly. Thus, regardless of whether the design was released prior to application for registration, a person having performed the acts complained of in section 20(1) and facing an infringement suit may rely on section 35(9) to raise the so-called Gillette defence against the action.

It is important to note that, when dealing with an assertion by the defendant that a registered design is not new, the registered design should be compared to other articles on a three-dimensional level. In *Sunsmart v. Flag and Flagpole Industries* [2007] SCA 50 (RSA), the court clarified this issue as follows: '*the fact that a two-dimensional representation of the shape of an article is similar to a two-dimensional representation of the registered design does not prove that the design of the article anticipated the registered design*'.

Section 17 provides an exception to the above in cases in which the proprietor of the registered design proves that the design was disclosed without his knowledge or consent, and that the knowledge acquired or the matter disclosed was derived or obtained from him, and that he applied for and obtained protection for his design with all reasonable diligence after learning of the disclosure, the registration of the design will be valid.

[2] Functional/Not Ornamental

In terms of section 14(5), no feature of an article in so far as it is necessitated solely by the function which the article is intended to perform, or method or principle of construction, affords the registered proprietor of an aesthetic design any rights in respect of such feature, method or principle. This provision is discussed above and applies if there is no component of the design that appeals to the eye. This clearly delineates the difference between functional and aesthetic design protection.

Where the element for which protection is sought is necessitated solely by the function of the article to which it applies, there will be no design protection on an aesthetic level. Further, where a method or principle of construction with no specific embodiment of the article is claimed, the design is not registrable, as this goes against the very nature of design protection.

In *Robinson v. Cooper* 1984 (3) SA (A), the court held that the features of the design in question did not have an aesthetic quality and they merely improved the closing mechanism of a trap door. In this way, the features were purely functional and as such the design was not registrable as an aesthetic design.

[3] Faults in Application

[a] 'Fraud'

In terms of section 14 of the Act, an application for the registration of a design must be made by the proprietor of such design. The defendant may argue that the applicant is not the true proprietor the design. Whether or not a person is the proprietor a design is a question of fact. Further, the registration of the design may not be in fraud of the rights of the applicant or of any person under or through whom he claims. The word 'fraud' in this respect is construed in the same manner as is discussed above in that the conduct on the applicant's part was deliberately designed to deprive the objector of vested rights as the proprietor of the design or as an assignee by operation of law. If this has happened, the court is entitled to mandate that the design be assigned to the appropriate party.

[b] False Statement

A defendant may argue that the application for the registration of a design contains a false statement or representation which is material and which the registered proprietor knew was false at the time when the statement or representation was made (section 31(1)(d)). A common example of this is where the applicant knowingly claims the incorrect priority date.

A notable case in this regard is that of *Gallagher Group Ltd v. IO Tech Manufacturing* (872/12) [2013] ZASCA 180. In this case, the court dealt with section 61(1)(g) of the Patents Act, which contains a provision similar to section 31(1)(d) of the Act. The court mentioned that a patent can be revoked if the declaration (form P3) signed by the applicant, in which the applicant states that, to the best of its knowledge, there is no lawful ground for the revocation of the patent, contains a false statement or representation which is material and which the patentee knew or ought reasonably to have known was false at the time when the statement or representation was made.

If this obiter dicta is confirmed there may be a significant number of South African patents and designs which are invalid due to being granted with such a false declaration on file at the South African Patent Office or Designs Office. Section 31(1)(d) of the Act has not yet been considered in this light, but it is highly likely that a court will turn to the decision in *Gallagher* if such an issue does arise.

[4] Other Possible Criteria

As a defence to infringement, a defendant may counterclaim for the revocation of a design registration on the ground that the article to which the design is applied is not intended to be multiplied by an industrial process (section 14(4)). In this respect, 'industrial' does not mean that the process must be automated and multiple handmade articles fulfil the requirement as indicated above.

A further defence to infringement may be raised in terms of section 35(1)(e) read with section 16 by alleging that the application for the design registration was not made in the prescribed manner and should have been refused by the registrar. The prescribed manner for applying for a design registration is discussed below in §17.07.

[B] Marking

Ignorance as to the registration of a design affords no defence to a defendant in a design registration infringement suit. However, section 35(6), as mentioned above, under 'Limitation on remedies', limits damages for infringing acts committed during the period commencing with the representation that a design is registered and terminating two months after the date on which such an alleged infringer was notified in writing by the first-mentioned person of the number of the registered design concerned.

Furthermore, section 35(7) provides that any person who, having made the request referred to above, expends during the period referred to in that subsection, any money, time or labour with a view to making, using or disposing of any article embodying the registered design or a design not substantially different from the registered design, as the case may be, may apply in the manner prescribed to the court for compensation in respect of the money, time or labour reasonably so expended, and the court may issue such order as it may deem fit.

It should be noted that in terms of section 47(2), if any person disposes of any article on which is stamped, engraved or impressed or to which is otherwise applied the word 'design' or 'registered design' or any other word expressing or implying that there is a registered design in respect of the article, or that the article is the subject of an application for registration of a design, or to which any marking is applied in any manner so expressing or implying, he shall be deemed for the purposes of this section to represent that there is a registered design in respect of the article or that the article is the subject of an application for registration of a design.

In terms of section 47(1), where a person falsely represents that there is a registered design in respect of any article or represents that any article is the subject of an application for the registration of a design, knowing that no such application has been made or that an application made in respect thereof has been refused or withdrawn or has lapsed, that person is guilty of an offence and liable on conviction to a fine, or to imprisonment for a period not exceeding one year.

[C] Non-infringement

The test for infringement, as described in §17.05 above, fundamentally involves determining whether or not the article alleged to infringe embodies the registered design or a design not substantially different from the registered design. The court is tasked with determining infringement through the eyes of the likely consumer, bearing in mind that the measure of novelty of the registered design indicates its ambit of monopoly (*Homecraft* and *Clipsal supra*).

A typical defence to infringement thus involves alleging that the design of the allegedly infringing party differs substantially from the registered design and therefore falls outside of the scope of protection afforded to the registered design. The plaintiff bears the onus to prove, on a balance of probabilities, that infringement has in fact occurred.

In terms of section 36(1), if a party proves that he has applied in writing to the registered proprietor or licensee for a written acknowledgement of non-infringement, and has furnished him with full particulars of the article in question, and the registered proprietor or licensee has failed to give such acknowledgement, the court may issue a declaration of non-infringement. A declaration of non-infringement is a declaration that the making, importing, using, disposing or reproducing of an article by any person does not or would not constitute an infringement of a registered design.

[D] Other Defences

Other more general defences to infringement include the following:

(a) acquiescence;
(b) waiver; and
(c) estoppel.

§17.07 PROCEDURES FOR FILING AN APPLICATION FOR REGISTRATION OF A DESIGN

[A] Priority and Application for Registration

An application for the registration of a design must be made by paying the relevant official fee and filing the following documents at the Designs Office:

(a) Form D1, in duplicate, one copy of which shall be returned to the applicant as a proof of filing document and on which appears all the relevant details of the applicant, any foreign application from which priority is claimed as well as

the Part (A or F) of the register in which the design is filed, and the classification of the design.
(b) Form D2 in duplicate, one copy of which actually becomes the official register page for that design.
(c) Form D3, which includes a declaration by the applicant that he is the proprietor of the design, and a Power of Attorney in the case in which an agent is appointed.
(d) If the applicant is a cessionary of a former proprietor, a document serving as proof of the cession of the design rights may be requested.
(e) A representation or representations, in the prescribed form.
(f) A definitive statement on Form D6, in duplicate.
(g) Publication particulars on Form D8, with an attached publication representation, all in duplicate, for purposes of publication.
(h) If the application is a convention application, a certified copy of the first application and any translation that may be necessary and, if the applicant in the convention country was different, then a document serving as proof of the cession of the priority rights must be filed.

An application for registration of a design must be signed by the applicant(s) or appointed agent.

Minimum filing requirements in non-convention applications in order to secure a filing date are the payment of the fee and the filing of forms D1, D2, and D6 as well as a copy (which can be informal) of the representations. The information as to whether the application is to be filed in Part A or Part F of the register can be delayed for up to fourteen days from the date of filing.

Minimum filing requirements in convention applications in order to secure a filing date are the payment of the fee and the filing of forms D1 (including the foreign application details from which priority is being claimed) and D2, provided that the form D6 and the representations are filed within fourteen days from the date of filing. The information as to whether the application is to be filed in Part A or Part F of the register can, as in the instance of non-convention applications, also be delayed for up to fourteen days from the date of filing.

For the purposes of classifying articles in accordance with Schedule 3 of the Designs Regulations, classification broadly follows the International Classification of Industrial Designs, originally adopted by the Locarno Union in 1971, as subsequently amended and put into force on 1 January 1989 including the explanatory notes and the lists of articles and goods contained therein. Separate applications for registration can be filed in respect of the same article in different classes.

[1] Convention Applications: Priority

A person or their cessionary who has applied for a registered design or a similar right in a convention country may apply for registration of the design in South Africa within the non-extendable six-month period claiming priority from the original (first) application in terms of the Paris Convention. For a functional design application, a similar right includes a utility model, a design patent, or the like in the convention country but the convention period remains at six months.

Priority documents must be filed within six months from filing the application in South Africa, but this period may be extended on application. If the priority documents are in a language other than an official South African language (typically, not in English), a verified translation thereof must accompany the priority documents.

[B] Drawings/Representations

South African design law does not prescribe a maximum number, or specific views to be contained in representations to be filed. At least one representation must be filed. Representations may be in the form of photographs; however, this may hinder the reproduction of the representations by the Designs Office and also in subsequent proceedings and line drawings are generally more acceptable. As far as possible, drawings should be without colouring and shading, and should preferably be in line format. Any words, letters or numerals that are not part of the design and do not serve as reference symbols must be removed or disclaimed.

[C] Design Part of Another Article/Environment of Design

The Act requires that a design must be a design as applied 'to an article'. 'Article' is defined as 'any article of manufacture and includes a part of such article if manufactured separately'. A design may therefore be registered in respect of an article that is manufactured on its own, or for an article comprising a number of parts manufactured separately. If an article includes a unique part which is not manufactured separately, the registered design should be applied for the entire article, and the part(s) which are not novel/for which protection is not sought should be disclaimed in the definitive statement (see below).

[D] Description of Design (Definitive Statement)

A 'definitive statement' has to be provided for every registered design application. The definitive statement sets out the features of the design for which protection is claimed, and is used to interpret the scope of protection afforded by the design. Although most definitive statements are fairly straightforward, it is possible to achieve wider or more specific protection by excluding (disclaiming) certain features of an article, or by focusing on or emphasizing particular features and disclaiming particular, usually variable, parts.

An 'explanatory statement' may also be provided for a registered design application and may refer to features of the article to which the design is applied, including its function or the method or principle of construction thereof. The explanatory statement will not necessarily be used to interpret the scope of protection afforded by the design, although it may be used in certain circumstances.

In the case of an application for the registration of a functional design for an integrated circuit topography, a mask work or a series of mask works, the definitive statement must be accompanied, on Form D6, by an explanatory statement that refers to the function and operation of the integrated circuit topography, mask work or series of mask works.

§17.08 EXPUNGING/CANCELLING OR VARYING REGISTRATION

[A] Revocation

The grounds for an application for revocation of a registered design are dealt with in section 31 of the Act and, as mentioned above in §17.06, are as follows:

(1) That the application for registration of the design was not made by a person entitled thereto under section 14.
(2) That the registration of the design is in fraud of the rights of the applicant or of any person under or through whom he claims.
(3) That the design in question is not registrable under section 14 of the Act.
(4) That the application for the registration of the design contains a false statement or representation which is material and which the registered proprietor knew was false at the time when the statement or representation was made.
(5) That the application for the registration of the design should have been refused in terms of section 16.

The grounds for revocation given in section 31 are not exhaustive. However, in terms of section 32, a person cannot apply for the revocation of a design on the basis that the registered design comprises more than one design.

[1] Not Made by a Person Entitled Thereto

Only a proprietor or joint proprietors may apply for a design – an employer will also be entitled to apply where the employee made the design in the course and scope of his/her employment (in terms of the common law).

[2] On Ground of Fraud

If the court revokes a design on the basis that it was registered in fraud of rights, or that the registration was fraudulently obtained then, in terms of section 33, the court may mandate that the registration of the design be granted to the applicant or an assignee. In such a case, the design so registered will have the same date of registration as the design which was revoked. The design is thus not simply assigned, but rather revoked and then registered again.

[3] Not Registrable under Section 14

Section 14 deals with the intrinsic characteristics which a design requires in order to be registrable, i.e., novelty, originality, not common place in the art in question, intended to be multiplied by an industrial process, not a spare part, and must be registered in the correct part of the register. A design may be revoked on any one of these grounds.

[4] False Statement or Representation

This ground, which is discussed in greater detail in §17.06 above, may be raised where the applicant intentionally made a false statement in the application. It must be shown that the applicant made the false statement intentionally, and this ground is therefore fairly difficult to prove. An example would be where the applicant intentionally claims the incorrect priority date.

[5] Section 16

Section 16 sets out that the registrar may refuse an application if it is not made in the prescribed manner. The effect of including the section as a ground for revocation means that years after a design has been registered, it may be revoked if the application, or the procedure contained an error or was incorrectly followed. It is therefore of utmost importance that an application is correctly made and that the application procedure is followed correctly.

[6] Procedure

The Act defines 'Court' to mean the division of the High Court of South Africa having jurisdiction in respect of the matter. In the usual case, the court will consist of a single judge of the North Gauteng High Court. Normal High Court Rules will apply.

In terms of section 31(2), any application for revocation must be made by way of notice of revocation on Form D14 and must set out the ground(s) on which the revocation is based. The notice must also be accompanied by an affidavit by the applicant setting out the facts upon which reliance is placed. The notice and all annexures must be served on the registered proprietor, and any person recorded in the register as having an interest in the design, and must also be lodged at court and with the registrar. The registrar, as the guardian of the register, is the true respondent in the proceedings for revocation and he must therefore be joined as a respondent to the proceedings.

If the registered proprietor intends to oppose the revocation application, then a notice of intention to oppose must be delivered within one month of service of the revocation application. If no such notice is delivered, the application may be set down for hearing on a date not less than ten days after expiry of the one month period. The notice must be served on all parties listed in the application for revocation and on the registrar and at court.

The registered proprietor may oppose the revocation application by delivering an answering affidavit and supporting documents within two months of delivering the notice of intention to oppose. Thereafter, and within one month of delivery of the answering affidavit, the applicant may deliver his replying affidavit. The replying affidavit may only reply to new facts or evidence raised by the respondent in the answering affidavit. The judge or registrar of the High Court may, in his discretion, permit the filing of furthcr affidavits.

After this, the process of filing documents is closed and the applicant may within ten days apply to the registrar of the High Court to allocate a date for the hearing.

The parties then file heads of argument with the registrar of the High Court which must be filed not later than two court days before the date for the hearing. The heads consist of concise and succinct statements of argument and the main points intended to be argued, as well as authorities for each point relied upon.

Where the judge determines that the matter cannot be decided on affidavit, he may refer the matter to trial.

[7] Revocation as a Defence to Infringement

The procedure outlined above is not the only way in which revocation of a design may be achieved. In any proceedings for infringement of a registered design, the defendant may counterclaim for the revocation of the design.

§17.09 OTHER

[A] Interface between Registered (Industrial) Designs and Trademarks

Trademark and design protection can overlap in the registration of the shape of containers for goods and the shape of goods themselves. Both the repealed Trade Marks Act of 1963 and the Trade Marks Act of 1993 currently in force include a 'container for goods' as a registrable mark. The 1993 Trade Marks Act broadened protection to include shapes of goods as trademarks. It is to be borne in mind, however, that the container, if it is distinctive, can only be registered as a container trademark in respect of the contents and not as a trademark in respect of the container itself. If registration is to take place of the container as a container then in effect the applicant will seek to register the shape of the article as the mark in terms of the 1993 Trade Marks Act. The court *a quo* in *Triomed (Pty) Ltd v. Beecham Group plc* [2001] 2 All SA 126 (T) 137h–138 quoted Aldous LJ in *Philips Electronics NV v. Remington Consumer Products Ltd* [1999] RPC 809 (CA) 817–818 'In my view a shape of an article cannot be registered in respect of goods of that shape unless it contains some addition to the shape of the article which has trade mark significance. It is that addition which makes it capable of distinguishing the trade mark owner's goods from the same sort of goods sold by another trader.'

Registration of a design in terms of the Act does not preclude it from also being registered as a trademark. A parallel trademark registration for the shape of a product or container has two benefits. First, the shape does not require strict novelty to qualify for a trademark registration (as opposed to a design registration). Rather, a trademark must be distinctive to be registrable. Second, unlike a design registration, a trademark registration is not time-limited and can be renewed indefinitely. Thus, a trademark registration could potentially provide an everlasting monopoly in the shape of a product. The mark must remain capable of distinguishing the goods or services of one person from those of another. In *Bergkelder Bpk. v. Vredendal Koöp Wynmakery and Others (105/05) [2006] 4 All SA 215 (SCA)* [Par 9], Harms JA references *Henkel KGaA v. Office for Harmonisation in the Internal Market (Trade Marks and Designs) (OHIM)* and states that 'a container mark must, in order to be able to fulfil a trade mark function, at least differ "significantly from the norm or custom of the sector"'.

The point of departure for trademarks is that any mark capable of distinguishing the goods or services of one person from those of another should be registrable. Therefore the shape of a design may be registrable as a trademark insofar as it is so capable of distinguishing. Important provisos, however, are firstly that a mark is not registrable if it consists exclusively of the shape, configuration or colour of the goods, where such shape, configuration or colour is necessary to obtain a specific technical result or results from the nature of the goods themselves. Second, a mark is not registrable where registration it is likely to limit the development of any art or industry.

Thus, as long as the shape of the design is: (a) not necessary to obtain a specific technical result, (b) is not determined by the nature of the goods themselves or (c) is not likely to limit the development of any art or industry and (d) is capable of distinguishing those goods from the goods of another, it may be registrable as a trademark.

The 1993 Trade Marks Act further provides that a trademark shall not be infringed by the bona fide use by any person of any utilitarian features embodied in a container, shape, configuration, colour or pattern which is registered as a trademark. As trademarks can be renewed perpetually, this provision ensures that no rights are gained in features that are more appropriately protected by a design or a patent.

[B] Relationship between Registered (Industrial) Designs and Copyright/Design Copyright

Generally speaking, copyright in a work may be described as an exclusive bundle of rights to do or to authorize others to do certain acts in relation to material manifestations of intellectual effort. The right to control the use of a work in all manners in which that work can be exploited for profit or gain is an essential right under the laws of copyright.

The protection afforded by the laws of copyright, specifically in relation to artistic works, overlays to a certain degree (and subject to certain qualifying exceptions and conditions) with the protection received by the registration of registered designs.

As indicated previously, the protection afforded by registered designs lasts for ten years in the case of functional designs and fifteen years in the case of aesthetic designs. This is contrasted with the duration of copyright protection which subsists for a period equal to the lifetime of the author plus fifty years. Therefore, copyright protection in respect of manufactured articles will continue to subsist long after design protection has lapsed. It is for this reason, among others, that the juncture between these two separate fields of intellectual property law is so important.

The troubled relationship between the two forms of protection has swung wildly since the first South African intellectual property legislation of 1916 to the current Act of 1998 as amended. This has occurred as the legislature has attempted to separate the two forms of protection from one another to prevent allowing dual protection under both forms of intellectual property protection. The exhaustive copyright protection that existed for functional articles under the 1916 Act has been curtailed to the present position of no effective protection under the 1988 Copyright Amendment Act.

The current standing between the two branches of law is that, in principle, manufactured products can be protected under both design and copyright laws, subject to certain qualifications and exceptions. Under the correct conditions, copying of manufactured articles can give rise to copyright and design infringement simultaneously.

[1] Copyright in South African Law

The Copyright Act No. 98 of 1978 grants the eligible holder of copyright a qualified monopoly in relation to the material manifestations of intellectual effort. In order for a work to be protected under the Copyright Act, certain minimum qualifying conditions must be met. No registration is necessary in order for copyright to subsist, although registration of copyright in relation to cinematograph films is possible. The work must be original in that it must be the product of the author's own independent skill, labour and effort. The work must be reduced to material form, and this flows from the premise that copyright does not subsist in ideas alone. Finally, the author must be a qualified person in terms of the Copyright Act, qualified meaning either a South African citizen or a citizen of a Berne Convention member country. Alternatively, should the author not be a qualified person, the work must have been first made or published in the Republic of South Africa or a Berne Convention member country.

The Copyright Act recognizes certain categories of works which may qualify for copyright protection. Most notably for the purpose of this discussion is the definition of an artistic work under the act. The Copyright Act defines an artistic work as:

> Irrespective of the artistic quality thereof –
>
> a) Paintings, sculptures, drawings, engravings and photographs;
> b) Works of architecture, being either buildings or models of buildings; or
> c) Works of craftsmanship not falling within either (a) or (b).

The terms used in this definition are, in turn, subject to their own definitions in the Act. Articles of manufacture and industrial products as well as prototypes of such goods may comprise 'drawings' and 'works of craftsmanship' under the Copyright Act. The Copyright Act specifies which acts may be done or authorized exclusively by the copyright holder in relation to each recognized species of work. The Copyright Act stipulates that the copyright in an artistic work vests in the holder thereof the exclusive right to do or to authorize the doing of any of the following acts in the Republic of South Africa:

(a) Reproduce the work in any manner or form.
(b) Publish the work.
(c) Include the work in a cinematograph film or a television broadcast.
(d) Cause a television or other programme, which includes the work, to be transmitted in a diffusion service, unless such service transmits a lawful television broadcast, including the work, and is operated by the original broadcaster.
(e) Make an adaptation of the work.
(f) Do in relation to an adaptation of the work, any of the acts specified in relation to the work in the paragraphs above.

Furthermore, the Copyright Act defines what constitutes a reproduction or an adaptation of an artistic work. The definition of reproduction is of particular importance in connection with the infringement of artistic works, particularly drawings of a technical nature. A reproduction of an artistic work includes a version produced by converting the work into a three-dimensional form, or if it is in three dimensions, by converting it to a two-dimensional form. An adaptation in relation to an artistic work includes a transformation of the work in such a manner that the original substantial features thereof remain recognizable. The definition of reproductions also, in relation to any work, includes a reproduction of a reproduction of that work. Thus it is possible to commit both direct and indirect copyright infringement.

Copyright infringement in relation to artistic works occurs, *inter alia*, when direct or indirect reproductions or adaptations are made without the permission or authorization of the copyright holder. Infringement may take place in relation to any substantial part of the work, and therefore the work as a whole need not be reproduced or adapted. In other words, making a copy of an industrial product which is derived either from an original drawing or an original prototype, can constitute infringement of the copyright in the underlying work if such copy is made without the consent of the copyright owner.

The nature of design protection as well as requirements for subsistence of design protection have been expanded upon above.

In 1983, an amendment to the Copyright Act brought about changes to the relationship between copyright and registered designs. The amendment allowed the so-called reverse engineering of industrial products but only ten years after the authorized derivative products of the artistic work had been placed on the market, or after 1993, whichever was later.

These provisions were amended in 1988 so as to render the reverse engineering exclusion operable immediately.

[2] Reverse Engineering

In its current format, the Copyright Act makes full provision for protection of industrially manufactured articles which are derived from artistic works. Full copyright protection is afforded to the work, including the rights to make reproductions of the work and the rights to restrain others from doing so. An exception lies in section 15(3A) of the Copyright Act, which deals with reverse engineering.

Copyright in an artistic work of which three-dimensional reproductions have been made available to the public, either in the Republic of South Africa or elsewhere, by or with the consent of the copyright holder (referred to as 'authorized reproductions') is not infringed if someone, without the consent of the copyright holder, makes three-dimensional reproductions or adaptations of the authorized reproductions, provided that the authorized reproductions primarily have a utilitarian purpose and are made by an industrial process. The copyright holder can still restrain the making of three-dimensional articles from the drawing itself or two-dimensional reproductions of it. In effect, all that is forfeited is the right to restrain the making of three-dimensional reproductions or adaptations by indirect copying from an authorized three-dimensional reproduction.

Whether or not an article has a primarily utilitarian purpose is determined through an objective test and the intention of the manufacturer of the product as to its intended use is irrelevant.

This section was introduced into the Copyright Act due to the large number of copyright infringement matters which were brought to restrain the reverse engineering of spare parts and other technical products. The objective of the section was to preclude copyright owners from effectively using their copyright as a form of unregistered design and thus interfering with the production and distribution of competing spare parts. To compensate for this removal of rights from the Copyright Act, the Designs Act makes provision for the registration of functional designs. The shapes of items of machinery, spare parts and the like were intended to be protected under the Designs Act while the shapes of non-utilitarian goods were to be protected under the Copyright Act.

[C] Relationship between Registered (Industrial) Designs and Unfair (unlawful) Competition

Liability on the basis of unfair or unlawful competition is delictual in nature and protection is based on the common law principles of the lex Aquilia derived from Roman law. The point of departure in South African law, is as established in the leading decision of *Schultz v. Butt*, and emphasized in *Aruba Construction (Pty) Ltd v. Aruba Holdings (Pty) Ltd*, remains that:

'...in the absence of statutory protection, copying by a person of a competitor's product does not *per se* constitute actionable unlawful competition in South African law.' To determine whether the conduct was unfair, regard must be had to the *boni mores* criterion and the general sense of justice of the community. Protection by way of a registered design remains the appropriate manner in which to protect the design of an article as unfair competition has a restricted scope and application. Unfair competition cannot be used as an alternative cause of action when intellectual property rights such as copyright, a trademark, a patent or a registered design have been infringed. Common law protection for the design of an article is unlikely to be available over and above these mechanisms unless there is some contributing illegality that moves it into the ambit of the law of delict (*Premier Hangers CC v Polyoak (Pty) Ltd* 1997 1 SA 416 (A)).

CHAPTER 18

Industrial Design Rights: Spain

Gonzalo Ulloa & Álvaro de Castro

§18.01 INTRODUCTION

[A] Highlights of New Developments in This Jurisdiction since Last Edition

Under the 2001 edition of this book, the Spanish Industrial Designs were ruled by the old Industrial Property Statute, dating back from 1929.

This statute has been replaced by more modern laws, each one relating to a specific industrial property right. A Patents Act was approved in 1986[1], and two successive Trademarks Acts were passed, the first one in 1988 and the second one in 2001.

As far as industrial designs are concerned, an Industrial Designs Act was approved in 2003, further developed by an Industrial Designs Regulation passed in 2004.

This new legal basis has provided substantial amendments to the former regulation contained in the Industrial Property Statute, such as: the suppression of the distinction between industrial models (three-dimensional designs) and industrial drawings (two-dimensional designs);[2] the suppression of the category of 'artistic models and drawings having an industrial application'; the introduction of a twelve-month grace period in which the applicant may disclose the design without losing its

1. This law has recently been replaced by a new *Patents Act* passed in July 2015 but not coming into effect until 1 Apr. 2017.
2. However, this purely terminological distinction persists in the Spanish version of the EU Regulation on Community Industrial Designs and in the Industrial Designs Directive, without any other repercussion, considering that both models and drawings (3D and 2D designs) are subject to the same legal regime.

novelty; the setting of the 'specialized circles in the European Union' as the relevant reference for assessing disclosure; the introduction of the concept of 'individual character'; the extension of the renewability of the designs up to twenty-five years instead of twenty years; etc.

§18.02 CONVENTIONS AND LEGISLATION

The Spanish legislation on industrial designs is influenced, not only by the EU legislation,[3] but also by the international treaties on this matter to which Spain is a signatory, namely:

- *Hague Agreement Concerning the International Registration of Industrial Designs,* of 6 November 1925. Spain adhered to the London Act of 1934, and ratified the Additional Act of Monaco of 1961 and the Geneva Act of 1999.[4]
- *Common Regulations Under the 1999 Act and the 1960 Act of the Hague Agreement* of 27 September 2011, as amended on 5 October 2011 and on 29 July 2014.
- *Locarno Agreement Establishing an International Classification for Industrial Designs,* signed at Locarno on 8 October 1968.
- *Convention establishing the World Intellectual Property Organization*, signed at Stockholm on 14 July 1967, and as amended on 28 September 1979.
- *Paris Convention for the Protection of Industrial Property*, of 20 March 1883, as revised at Brussels on 14 December 1900, at Washington on 2 June 1911, at The Hague on 6 November 1925, at London on 2 June 1934, at Lisbon on 31 October 1958, and at Stockholm on 14 July 1967, and as amended on 28 September 1979.
- *Agreement establishing the World Trade Organization,* made on Marrakech, on 15 April 1994 (including its *Annex 1c: Agreement on Trade-Related aspects of Intellectual Property rights).*

As mentioned above, the current regulation on Spanish industrial designs is contained primarily in *Law No. 20/2003 of 7 July 2003 on the Legal Protection of Industrial Designs.* As expressed in its Preamble, this statue was born with a double goal: incorporating the *Directive 98/71/EC of the European Parliament and of the Council of 13 October 1998 on the legal protection of designs* (hereinafter, the 'Industrial Designs Directive') into the Spanish legal system, and adapting the legal protection of industrial designs to the current needs.

This act was later amended by *Law No. 19/2006, of 5 June 2006 expanding the means for the enforcement of intellectual property rights and providing procedural*

3. See, specifically, Directive 98/71/EC, Regulation (EC) No. 6/2002 of the Council, Regulation (EC) No. 2245/2002 of the European Commission.
4. On 18 Sep. 2012, Spain notified acceptance of the termination of the London Act (1934) and the Additional Act of Monaco (1961).

provisions to facilitate the implementation of several Community Regulations.[5] Further, the General State Budget Acts modify periodically the Annex of the Law relating to the applicable taxes and fees in connection with industrial designs.

Unless otherwise stated, the terms 'Industrial Designs Act' will be used hereinafter to refer to the consolidated version of Law 20/2003 as currently in force.

The Industrial Designs Act is further developed in *Royal Decree No. 1937/2004 of 27 September 2004, approving the Regulation implementing Law No. 20/2003 on Legal Protection of Industrial Designs,* later amended by *Royal Decree No. 1431/2008 of 20 August 2008, amending several intellectual property regulations.*[6]

Unless otherwise stated, the terms 'Industrial Designs Regulation' will be used hereinafter to refer to the consolidated version of Royal Decree 1937/2004 as currently in force.

Finally, and although the present work is focused on the civil protection of industrial designs, Article 273 of the Criminal Code punishes with a sentence of imprisonment of six months to two years, and a fine from twelve to twenty-four months, whoever, for industrial or commercial purposes, without the consent of the holder of a design, and being aware of its registration, manufactures, imports, possesses, uses, offers or markets objects protected by such design.

§18.03 SUBJECT MATTER

[A] Definition/What Constitutes a Protectable Design?

According to Industrial Designs Act's Preamble, industrial designs are envisaged as a type of formal innovation referring to the appearance of the product per se or of its ornamentation.

An essential element for the protection of the design under the Industrial Designs Act is the principle of registration. The Industrial Designs Act does not confer protection to unregistered designs as industrial designs. However, unregistered designs may be protected in Spain under the national Copyright Law and Unfair Competition Law (provided that the relevant requisites of these other pieces of legislation are met), and also under the EU Regulation on Community Industrial Designs (Regulation [EC] 6/2002). In fact, the Spanish lawmaker understood (as is expressed indeed in the Preamble of the Industrial Designs Act) that it was superfluous to add a national protection to unregistered designs when there is an EU Regulation granting such protection over those designs which is directly enforceable by the Spanish Courts.

5. This amending law modified two provisions of Law 20/2003 relating to remedies and damages with the purpose of adapting the Spanish legislation to *Directive 2004/48/EC, of the European Parliament and of the Council, of 29 Apr. 2004, on the enforcement of intellectual property rights* (hereinafter, the 'Enforcement Directive').
6. The aim of this amendment was the suppression of the requisite to have a domicile in Spain in order to act before the Spanish Patent and Trademark Office. This requisite was declared by the European Commission incompatible with Art. 49 of the European Union Treaty, insofar as it created a restriction on the freedom to provide services within the European Union.

[1] Shape and Configuration

In accordance with the Industrial Designs Directive, Article 1 of the Industrial Designs Act defines '*design*' as the appearance of the whole or a part of a product resulting from the features of, in particular, the lines, contours, colours, shape, texture and/or materials of the product itself and/or its ornamentation.

'*Product*', in turn, is defined in the law as any industrial or handicraft item, including, *inter alia,* parts intended to be assembled into a complex product, packaging, get-up, graphic symbols and typographic typefaces, but excluding computer programs.

'*Complex product*' means, in the context of the law, a product which is composed of multiple components which can be replaced permitting disassembly and reassembly of the product.

[2] Ornamental or Functional

As the Industrial Designs Act's Preamble states, the legal protection afforded to industrial designs is based on the commercial value added by the design to the product, irrespective of its degree of originality, aesthetic merit or artistic value.

Both purely ornamental designs and functional designs are admitted to registration. What is banned from registration are those designs whose features are *exclusively* dictated by their technical function.

In terms of the Industrial Designs Act, the registration of a design shall not confer any right over the features of appearance of a product which are *solely* dictated by its technical function.

The word '*solely*' has been construed by Courts and scholars in two ways:

(a) Doctrine of the *designer's intent*: the design can subsist if it was conceived not only to serve a technical function, but also other (aesthetical) functions, provided that these are essential or relevant. This criterion was imported from the UK *Amp v. Utilux* case.
(b) Doctrine of the *multiplicity of forms:* the design can subsist if there are other alternative aesthetical forms which can perform the same technical function and achieve the same technical result. This is the criterion predominantly followed by the Spanish Community Design Courts (which consistently apply the EU Regulation on Community Designs, which is in many aspects identical to the Industrial Designs Directive, and thus to the Industrial Designs Act).[7]

7. Although the theory of the *multiplicity of forms* has been expressly rejected by the European Court of Justice ('ECJ') in the field of trademarks (see cases *Philips v. Remington* –case C-299/99- *and Lego* –case C-48/09-), supporters of this theory say that the conclusions of the ECJ in these cases are not applicable to design cases, and they precisely quote the following reflection made by the Advocate General Ruiz Jarabo in the *Philips v. Remington case* (emphasis added):

> 34. The wording used in the Designs Directive for expressing that ground for refusal does not entirely coincide with that used in the Trademarks Directive. That discrepancy is

More particularly, the registration of a design shall not confer any right over features of appearance of a product which must necessarily be reproduced in their exact form and dimensions in order to permit the product in which the design is incorporated or to which it is applied to be *mechanically connected* to or placed in, around or against another product so that either product may perform its function. This is informally known as the '*must-fit clause*', which excludes from registrability the designs of interconnections.

However, the rights arising from registration shall be recognized to designs serving the purpose of allowing the multiple assembly or connection of mutually interchangeable products within a *modular system*. This is informally known as the '*Lego clause*'.

[B] Originality/Novelty

[1] Test for Originality Generally

In order for a design to qualify for registration, it must be new and have individual character.

The *novelty* of a design requires that no identical design has been made available to the public before the date of filing of the application for registration of the design for which protection is claimed (or, if priority has been claimed, before the date of priority). In this regard, designs will be deemed to be identical if their features differ only in immaterial details.

---

not capricious. Whereas the former refuses to recognise external features 'which are *solely* dictated by its technical function', the latter excludes from its protection 'signs which consist exclusively of the shape of goods which is *necessary* to obtain a technical result. In other words, the level of 'functionality' must be greater in order to be able to assess the ground for refusal in the context of designs [...] *This means that a functional design may, none the less, be eligible for protection if it can be shown that the same technical function could be achieved by another different form.*

35. The Trademarks Directive excludes all shapes necessary (in the sense of *ideally suited*) to achieve a technical result. That is to say, in so far as the essential features of a shape are necessary in order to fulfil a function, trademark protection must not be granted without investigating whether that function could also be achieved by other features.

36. It is logical that the bar for assessing whether a ground for excluding a functional form applies is set higher for designs than for trademarks: the nature and scope of their protection are completely different from one another.

37. First, a trademark seeks to protect the identity of the origin of the goods and, therefore, indirectly, the goodwill which the goods attract, whereas designs - like patents - seek to protect the goods, in their own right, as an economic factor: their substantial value (in the case of designs) or the value which derives from their technical performance (in the case of patents). In that sense, it is entirely logical that the legislature is less concerned by the strict delimitation between designs and patents than by that which ought to exist between the latter and trademarks. Moreover, this makes it easier to give protection to designs that combine functional and aesthetic features.

38. Secondly, whereas trademarks enjoy protection unlimited in time, rights in designs - like rights in patents - are limited in time. From that viewpoint, too, it is appropriate to use a stricter test for excluding functional or ornamental shapes from registration as trademarks than that to be used in separating designs from patents.

As for *individual character,* a design will be considered to have individual character if the overall impression it produces on the informed user differs from the overall impression produced on such a user by any design which has been made available to the public before the date of filing the application (or, if priority has been claimed, before the date of priority). The comparison here must also be made individually between, on one hand, the overall impression given by the design, and, on the other hand, the overall impression given by each one of the prior designs existing in the design corpus and invoked in the case at hand. In assessing individual character, the degree of freedom of the designer in developing the design must be taken into consideration. The concepts of 'informed user', 'overall impression' and 'designer's freedom' are the same concepts which are used in the infringement test (see §18.05[B] below).

Further, a design applied to or incorporated in a product which constitutes a *component part of a complex product* will only be considered to be new and to have individual character:

(i) if the component part, once it has been incorporated into the complex product, remains visible during normal use of the latter; and
(ii) to the extent that those visible features of the component part fulfil in themselves the requirements as to novelty and individual character.

'Normal use' within this context means use by the end user, excluding maintenance, servicing or repair work.

The applicant does not need to demonstrate that his design is new and has individual character, nor will the office examine these circumstances ex officio, but only as a result of a post-grant opposition filed by a third party.

[2] Novelty

(i) Prior Use.
(ii) Prior Publication.
(iii) Prior Application or Registration.

For the purpose of assessing novelty (and also for assessing individual character), a design will be deemed to have been made *available to the public* if it has been published (following registration or otherwise), or exhibited, used in trade or otherwise disclosed. That being said, the Industrial Designs Act (following also the Industrial Designs Directive) lays down a number of exceptions where a design shall not be deemed to have been made *available to the public,* namely:

(a) where the disclosure could not reasonably have become known in the normal course of business to the *circles specialized* in the sector concerned, operating within the European Union;
(b) when the design has only been disclosed to a third person under *explicit or implicit conditions of confidentiality*;

(c) when the design has been made available to the public by the designer, his successor in title, or a third person (as a result of information provided or action taken by the designer or his successor in title) during the *twelve-month period preceding the date of filing* of the application; or
(d) if the design has been made available to the public as a consequence of an *abuse* in relation to the designer or his successor in title.

§18.04 DURATION OF PROTECTION

Registered designs have a five-year initial term of protection as from the date of filing of the application. The registration is renewable for one or more periods of five years each, up to a total term of twenty-five years from said date.

§18.05 INFRINGEMENT

[A] Scope of Infringement Action

According to the Industrial Designs Act (and consistently with the Industrial Designs Directive), the registration of a design confers on its holder the exclusive right to use it and to prevent any third party not having his consent from using it.

The aforementioned *use* covers, in particular, the making, offering, putting on the market, importing, exporting or using of a product in which the design is incorporated or to which it is applied, or stocking such a product for those purposes.

[B] Test for Infringement/Criteria to Constitute Infringement

The scope of the protection conferred by a design includes any design which does not produce on the *informed user* a different *overall impression*. In assessing the scope of protection, the *degree of freedom* of the designer in developing his design must be taken into consideration.

Following these parameters, the Spanish Courts have established (often following previous judgments from the European Court of Justice or from the EU General Court) some criteria of assessment, such as the following:

- The comparison must be made between the asserted design and the contested embodiment. Any difference or similarity between the defendant's product and the plaintiff's product as actually marketed are irrelevant for the assessment if the same similarities or differences do not exist with regards to the registered design, which is what defines the scope of protection.[8]

8. The comparison product versus product is relevant in the context of unfair competition, not in the context of infringement of intellectual property rights.

- The 'impression' to be considered is the general or global sensation or effect, as opposed to the mere comparison of individual elements.
- Such impression is to be considered bearing in mind that the users will not always have both designs at hand simultaneously, in order to be able to discriminate between one and the other; whenever possible, he would do so, but the usual situation is that the two are not available simultaneously and that the user has to rely on the imperfect image which is stored in his/her memory.
- A mere variation in size does not appear to imply a change which transcends the overall impression produced; the decisive factor for the informed user in this case appears, rather, to be the proportions between the different magnitudes, since it would otherwise be very easy to get around the protection afforded by registration.
- The 'informed user' is an indeterminate juridical concept which is defined as the person concerned who uses the product into which the design is incorporated in conformity with the purpose for which said product is intended. The informed user is thus a person possessing a certain level of knowledge of the designs published in the sector, but without being a design specialist. As a result, the informed user is an intermediate person who is neither an average consumer with no specific knowledge (a concept used in relation to trademarks), nor an expert in the field (a concept related to patents), nor a member of the 'sector's specialized circles' (a concept used in designs law, but which is closer to the expert in the field). It must be defined on a case-by-case basis depending on the sector and market segment involved.
- The informed user pays more attention to similarities between non-necessary elements and to differences between essential elements.
- The more the designer's freedom in developing the challenged design is restricted, the more likely minor differences between the designs at issue will be sufficient to produce a different overall impression on an informed user. And on the opposite, if the designer enjoys a high degree of freedom in developing a design, the designs which do not have significant differences will be more likely to produce the same overall impression on an informed user.

[C] Standing to Sue

Aside from the design owner, licensees also have standing to sue when their licence agreement so stipulates. Otherwise, licensees can only file a lawsuit in their own name with the express authorization of the design owner. However, an exclusive licensee is entitled to file an infringement lawsuit on his own behalf if he proves that he has formally requested the design owner to enforce his rights against the alleged infringer, and the design owner has refused to take any action or has not enforced his rights in the three months following such request. Even before these three months have elapsed, the licensee is entitled to ask for urgent precautionary measures in order to avoid an important damage.

Both the licensor and the licensee are obliged to notify each other about the legal action taken. The design owner is entitled to appear in the proceedings initiated by the licensee to defend his rights, and the licensee is entitled to appear in the proceedings initiated by the design owner to claim for an award of the damages caused to him.

[D] Remedies for Infringement

The holder of a registered design can exercise the civil or criminal actions available and can apply for any remedies which are necessary to safeguard his rights.

In civil proceedings, the remedies normally granted are those which (in a non-limiting manner) are provided in the Industrial Designs Act. These are the following:

- Injunctions (orders of cessation of infringing acts).
- Compensation for damages.
- Withdrawal of the infringing products from the market, at the infringer's expense.
- Seizure or destruction of the means principally aimed at performing the infringing activity, or assignment of the same to the claimant (valued at cost) as part of the payment of the damages, at the infringer's expense and at the claimant's choice.
- Destruction of the infringing products, or their assignment for humanitarian purposes, or to the claimant (valued at cost) as part of the payment of the damages, at the infringer's expense and at the claimant's choice.
- Publication of the judgment in the media, or its notification to interested persons.

The claimant is entitled to request one, several or all of these remedies, but these will be granted or not by the Court on a necessity and proportionality basis. For example, the Court is not expected to order the destruction of the goods if the infringing design can be easily removed from the good without damaging the product. Courts are not expected to grant punitive remedies, but just the necessary remedies to repair the damage caused and to avoid further damage in the future.[9]

The above remedies do not apply to goods which have been acquired in good faith for personal use.

The injunctions and preventive remedies may also be imposed to intermediaries (i.e., third parties rendering services used by the infringer to perform his infringing activities) even when these intermediaries are not liable for infringement (without prejudice to the specific provisions set forth in the Internet and e-commerce legislation). The remedies requested must be objective, proportionate and non-discriminatory.

9. However, if the defendant fails to comply with an injunction, coercive fines of a minimum of EUR 600 will be imposed for each day elapsed until the cessation of the infringing activity, without prejudice to the criminal consequences that this behaviour of Court disobedience could have in the most severe cases.

The compensation for damages can be imposed only to those who have been previously warned by means of a cease-and-desist letter, or where recklessness can be established. However, these requisites are not necessary in the case of manufacturers and first sellers of the infringing goods, who will be liable for the damages caused in any case.

The claimant can opt between two criteria of assessment of the damages:

(a) the negative economic consequences (which are normally quantified in either the infringer's profit or the claimant's lost profit plus, if applicable, other damages suffered such as the moral prejudice –e.g., harm caused to the design's reputation as a result of a faulty manufacturing of the infringing products, or of the conditions and channels under which such products are marketed);
(b) a reasonable royalty (i.e., the amount that the infringer would have reasonably had to pay to the design holder if it had acted with his consent).

The damages must be assessed by taking into account the economic importance of the design and its contribution to the demand of the product into which it is incorporated, its repute and the number and types of licences granted at the moment when the infringing activity commenced. The moral prejudice must be assessed by taking also into account the circumstances of the infringement, the importance of the harm and the spreading of the infringing products in the market.

The plaintiff may claim for the damages caused from the date of publication of the design. Additionally, he is entitled to claim for a reasonable compensation for infringement acts occurred between the filing of the design application and the publication of the design, provided that he had formally notified the existence and content of the application to the alleged infringer. Such provisional protection can only be requested once the design has been published, and provided that it has not been withdrawn, refused, cancelled or revoked.

In order to quantify the damages, the plaintiff can ask the Court to order the infringer to disclose the necessary documents (invoices, books and accounts, etc.). This does not mean a general discovery, but only a limited disclosure of the documents which are strictly necessary to quantify the damages and which cannot be legitimately obtained by the plaintiff otherwise. For confidentiality reasons, often this measure is replaced by the designation of a Court-appointed accounting expert, who will inspect the defendant's financial and accounting records and will issue a report to be explained at trial.

If, despite the above, the damages cannot be quantified, or in case they amount to less than 1% of the turnover made by the infringer with the sale of the infringing goods, the plaintiff can ask for such 1% as the minimum compensation for damages established in the law.

Finally, the plaintiff can also claim for the reimbursement of the investigation expenses (i.e., costs incurred in obtaining evidence of the infringement) and, on a different note, he can claim (up to certain limits) for the reimbursement of the legal costs incurred during the proceedings (attorney's fees, Court agent fee, expert fees, etc.).

§18.06 DEFENCES TO INFRINGEMENT

[A] Validity

The infringer may challenge the validity of the design via defence or counterclaim. If the nullity of the design is declared as a result of a defence, this declaration will only have effects *inter partes*, and will only entail the dismissal of the claim for infringement and the acquittal of the defendant, but not the cancellation of the design at the registry. Conversely, if the nullity is declared as a result of a counterclaim (or as a result of an independent nullity claim filed by the alleged infringer), the effects of this pronouncement will be *erga omnes* and will entail the revocation of the design and its cancellation from the registry. The counterclaim is considered an independent claim, and thus normally entails an additional/separate award of the legal costs against the losing party.

When the ground for nullity only affects part of the design, it is possible to declare null and void only that part of the design and maintain the rest, provided that such modification does not substantially alter the identity of the design and the remaining part still qualifies for registration.

The nullity declaration will be retroactive, and the design and the design application will be deemed to have never produced their effects. However, without prejudice to the compensation for damages that the design holder will have to pay if he acted in bad faith, the nullity of the design will not affect the Court decisions on design infringement which have become final and have been already enforced by the time the nullity declaration is issued, nor the contracts concluded before that date. Nonetheless, the Industrial Designs Act leaves the doors open to the licensee to be able to obtain a reimbursement of the amounts paid when this is justified from the circumstances for equity reasons.

The grounds for nullity are the same grounds as the grounds for refusal of the design during the registration proceedings.

[1] Not Novel/Original

See §18.03[B] above.

[2] Functional/Not Ornamental

See §18.03[A][2] above.

[3] Other Possible Criteria, such as Non-obviousness

Other additional grounds for invalidity are:

(a) The design does not meet the requisites to be considered a design under the legal definition provided in the Industrial Designs Act (see §18.03[A][1] above).
(b) The design does not qualify as protectable because it lacks one of the legal requisites (novelty, individual character, designs dictated exclusively by their technical function, designs contrary to the public order or the public morals).

Further, if the alleged infringer holds one of the followings rights, he can argue:

(c) The design holder is not the author or does not have entitlement to the design, and there is a final Court decision declaring so.
(d) The design conflicts with an earlier registered design which was applied for before, but published after, the filing/priority date of the later design.
(e) The design reproduces without authorization an official flag, badge or emblem.
(f) The design conflicts with an earlier trademark or distinctive sign.
(g) The design entails an unauthorized use of a copyrighted work.

[4] Faults in Application

Other possible faults in the application or in the registration procedure can only be invoked by challenging the administrative decision to grant the design through the administrative or contentious-administrative channels.

[B] Marking

An infringer cannot exempt himself from liability by pleading that the original products did not include the mention that they incorporate a protected design. However, if the infringer is not the manufacturer nor the introducer of the infringing goods in the market, he will not have to pay damages for acts performed prior to receiving a warning letter from the design holder making him aware of the infringement, unless negligence can be attributed to him (e.g., if it results obvious from the circumstances at hand that the infringer should be aware of the existence of the infringed design).

[C] Non-infringement

See §18.05 above.

[D] Other Defences

The Industrial Designs Act contemplates other defences on the merits (i.e., aside from the general defences based on procedural or formal faults) which may be invoked by the alleged infringer:

(a) Exceptions to the right conferred by the design:
  - Private use without commercial intent.
  - Experimental use.
  - Education use (i.e., acts of reproduction of the design for the purpose of making citations or for teaching, provided that such acts are not contrary to fair trade practice and do not unduly damage the normal exploitation of the design, and that mention is made of the relevant source).
  - The equipment and repair works on foreign ships and aircrafts, when these temporarily enter Spain, or the import of spare parts and accessories for the purpose of repairing such ships or aircrafts.

(b) Exhaustion of rights (the rights conferred by the registered design shall not extend to acts relating to a product incorporating a design included within the scope of protection of the registered design, when the product has been put on the market in the European Union by the holder of the design right or with his consent; in other words, the registered design cannot be enforced against resalers who acquired the product from authorized wholesalers).

(c) Prior honest use (use, or serious and effective preparations for the use, in good faith by a third party of a design falling within the scope of protection of the registered design, provided that said use had started before the filing date - or, when applicable, the priority date - of the registered design, and that the exploitation is made in the same way and with the same purpose as the one which took place before the filing/priority date of the registered design, with no possible licensing or separate assignment; besides, prior honest users are deemed to be 'authorized users' for the purpose of assessing the exhaustion of rights, so any subsequent resales of these products would also be excluded from the scope of protection of the later registered design).

(d) Limitations to the exercise of the right conferred by the design (the exploitation of the registered design can be limited by reasons of law, moral or public order, and use of the registered design cannot be used as a defence against the infringement of other prior intellectual property rights).

(e) Limitation to the exercise of actions (the design cannot be enforced against third parties making use of products introduced in the course of trade by persons or entities which have already compensated the design holder for the damages caused).

(f) Prescription of actions (the deadline to exercise the civil actions arising from the design is five years since the moment they could be exercised - i.e., normally, the moment when the design holder became aware, or should reasonably have become aware, of the infringement, although in cases of continuing acts it has been stated that the term does not commence until the cessation of the infringing acts takes place –; this makes it possible in many cases to file the lawsuit more than five years after the starting of the infringement; however, the compensation of damages has an additional limitation whereby only the damages arising from infringement acts taking place in the five years preceding the filing of the suit can be claimed).

(g) Special exception related to the use of design of components for repairing complex products in order to restore their original appearance (see §18.07[C] below).

§18.07 PROCEDURES FOR FILING APPLICATION FOR REGISTRATION

[A] Priority

Those who have regularly submitted a first application for design registration in or for any of the Member States of the Paris Convention or the Agreement establishing the World Trade Organization, or their successors in title, enjoy, for the filing in Spain of an application for registration of the same design, a right of priority of six months under the conditions laid down in Article 4 of the Paris Convention. The same deadline applies when a right of priority based on a previous application for a utility model is invoked.

Those who have submitted a first application for protection of the same design in a State or international organization other than those mentioned in the preceding paragraph enjoy the same right of priority when such State recognizes design registration applications filed in Spain a right of priority under equivalent conditions and effects to those provided under the Paris Convention.

An applicant who claims the priority of an earlier application must file a declaration of priority and a copy, certified by the office of origin, of the earlier application, accompanied by a translation into Spanish, when that application is drafted in another language.

The Industrial Designs Act also contemplates an *exhibition priority*, whereby the applicant for registration of a design who had exhibited products incorporating the design in an official or officially recognized exhibition may claim the priority date of the first presentation of the products in the exhibition, provided that the application for registration of the design is filed within six months from that date.

An applicant who wishes to claim the exhibition priority must justify that the products incorporating the design were presented at the exhibition on the alleged date.

The effects of the right of priority are that the priority date (i.e., filing date of the earlier application whose priority has been validly claimed) will be deemed to be the filing date of the later application for the purposes of assessing novelty, individual character, accessibility to the public, date of filing of unpublished designs, date of filing in cases of conflict of authorship, and prior honest use.

[B] Drawings

According to the Industrial Designs Regulation, the application for an industrial design before the Spanish Patent and Trademak Office must be accompanied by a graphic (or photographic) representation of the design, in colour or in black and white, which is suitable to be reproduced.

The representation must not include expressions or explanatory symbols except the indication 'top', if deemed necessary. The representation of each design can range from one to seven different views of the design requested.

[C] Design Part of Another Article/Environment of Design

As is explained in the Industrial Designs Act's Preamble, the adoption of the Industrial Designs Directive was possible only through a transitional solution of maintaining the 'status quo' on the contentious issue of using designs of components for repair of complex products to restore their original appearance. The compromise reached was the establishment of a transitional regime summarized in the formula 'freezing plus liberalization', which essentially requires Member States to maintain the legal regime applicable to the use of these components designs for repair purposes into force, and only authorizes changes aimed at liberalizing the market for such components.

The Industrial Designs Act, following the Industrial Designs Directive, admits the registration of these components as designs provided that the component, once incorporated into the complex product, remains visible during its normal use. The applicant has the option to register the design of the complex product as a whole, without prejudice to also being able to protect the design of the components separately, and such a request can be made in several applications or in a multiple application. However, until the Industrial Designs Act is modified to implement new EU provisions in this regard, the rights resulting from the designs of components cannot be exercised to prevent the use of the components design for repair purposes, when the component design needs to be reproduced in its exact form and dimensions to restore the original appearance of the complex product.

Unlike what happens with the interconnections (subject to the *must-fit clause*), the components exception is configured as a limit on the exercise of the right, and not as an exception to the registration, since it only applies to the use of these components for repair purposes, and not in other cases. But the foundation is the same: to prevent the creation of captive markets and to avoid that the *monopoly over the form* granted by the design becomes a *monopoly over the product* itself, which is what happens when there is no alternative to restore the original appearance of the complex product.

[D] Description of Design (Where Required)

The description of the design is not mandatory, but optional for the applicant. It must only refer to the features shown in the graphic representation, and cannot include statements regarding the alleged novelty, individual character or functionality of the design.

§18.08 EXPUNGING/CANCELLING OR VARYING REGISTRATION

The application for registration may be refused, or deemed to be withdrawn, if it has formal defaults which have not been corrected in time during the admissibility stage.

If the formal examination of the admissibility stage has been overcome (which does not entail any substantial examination of requisites such as novelty or individual character), the design will be granted and published.

Once published, the design can be cancelled or modified as a result of an opposition filed in the two months following its publication. The modification has retroactive effects as to the determination of the scope of protection of the design; it cannot substantially alter the identity of the design, and the design after the modification must keep on fulfilling the necessary requisites for registration. The cancellation of a design as a result of an opposition has the same retroactive effects as the declaration of nullity (see §18.06[A] above).

In addition to the above, the applicant may at any time withdraw his application, renounce some of the designs applied for, or limit the goods specified in the application. He can also introduce corrections to the name and address, defects of expression or transcription, or evident mistakes, provided that the amendment does not affect the identity of the design or replace or expand the products listed in the application.

The applicant or holder of a multiple registration may also request the division of the application or registration in two or more divisional applications or registrations, by splitting from the original file one or more designs in order to overcome certain objections during the registration proceedings, or when a change in ownership affects part of the design of an application or multiple registration. Divisional applications or registrations retain the filing date of the initial application and the benefit of the right of priority, if any.

Aside from the administrative stage, the cancellation of the design can be sought by interested third parties at Court, by appealing the decisions rendered by the Spanish Patent and Trademark Office before the contentious-administrative Courts, or in the context of invalidity/infringement proceedings substantiated in Civil Courts. However, it is not possible to invoke before a Civil Court the same grounds of invalidity which have already been resolved on the merits in a contentious-administrative judgment issued between the same parties and over the same facts invoked as grounds for invalidity.

Finally, the design may lapse due to:

(a) non-renewal;
(b) waiver of rights by the design holder (which will not be possible if it is made without the authorization of the holders of rights, liens or encumbrances over the design, or when the entitlement to the design is *sub judice* and the claimant does not give his consent to the waiver); or
(c) lack of legal standing to be a design holder (for not having a serious and effective establishment in Spain, or in another country member of the Paris Convention or the World Trade Organization, or in any country with reciprocal conditions regarding standing to own design registrations).

The first two cases are to be declared by the Spanish Patent and Trademark Office, and the third case is within the competence of the Court. The lapse of the design shall be retroactive with the same effect as the nullity (regarding past contracts and lawsuits) but the effect of the retroactivity will only be backdated to the date when the cause originating the lapse took place.

§18.09 OTHER

[A] Interface between Industrial Design and Trademarks

Industrial designs and trademarks play a different role in the world of industrial property. Whilst trademarks are primarily aimed at ensuring the correct identification of the origin of the products placed in the market, industrial designs are primarily aimed at protecting a formal innovation consisting of the appearance of the product or its ornamentation. Therefore, both institutions are *a priori* fully compatible and complementary to each other.

This does not mean that in some circumstances there may be a conflict between trademarks and designs. In fact, a trend nowadays is for companies to actually register as trademarks the design or trade-dress of their products, due to the advantages that the trademark protection confers in terms of duration, *inter alia* (trademarks can *a priori* be renewed unlimited times, while industrial designs have a maximum life of twenty-five years)[10]. These figurative (device) trademarks and tridimensional trademarks may destroy the novelty or individual character of future design registrations, and may prevent future design registrations incorporating such marks from being registered, as the Industrial Designs Act clearly states. And likewise, the Trademarks Act entitles the holder of an industrial design to prevent the granting of a trademark reproducing or imitating such industrial design.

[B] Relationship between Industrial Design and Copyright/Design Copyright

The legal protection afforded to industrial designs is based on the commercial value added by the design to the product, irrespective of its degree of aesthetic value, artistic merit or originality. If the latter concurs, then the design may also be protected simultaneously by Copyright Law by the mere fact of its creation and its fixation in any form, and without need of registration. The protection granted by the Industrial

10. In fact, it is not unreasonable to consider that, when a design has survived more than twenty-five years in the market, it has actually become a distinctive sign, a 'sign of identity' of the company holding the rights over such a design. In parallel, the traditional functions of the trademarks are broadening to incorporate other additional functions (such as the investment function or the advertising function) as results from the doctrine of the European Court of Justice. All these circumstances contribute to diluting the frontier between the suitability of a drawing or model to be registered as a trademark or as an industrial design.

Designs Act is independent, supplementary and compatible with the protection granted by the Copyright Act.

Similarly to what has just been said in the previous section in regards to trademarks, this compatibility does not avoid the existence of conflicts when different authors claim different rights (industrial design rights/copyrights) over the same design. In principle, the earlier right should be the one which would prevail in such a conflict, considering that an earlier copyright constitutes a ban on registration of a later industrial design according to the Industrial Designs Act, and that on the other hand, the existence of an earlier industrial design seems to exclude that a later identical work can be considered 'original' under Copyright Law (unless by adopting a wide, *subjective* construction of the word 'original', which is not the predominant one among Courts, and which would require to establish that the author of the later work created it by himself independently and could not be aware of the pre-existing design prior to the creation of such work – in such a case, it might be arguable, though it would be subject to controversy, to defend a coexistence between both intellectual property rights)[11].

[C] Relationship between Industrial Design and Unfair Competition

Courts are very reluctant to examine under unfair competition law acts which qualify to be examined under industrial property law. This is due to the so-called *doctrine of the concentric circles* (Bercovitz[12]), which envisages the scope of industrial property law as an inner circle between the broader circle which delimitates the protection granted by unfair competition law, in such a way that when both circles overlap, the industrial property must prevail, and the unfair competition is left to repress anticompetitive practices where there is no intellectual property right preventing such act.

This doctrine, supplemented and nuanced by another, more recent construction known as the *relative complementarity principle* (Massaguer),[13] implies that the unfair competition law cannot be used to punish acts which fall within the scope of the industrial property law and are considered licit under such branch of law. If, for instance, a challenged design is considered not to produce the same overall impression in the informed user as a prior registered design owned by a third party, then this third party cannot claim that the later design imitates his prior design right and therefore incurs in an unfair competition act of imitation. And on the opposite, if the defendant's product is deemed to infringe the claimant's registered design, then the claimant will be

11. This does not mean, however, that the later 'independent' author is entitled to register his work as an industrial design as well. The Industrial Designs Act provides that if the same design has been created by different persons independently, the right to register the design belongs to the person whose application for registration has an earlier filing date in Spain, provided that such registration is eventually granted. However, some recent Supreme Court judgments declared that, in order for an industrial design to be protectable by the copyright laws, it must be original *in addition to being novel and having individual character;* this may be interpreted as reinforcing the idea that an objective novelty must exist in any case for an industrial design to be protected by copyright laws.
12. *La formación del Derecho de la Competencia*, Actas de Derecho Industrial, 1975.
13. *Comentario a la Ley de Competencia Desleal*, Civitas, 1999.

able to obtain the remedies sought under the Industrial Designs Act, but will not be able to claim, for instance, an additional compensation for damages under the Unfair Competition Act.

A different thing is that the unfair competition is attributed to a different, independent act (though related to the act examined by the industrial design law) or to a different anticompetitive effect arising from a different aspect or dimension of the conduct. For instance, it may be concluded that the contested product which apparently infringes a registered design does not fall within the scope of protection of such design, and that, as a result, its marketing is licit; however, if the rest of circumstances surrounding the marketing of the product (i.e., packaging, accessories, advertising, etc.) are a copy of the competitor's ones, it might then be established that such presentation of the product in the market constitutes an act of unfair competition, provided that the remaining requisites of the Unfair Competition Act are met in the case at hand.

CHAPTER 19

Industrial Design Rights: Taiwan

Yu-Jiunn Kao & Frank Liu

§19.01 INTRODUCTION

[A] Highlights of New Developments in This Jurisdiction since Last Edition

In the past years, the Intellectual Property Laws in Taiwan have undergone several revisions. All the revisions may well demonstrate the trend that Taiwan is improving its protection of IP rights.

There is not a separate Industrial Design Protection Law in Taiwan. While Patent Law is the primary legislation protecting designs, Copyright Law and Fair Trade Law also contain relevant provisions for the protection of industrial designs. In this chapter, the principle focus will be placed on Patent Law with a secondary emphasis on Copyright Law and Fair Trade Law.

As a result of the most recent extensive amendments to the Patent Law implemented since 1 January 2013, partial designs, computer-generated icons, graphic user interfaces (GUIs), and sets of articles are able to receive design patent protection. Moreover, the associated design system has been abolished and replaced by the derivative design system. According to this new system, the filing date of a derivative design application shall not be earlier than that of the original design application, and no derivative design application can be filed after the publication of the original design application. Unlike an associated design patent which is without a separate Letters Patent and the patent right of which expires or extinguishes with its parent design patent, a derivative design patent has its own Letters Patent and independent patent right, which also comes with a necessity to pay annuities therefore.

§19.02 CONVENTION AND LEGISLATION

Taiwan is neither a member of the Paris Convention nor a member of the Patent Cooperation Treaty (PCT).[1] However, the Patent Law in Taiwan is generally in harmony with the spirit of the laws prevalent in other leading countries.

§19.03 SUBJECT MATTER

[A] Definition/What Constitutes a Protectable Design?

'Designs' as referred to in the Patent Law means any novel design created with respect to the shape, pattern or colour of a portion of an article of manufacture, or the entire article, or any combination thereof, thereby creating eye-appealing aesthetic effect.

A new design application can only be filed to seek protection of the aesthetic outer appearance of an article that is capable of industrial application. In other words, the mechanical function and construction of an article of manufacture, as well as the operation processes involved in the use of the article, are excluded from design patent protection. When filing a design application, an applicant must specify the article to which the design applies.

In general, 'an entire design of the appearance of an article of manufacture', 'a partial design of the appearance of an article of manufacture', and 'a computer generated icon and graphic user interfaces applied to an article UI design applied to an article of manufacture' are statutory subject matters. However, exceptions to design patent protection include: (a) the configuration of an article of manufacture which is purely functional; (b) a purely artistic creation; (c) an integrated circuit layout or an electronic circuit layout; and (d) an article, the use of which is detrimental to public order or morality.

In view of the strict unitary-of-design requirement, a design application can only cover one embodiment; however, when two or more articles belong to the same class and are customarily sold or used as a unit, filing one single design application is permitted.

Where a person owns two or more similar designs, he/she may choose to file a design application and derivative application(s).

[1] Shape and Configuration

'Design patents' offers protection to any design created with respect to the 'shape', 'pattern' or 'colour' of a portion of an article of manufacture, or the entire article, or any combination thereof, thereby creating eye-appealing aesthetic effect. On this note, patentable designs may be two-dimensional or three-dimensional.

'Shape' refers to the form of an article or its external boundary, outline, or external surface in three-dimensional space. 'Pattern' is an arrangement of dots, lines,

1. See http://www.wipo.int/export/sites/www/pct/en/texts/pdf/pct_paris_wto.pdf.

planes, or colours that are presented in a recurring manner. 'Colour' is a visual perceptual property that derives from the spectrum of light interacting with human eyes. A shape, pattern or colour of an article of manufacture or any combination thereof can be a proper statutory subject matter.

Computer-generated icons and graphical user interfaces (GUIs) used in computer and electronic device environments are statutory subject matters since they are the combination of patterns and/or colours to form a virtual presentation that allows users to interact with electronic devices.

[2] Ornamental or Functional

A design patent in Taiwan protects only the visually appreciable appearance of an article of manufacture. The mechanical function and construction of an article of manufacture, as well as the operation processes involved in the use of the article are excluded from design patent protection.

[B] Originality/Novelty

[1] Test for Originality Generally

'Test for originality' is applied in determining the patentability of a design. A design to be patentable must be 'original'. In other words, a design that simulates a known object is not original as required by the Patent Law.

[2] Novelty

Taiwan adopts a standard of absolute novelty. Any public use or publication of a design before the filing date of the design in any place around the world would defeat the novelty of the design. Moreover, in view of the 'first to file' principle, where two or more design applications are filed for the same design, only the applicant whose application is filed first shall be granted a design patent.

As the objective of a design patent is to protect only the visual aspect of a design, consideration of the similarity of a design to a prior art shall rest on naked eyes without the aid of any instrument. In general, two prongs of test will be addressed in determining whether a design is novel, namely: (1) the degree of similarity between the applied-to articles; and (2) the degree of similarity between the designs per se.

A comparison based on these tests could produce the following results:

(a) Where the applied-to articles are similar and the designs are identical, the later filed design shall be devoid of novelty.
(b) Where the applied-to articles and designs are both similar, the later filed design is vulnerable to rejection for want of novelty, provided that the

differences between the two designs are readily perceivable by an ordinary observer.

(c) Where the applied-to articles are the same and the designs are similar, the later filed design shall be devoid of novelty, provided that the differences between the two designs are readily perceivable by an ordinary observer.

[a] *Prior Use*

As provided in the Patent Law and the Patent Examination Standards and Guidelines, where a design is 'readily accessible' to the public, i.e., nonspecific persons, the design shall be deemed to fall within the definition of 'prior use'. In addition, 'prior use' refers to the making, offering for sale, selling, or using an article to which a design is applied, or the importing of an article to which a design applied for such purposes prior to the filing date of the design, and does not necessarily mean that the public has already practiced the design or had good knowledge of the design.

[b] *Prior Publication*

'Publication' referred to in the Patent Law means any document or data storage medium carrying information that is open to the public in any language and in any place around the world, and is not limited to one that is printed on paper. It also refers to a medium that renders a design readily accessible to the public by means of texts, figures, or any other data carrying media, through transcribing, photographing, or replicating, or disseminating online.

[c] *Prior Application or Registration*

A prior design application published before the filing of a design application is definitely a valid prior art. Article 123 of the Patent Law further provides that where the design in a design application is identical with or similar to that disclosed in a prior design application published after the filing of the former, a design patent shall not be granted. This provision does not apply, however, where the later filed design application and the earlier filed design application were filed by one and the same person.

[C] Unobviousness (if Appropriate)

A design is devoid of unobviousness if it is readily perceivable by an ordinary observer based upon a prior art reference in combination with other references, in part or in whole.

§19.04 DURATION OF PROTECTION

A design patent becomes effective as of the publication date and is granted for a term of twelve years from the filing date. A derivative design patent expires simultaneously with its parent design patent.

§19.05 INFRINGEMENT

[A] Scope of Infringement Action

Article 136.1 of the Patent Law mandates that '[u]nless otherwise prescribed in this Law, a design patent owner shall have the exclusive right to exclude others from practicing the patented design or a similar design, without his/her consent'. According to Article 58.2 of the same law applying *mutatis mutandis* to design patents, 'practice of a patented design or a similar design' refers to the making, offering for sale, selling, or using a design or a designed article that is similar to the patented design, or the importing of a design or a designed article that is similar to the patented design for such purposes. Moreover, the right of a derivative design patent may be enforced independently of its parent design patent, and the right can be enforced against similar designs.

[B] Test for Infringement/Criteria to Constitute Infringement

The Taiwan Intellectual Property Office (TIPO) has published 'Patent Infringement Test Guidelines' which discloses, *inter alia*, that in the test for design patent infringement, the scope of the allegedly infringed design patent needs to be interpreted and ascertained before a comparison between the two designs concerned is conducted to determine whether they are similar.

A design patent aims at protecting the shape, pattern or colour(s) of an article of manufacture, or any combination thereof. Therefore, in the process of interpreting the scope of a design patent, the drawings, photos or colour charts of record shall all be consulted with reference to the state of art, such as the drawings of the relevant prior art, reference books, dictionaries, publications, and the like. In other words, where a specific portion of the design embodied in a design patent falls into the status of public domain, it shall be excluded from the scope of the design patent and should not be given much weight in interpreting the scope of the design patent. Moreover, the doctrine of 'file wrapper estoppels' may be applied in interpreting the scope of a design patent.

In the similarity test, the two designs shall be viewed side-by-side, as well as apart in the eyes of an ordinary observer to determine whether they have common or similar ornamental features and whether they are identical or similar in overall outer appearance. The designs cannot be segregated into individual elements for comparison in a piecemeal manner, and the functional features dictated by the intended use should be excluded from consideration. In case of an article having a visible internal structure, the visual effect achieved via light reflection or deflection shall be taken into account.

[C] Standing to Sue

A design patent owner is entitled to enforce his/her right against an infringer and claim damages for his/her loss of business reputation as a result of infringement. An exclusive licensee may make substantially the same claims within the scope of the licence, unless otherwise provided in the licence contract.

[D] Remedies for Infringement

Remedies available to a design patent owner are: (1) injunctive reliefs; (2) damages; (3) destroying infringing products and the materials and tools for making the infringing products; and (4) necessary measures to recover the design patent owner's goodwill.

A design patent owner could select one of the following methods to calculate damages: (1) the design patent owner's lost profit as a result of the infringement; (2) the profit gained by the infringer as a result of his/her infringement; or (3) an amount equivalent to the compensation that can otherwise be collected by licensing to others the right to practice the patented design.

Where an infringement is found to be intentional, the Court may, upon request and depending on the level of severity of infringement, increase the award of damages up to an amount not to exceed three times the proven loss.

Further, where a design patent owner suspects infringement of his/her design patent right by imported goods, he/she may request Customs to detain the suspected goods with the fact of infringement explained and a bond equivalent to the customs value of the imported goods assessed by Customs or an equivalent security deposited.

§19.06 DEFENCES TO INFRINGEMENT

[A] Validity

An alleged infringer may assert design patent invalidity as a defence to infringement, and the court is empowered by law to rule on the validity of the design patent at issue in determining infringement or non-infringement. However, since only the TIPO has the right to revoke a design patent, the alleged infringer needs to file an invalidation action before the TIPO with submission of evidence and grounds in order to cancel the design patent, if intended.

[1] Not Novel/Original

A design to be patentable must be 'original'. In other words, a design that copies a known object is not original as required by the Patent Law. Basically, a design shall be deemed to be novel unless:

(1) prior to its filing date, an identical or similar design has appeared in any printed publication;

(2) an identical or similar design has been made known to the public; or
(3) the design had been publicly known prior to its filing.

[2] Functional/Not Ornamental

A design for an article of manufacture that is dictated primarily by the function of the article lacks ornamentality and is not a proper statutory subject matter.

[3] Other Possible Criteria

A design for an article of manufacture is not patentable if the article is not capable of industrial application.

The following types of design are all unpatentable subject matters: (1) a purely artistic creation; (2) an integrated circuit layout or an electronic circuit layout; and (3) an article, the use of which is detrimental to public order or morality.

Because design patents protect only the appearance of an article of manufacture, slight differences between similar designs can render each design to be patentable. Accordingly, the protection afforded by design patents in Taiwan may be limited.

[4] Faults in Application

According to the Patent Law, any of the following procedural faults could render a design patent to be invalid:

(a) the design patent owner is not entitled to file the design application concerned;
(b) where the right to file a design application is jointly owned, the application is not filed by all of the owners jointly; and
(c) a derivative design application is filed after the publication of the parent design patent.

[B] Marking

A design patent owner shall mark the design patent number on all his/her patented articles; if it is unable to so mark the article, the owner may mark the labels, packaging, or other areas noticeable and obvious to an observer, according to the Patent Law. In the absence of such markings, the owner when claiming damages shall provide evidence proving that the alleged infringer had knowledge or had access to knowledge of the patented articles.

[C] Non-infringement

An alleged infringer may argue that his/her product is dissimilar to either the design embodied in the design patent or the article to which the design is applied. To support such contention, the alleged infringer may submit a non-infringement analysis report prepared by a competent patent attorney. Moreover, the infringer may consider whether the doctrine of 'file wrapper estoppels' may be applied or whether the design embodied in the design patent concerned should be excluded from protection for falling within the scope of public domain.

[D] Other Defences

In addition to non-infringement or invalidity, the alleged infringer may raise the following defences: (1) no damages should be awarded due to the design patent owner's non-compliance with the marking requirements; (2) no damages should be awarded because the alleged infringer lacks the subjective intention or negligence on which an award of damages must be based; (3) the design patent has been exhausted; (4) the design patent owner is an exclusive licensee not recorded with the TIPO; and (5) the design patent owner's claim is time-barred.

As the Patent Law allows in some instances for fair or bona fide use of others' intellectual property rights, when being accused of design patent infringement, the defendant may argue that the practice of the design patent is non-public or for non-commercial purposes or the practice is necessary for research or experimental purposes.

§19.07 PROCEDURES FOR FILING APPLICATION FOR REGISTRATION

An entitled Applicant of a design application may be a designer of the design, his/her assignee or heir, an employer of the designer, or a party responsible for the research and development costs of the design. The documents and information required for filing a design application in Taiwan are as follows:

(1) the title of the claimed design which, basically, need be named based upon the Locarno International Classification System or the commonly used terms in the related art;
(2) drawings or photos illustrating the six views (i.e., front view, left side view, right side view, bottom view, top view and rear view) and the perspective view of the claimed design;
(3) a Power of Attorney duly signed by the applicant; and
(4) the priority document, if applicable.

Assignment is not a prerequisite document for filing purposes under the current practice. An applicant having no domicile or place of business in Taiwan shall appoint a registered agent to act on his/her behalf in connection with design application and design-related matters.

According to the Patent Law, a design patent shall not be granted to a design that has appeared in any printed publication, been publicly practicing prior to filing, or was publicly known prior to filing. These limitations, however, shall not apply if the publication, public use, or disclosure of a design was due to the following reasons and a patent application has been filed within six months from the date of publication, public use or disclosure:

(1) the design was disclosed in a printed publication;
(2) the design was displayed at an exhibition sponsored or approved by the government; or
(3) the disclosure was against the applicant's intent.

According to an official announcement, the 'exhibition sponsored or approved by the government' referred to in the Patent Law, with reference to the stipulations of the Paris Convention, should be an international exhibition sponsored or approved by the government having jurisdictions over the place where the exhibition is held.

[A] Priority

Any foreign applicant, whose country of origin is a World Trade Organization (WTO) member or who has a domicile or a place of business within the territory of a WTO member country, is entitled to claim priority in Taiwan based on his/her design application first filed in any member country of the WTO within six months of the filing date in Taiwan.

[B] Drawings

To file a design application for a three-dimensional design, the applicant needs to submit drawings or photos illustrating the six views (i.e., front view, left side view, right side view, bottom view, top view and rear view) and the perspective view of the design. To file a design application for a two-dimensional design, drawings showing the front view and the recurring unit of the design (if applicable) are needed. The applicant may additionally submit auxiliary figures, if necessary.

The drawings shall be ink drawings composed of solid black lines drawn in accordance with industrial drafting method, or photographs or computer printouts clearly showing the claimed design. When the drawings are reduced in size to two-thirds, they shall clearly demonstrate details of the design. Where colours are claimed, the colours should be clearly illustrated in the drawings.

Each figure in the drawings shall be annotated with a title. The perspective view or a figure that can best characterize the claimed design shall be designated as the representative figure.

[C] Design Part of Another Article/Environment of Design

The current Patent Law permits the filing of 'partial design' applications to protect the partial appearance of an article of manufacture. If the applicant wishes to seek protection for a portion of a design: (i) the portion of the design to be claimed shall be clearly stated in the description accompanying the drawings; and (ii) 'the claimed design portion' and 'the non-claimed design portions' shall be distinguishable in the perspective view and six side views. In this regard, solid lines can be used to illustrate 'the claimed design portion', and dashed lines, grey scale colouring, or semi-transparent colouring can be used to illustrate 'the non-claimed design portion'.

[D] Description of Design

The description of a design is required only upon request. It shall contain: (1) the intended use of the applied-to article; and (2) the ornamental features of the claimed design.

In case of any of the following circumstances, the applicant would be required to explain or state as such in the description:

(1) the drawings contain non-claimed design portions;
(2) the change sequence of the claimed design, the visual effect of which contains continuous dynamic changes; or
(3) any omission of the required views for being identical or symmetrical to other views or for other reasons.

In case of any of the following circumstances, the applicant should briefly state as such in the description where necessary:

(1) the characteristics of the materials used and variations in the spatial arrangement or in the state of use have resulted in any change of the appearance of the claimed design;
(2) any auxiliary figures or referential figures; or
(3) the title/name of each article when the claimed design is applied to multiple articles of the same category and customarily sold as a set.

§19.08 EXPUNGING/CANCELLING OR VARYING REGISTRATION

After a design application matures into a design patent, any party may file an invalidation action there against, except that only an interested party may file an invalidation action against a design patent based on the reasons that the design patent owner is other than the person entitled to file the design application.

Invalidation proceedings in Taiwan are conducted on a pleading-and-defence basis under which the two parties are allowed to alternately present their contentions in written form and the TIPO, upon receipt of the brief/counter-statement lodged by either party, will serve a copy on the other party along with a notification for filing a response and/or amendment (the amendment shall not go beyond the original disclosure) before a designated deadline. The invalidation procedure will continue until the two parties have exhausted their views and have no new observations to file. The TIPO will then, based upon all the documents/materials on file, render a decision.

After a design application matures into a design patent, the design patent owner may apply for amendment/correction of the drawings and/or description only for the following purposes:

(1) to correct clerical errors; or
(2) to clarify uncertain statements.

The amendment/correction shall not substantially alter the original scope of protection.

The TIPO shall publish the amendment/correction after approval, and the amendment/correction, once published, shall be retroactively effective from the filing date.

§19.09 OTHER

[A] Interface between Industrial Design and Trademarks

'Design patents' can offer protection to any design created with respect to the 'shape', 'pattern' or 'colour' of a portion of an article of manufacture, or the entire article, or any combination thereof, thereby creating eye-appealing aesthetic effect.

Trademark Law can offer protection to any distinctive indication composed of a word, character, device, symbol, colour, three-dimensional shape, motion, hologram, sound, or any combination thereof, which is capable of enabling the customers of related goods/services to identify the origin of the applicant's goods/services, thereby distinguishing such goods/services from those of others.

Therefore, when a design, either two-dimensional or three-dimensional, serves the trademark function of identifying the source of origin, a trademark application can be filed therefor to seek protection. However, when a device, symbol, colour, three-dimensional shape, hologram, or any combination thereof meets the patentability requirements of a design patent, both trademark and design protection can be sought.

[B] Relationship between Industrial Design and Copyright/Design Copyright

Copyright Law in Taiwan offers protection to artistic works, including painting, plate painting, caricature, comic strip (cartoon), sketch, masterpiece of calligraphy (calligraphy), letter form drawing (typeface), sculpture, craftwork, and pictorial and graphical works, such as any map, chart, scientific or engineering design drawing.

Where a novel design created with respect to the shape, pattern or colour of an article of manufacture, or the combination thereof falls within the definition of 'artistic work' or 'pictorial and graphic work' referred to in Copyright Law, it may be protected under the Copyright Law upon completion.

The duration of copyright is the life of the author plus fifty years or seventy years.

[C] Relationship between Industrial Design and Unfair Competition

Fair Trade Law in Taiwan is enacted to maintain trading order, protect consumers' interests, ensure fair competition, and promote economic stability and prosperity and to prohibit unfair competition acts, including passing-off, i.e., manufacturing and selling of counterfeits.

According to Fair Trade Law, no one is allowed to use in the same or similar manner, the appearance of another's article of manufacture, or any other symbol that represents such person's article commonly known to the relevant consumers. On this score, a design patent owner may additionally resort to Fair Trade Law for protection of his/her design if the design has become well-known. In case of a design whose popularity has not yet been established, the design owner may, based on Fair Trade Law, claim that an alleged infringer has plagiarized his/her design and such deceptive or obviously unfair act would adversely affect market order.

When wilful infringement is found, a design patent owner resorting to Fair Trade Law is entitled to request the Court to increase, within its discretion, total damages awarded up to triple the original amount.

CHAPTER 20

Industrial Design Rights: United Kingdom

David Charles Musker

§20.01 INTRODUCTION

[A] Highlights of New Developments in This Jurisdiction since Last Edition

The most significant changes in UK design legislation have been the amendment of the (already much-amended) Registered Designs Act 1949[1] (the 'RDA') to effect harmonization of substantive design law to match the Community Design Directive ('CDD'),[2] and the entry into effect of Council Regulation (EC) No. 6/2002 on the Community Design (Community Design Regulation, 'CDR'),[3] in 2001 and 2002 respectively.

The validity of the remaining UK designs filed before December 2001 (distinguishable by beginning with 1- or 2-), continues to be governed by the previous UK law, for which the reader should consult the last edition of this work.

Other significant changes include:

- The introduction of multiple design applications.
- The introduction of criminal offences of registered design infringement.
- The forthcoming removal of the twenty-five-year cap on the lifetime of design copyright.

1. 1949 c. 88.
2. Directive 98/71/EC of the European Parliament and of the Council of 13 Oct. 1998 on the legal protection of designs, OJ L 289/28, 28/10/1998.
3. Council Regulation (EC) No. 6/2002 of 12 Dec. 2001 on Community designs OJ L 003/1, 5/1/2002, entering into force on 6 Mar. 2002, as amended by as amended by Council Regulation No. 1891/2006 of 18 Dec. 2006 amending Regulations (EC) No. 6/2002 and (EC) No. 40/94 to give effect to the Accession of the European Community to the Geneva Act of the Hague Agreement concerning the international registration of industrial designs, OJ L 386/14 29/12/2006.

- The alteration of the appeal routes from decisions of the Intellectual Property Office (the 'IPO'), who register UK designs.
- The forthcoming introduction of an official Design Opinion Service.
- Facilitation of accession to the Hague Agreement Concerning the International Registration of Industrial Designs[4] (the 'Hague Agreement').

Finally, in relation to enforcement, the entire structure of civil litigation in England and Wales was reformed in 1999 to implement Lord Woolf's reforms. It now operates under the Civil Procedure Rules[5] (the 'CPR') with the possibility of litigating low-value cases in the Intellectual Property Enterprise Court (the 'IPEC'), formerly the Patents County Court, which also has a Small Claims Track. Scotland and Northern Ireland retain their independent legal systems which are relevant to enforcement of designs.

[B] Overview

One of the striking features of UK design law is the plethora of available rights. Setting aside trademarks and unfair competition (the overlaps with which are considered below) and patents, the set of design rights to be considered in the UK consists of:

(i) UK national registered designs.
(ii) International designs designating the UK (coming in the near future).
(iii) Registered Community Designs ('RCD's).
(iv) International designs designating the EU.
(v) Unregistered Community Designs ('UCD's).
(vi) UK unregistered Design Right ('UK UDR').
(vii) UK copyright.

The infringement and validity criteria for the first five of these are all essentially the same, as they are harmonized by EU law. The last two are considerably less similar to the first five, but are related to each other since UK unregistered design right is derived from, and to some extent complementary to, UK domestic copyright.

[1] Qualification

UK and Community Registered designs are open to any person regardless of nationality. Since most countries are members of a relevant copyright treaty, entitlement to copyright protection in the UK is also in practice unlimited by national qualification concerns.

4. The Hague Agreement Concerning the International Registration of Industrial Designs, Geneva Act (1999) entering into force on 23 Dec. 2003.
5. Originally The Civil Procedure Rules 1998, but for the current version of these everchanging rules, consult the Ministry of Justice website.

However, the International Design route is available only to persons with an entitlement arising from a connection with a Hague Agreement member country (by nationality, residence or domicile for example).

The qualifications for UK UDR are considerably more restrictive, extending only to those with a qualifying connection to Europe or, beyond that, to countries providing similar protection for functional designs to UK nationals.[6] New Zealand and Hong Kong feature on this list, but most major economies (outside Europe) do not.

If the designer has the necessary nationality or residence connection with a qualifying country, and is not employed, the design may be protected. If he is employed, then the connection considered is that of the employer not the designer. If it does not qualify in any other way, a design qualifies if its first marketer is a qualifying person and the territory in which it is first marketed is a qualifying country such as the EU.

Finally, although Unregistered Community Designs do not explicitly involve qualification by reference to nationality or residence, the right only comes into being following a disclosure (marketing or otherwise) within the EU.[7] Some national courts have held that this requires the first disclosure to be within the EU (*Thane Ab Swing-Hometrainer)*.[8] If correct, this has the practical effect of disentitling from UCD protection the many non-European entities who first disclose in their own home market rather than in Europe.

§20.02 CONVENTIONS AND LEGISLATION

[A] International Conventions

International Conventions are not self-executing in the UK, but the UK is a member of, and has generally implemented, the following:

- The Paris Convention for the Protection of Industrial Property ('Paris Convention')[9] revision of Lisbon in 1958, which requires protection for industrial designs and provides for priority claims.
- The Berne Convention for the Protection of Literary and Artistic Works[10] ('Berne Convention') which requires protection for works of applied art (Articles 2(7) and 7(4) Berne Convention).

6. See The Design Right (Reciprocal Protection) (No. 2) Order 1989 SI 1989 No. 1294.
7. Articles 11 & 110a CDR.
8. 3/12 O 5/04 *Ab Swing-Hometrainer,* [2006] ECDR 8, GRUR-RR 2005 Heft 1 (Frankfurt Regional Ct.); see also I ZR 126/06 *Gebäckpresse II/Pastry press II,* GRUR 2009, p. 79 (German Supreme Court).
9. Paris Convention for the Protection of Industrial Property of 20 Mar. 1883, as revised at Brussels on 14 December 1900, at Washington on 2 Jun. 1911, at The Hague on 6 Nov. 1925, at London on 2 Jun. 1934, at Lisbon on 31 Oct. 1958, and at Stockholm on 14 Jul. 1967, and as amended on 28 Sep. 1979.
10. Berne Convention for the Protection of Literary and Artistic Works (as amended on 28 Sep. 1979).

- The Agreement on Trade-Related Aspects of Intellectual Property Rights ('TRIPs'),[11] which provides for a minimum protection threshold for industrial designs (Articles 25 and 26 TRIPs).
- *[The Hague Agreement. Implementing provisions to permit accession were passed in the Intellectual Property Act 2014*[12] *introducing RDA Section 15ZA. The UK had not yet acceded at the time of writing, but will do so shortly. When operative this will allow the national designation of the UK in International Designs].*

[B] Domestic Legislation

There is a system of domestic design registration in the UK under the Registered Designs Act 1949[13] as amended by, *inter alia*:

- The Copyright, Designs and Patents Act 1988 ('CDPA 1988').
- The Registered Designs Regulations 2001.[14]
- The Registered Designs Regulations 2003.[15]
- The Community Design Regulations 2005.[16]
- The Regulatory Reform (Registered Designs) Order 2006.[17]
- The Intellectual Property (Enforcement etc.) Regulations 2006.[18]
- The Designs (International Registrations Designating the European Community) Regulations 2007.[19]
- The Intellectual Property Act 2014.[20]

Certain designs (mainly surface decoration) attract automatic copyright protection under Part I of the CDPA 1988 as amended by, *inter alia,* the Enterprise and Regulatory Reform Act 2013.[21] These will be discussed at the 'overlap with copyright' section below.

A further category of designs (essentially, three-dimensional designs meeting certain additional criteria) are protected automatically and without the necessity to register by national unregistered 'design right' ('UDR') under Part III of the CDPA 1988 as amended by The Intellectual Property Act 2014. The protection is for a relatively short period (ten to fifteen years) and there is essentially no overlap with copyright.

11. The TRIPs Agreement is Annex 1C of the Marrakesh Agreement Establishing the World Trade Organization, signed in Marrakesh, Morocco on 15 Apr. 1994, and entering into force on 1 Jan. 1995.
12. 2014 c. 18.
13. 1949 c. 88.
14. SI 2001 No. 3949.
15. SI 2003 No. 550.
16. SI 2005 No. 2339.
17. SI 2006 No. 1974.
18. SI 2006 No. 1028.
19. SI 2007 No. 3378.
20. 2014 c. 18.
21. 2013 c. 24.

Many aspects of civil design litigation in England and Wales are codified in the CPR at Part 63 – Intellectual Property Claims (and the associated Practice Direction PD63).

[C] European Legislation

The CDR provides unified Community-wide design protection; both registered (registered Community design, 'RCD') and unregistered (unregistered Community design, 'UCD'). For more information, see the EU chapter elsewhere in this volume. As a consequence, a single application for an RCD (or an equivalent International application designating the EU) results in design protection in, *inter alia,* the UK.

The CDR[22] is essentially self-implementing in the UK but some minor aspects of UK implementation are provided by:

- The Community Design (Fees) Regulations 2002.[23]
- The Community Design Regulations 2005.[24]
- The Community Designs (Designation of Community Design Courts) Regulations 2005.[25]
- The Designs (International Registrations Designating the European Community) Regulations 2007.[26]
- The Community Design (Amendment) Regulations 2014.[27]

Enforcement of Community Designs is through the UK's Community design courts (Article 80 CDR implemented by The Community Designs (Designation of Community Design Courts) Regulations 2005).[28] These are essentially the same as those having jurisdiction over domestic design litigation, namely:

(1) In England and Wales at first instance:
   (i) The High Court (specifically the Patents Court).
   (ii) The Intellectual Property Enterprise Court (IPEC).

(2) In Scotland at first instance:
   (i) The Court of Session.

(3) In Northern Ireland at first instance:
   (i) The High Court of Northern Ireland.

(4) On appeal, the normal appeal routes from each of the above.

22. And its various implementing Regulations, though these mostly concern pre-grant matters, see the relevant chapter in this volume.
23. SI 2002 No. 2942.
24. SI 2005 No. 2339.
25. SI 2005 No. 696.
26. SI 2007 No 3378.
27. SI 2014 No. 2400.
28. SI 2005 No. 696.

Jurisdiction and enforcement of judgments across the EU is dealt with by Regulation (EU) 1215/2012[29] (the 'Brussels I Regulation recast', implemented to some extent in the UK by the various Civil Jurisdiction and Judgments Acts[30] and Orders[31]), and for Community designs by the additional rules in the CDR.

§20.03 SUBJECT MATTER

[A] Definition/What Constitutes a Protectable Design?

[1] Shape and Configuration

RDA section 1(2), based on Directive Article 1(a) (and equivalent to Article 3(a) CDR) defines a design as 'the appearance of the whole or a part of a product resulting from the features of, in particular, the lines, contours, colours, *shape*, texture and/or materials of the product itself and/or its ornamentation'. A '"*product" means any industrial or handicraft item*': RDA section 1(3), based on DD Article 1(b) and equivalent to Article 3(a) CDR. Shape and configuration designs are therefore protectable by UK Registered Designs and Community Designs.

For unregistered UK Design Right, a 'design' (section 213(2) CDPA) is: '*the design of the shape or configuration (whether internal or external), of the whole or part of an article*'.

[2] Ornamental or Functional

As noted above, the definition of a design for UK Registered Designs and Community Designs includes, but is not limited to, the 'ornamentation' of a product.

Functional designs are not excluded by the definition of 'design', but there is an exclusion from protection of features of a product '*which are solely dictated by its technical function*' (RDA section 1C(1) (based on Directive Article 7(1) and equivalent to Regulation Article 8(1)). If all features are 'dictated' in this sense, the design as a whole is invalid. This is not equivalent to a requirement for ornamentality, as the

29. Regulation (EU) 1215/2012 of the European Parliament and of the Council of 12 Dec. 2012 on jurisdiction and the recognition and enforcement of judgments in civil and commercial matters, OJ L 351/1, 20/12/2012, as amended by Regulation (EU) No. 542/2014 of the European Parliament and of the Council of 15 May 2014 amending Regulation (EU) No. 1215/2012 as regards the rules to be applied with respect to the Unified Patent Court and the Benelux Court of Justice, OJ L 163/1 29/5/2014, which recast Council Regulation (EC) No. 44/2001 of 22 Dec. 2000 on jurisdiction and the recognition and enforcement of judgments in civil and commercial matters, OJ L-12/1. The equivalent is provided for EEA states by the Lugano Convention.
30. Civil Jurisdiction and Judgments Act 1982 (1982 c. 27) and the Civil Jurisdiction and Judgments Act 1991 (1991 c. 12).
31. Civil Jurisdiction and Judgments Order 2001 (SI 2001 No. 3929).

corresponding Recitals (Recital 14 DD and Recital 10 CDR) indicate that '*this does not entail that a design must have an aesthetic quality*'.

There is currently no consensus about the meaning of 'solely dictated by technical function'. According to the Court of Appeal in *Landor & Hawa v. Azure Designs*[32] and *Procter & Gamble v. Reckitt Benckiser*[33] this ground bites only if a design can be made in only one shape. In other words, even if the designer only considered functional aims, the design is protectable if it can be made in different shapes (the 'multiplicity of forms' test). They followed the opinion of the Advocate General of the European Court of Justice.[34] The IPO policy was based on that approach (DPN 5/03: Designs dictated by their technical function[35]).

However, in *Samsung v. Apple,*[36] the Court of Appeal reverted to an approach based on that taken under the previous law by the House of Lords in *Amp v. Utilux;*[37] that designs were 'dictated' by function if they were created for functional reasons, regardless of whether they could be made in other shapes, thus introducing what is essentially a requirement for some ornamental character.

UK UDR, by way of contrast, was intended to protect functional designs (*Landor & Hawa v. Azure Designs*[38]), but a 'method or principle of construction' is excluded from design right section 213(3)(a) CDPA). This exclusion is normally used as a 'squeeze' to prevent overbroad claims (e.g., *Rolawn v. Turfmech,*[39] *Landor & Hawa v. Azure Designs*[40]). Also excluded from UK UDR protection is 'surface decoration'. This may instead be protected by copyright, but there is no rule that every design must be protected by one or the other of these (*Lambretta v. Teddy Smith*[41]).

[a] 'Must Fit' Features

Closely allied is the exclusion of protection for features which '*permit the product in which the design is incorporated or to which it is applied to be mechanically connected to or placed in, around or against another product so that either product may perform its function*'. For UK Registered Designs (RDA section 1C(2) based on Directive Article 7(2)) and Community Designs (Article 8(2) CDR), this applies where the features 'must *necessarily* be reproduced in their *exact* form and dimensions'; these additional restrictions do not apply to UK UDR (CDPA section 213(3)(b)(i)).

32. *Landor & Hawa International Ltd v. Azure Designs Ltd* [2006] ECDR 31, [2006] EWCA Civ 1285, [2007] FSR 9.
33. *Procter & Gamble Co v. Reckitt Benckiser (UK) Ltd* [2007] EWCA Civ 936, [2008] FSR 8 at [35(ii)].
34. *Koninklijk Philips Electronics v. Remington Consumer Products,* Case C-299 (AGO) [2001] RPC 38, [2001] ETMR 48.
35. Designs Practice Notice, DPN 5/03: Designs dictated by their technical function, 18 Aug. 03.
36. *Samsung Electronics (UK) Ltd v. Apple Inc* [2012] EWCA Civ 1339, [2013] FSR 9.
37. [1971] FSR 572, [1972] RPC 103 (HL).
38. *Landor & Hawa International Ltd v. Azure Designs Ltd* [2006] ECDR 31, [2006] EWCA Civ 1285, [2007] FSR 9.
39. [2008] EWHC 989.
40. *Landor & Hawa International Ltd v. Azure Designs Ltd* [2006] ECDR 31, [2006] EWCA Civ 1285, [2007] FSR 9.
41. *Lambretta Clothing Company Ltd. v. Teddy Smith (UK) Ltd. and Next Retail Plc* [2004] EWCA Civ 886 [2005] RPC. 6.

The shape of the features must result in a product being enabled to perform its function. It is not enough that, if shaped differently, a feature would block proper function (*Dyson v. Qualtex*[42])

Further, for UK Registered Designs (RDA section 1C(3) based on Directive Article 7(2)) and Community Designs (Article 8(3) CDR), there is a saving of features of designs of 'modular products' such as Lego™.

[b] Must-Match Features (for UK UDR)

Additionally, for UK UDR, features of designs of products which '*are dependent upon the appearance of another article of which the article is intended by the designer to form an integral part*' are excluded from protection. Typically these are features of spare parts of a larger product, which have some aesthetic appeal (or otherwise it would be unnecessary to wish to copy them at all), but where the appeal is really that of the whole product not just the part in isolation. The customer for such spares is usually involved in a repair, and for many repairs it is necessary to use a replacement part which matches its predecessor, so as to keep the original appearance of the product. Design protection in the aftermarket for repair can therefore give rise to considerable market power, which the 'must-match' exclusion is intended to curb.

To determine whether it applies, it is necessary to consider whether there is a dependency of the kind, or to the extent, which would make the overall article in question radically different in appearance if the first article was shaped differently (*R. RDAT ex p. Ford,*[43] *Dyson v. Qualtex*[44]). The more there is freedom to vary the shape of the product, the less the market power conferred by design protection of the spare part, and the less is there need, or room, for the exclusion.

The 'must match' exclusion has no direct counterpart for UK Registered Designs and Community Designs, but the 'repair' exclusion from infringement discussed below has a similar effect in practice (*BMW v. Round)*[45].

[B] Originality/Novelty

[1] Test for Originality Generally

UK Registered Designs and Community Designs have no explicit requirement for originality.

For UK UDR, a protectable design must be both *original* and *not commonplace* (section 213(1), (4) CDPA). 'Original' is, as for UK copyright, 'not copied' (*C&H*

42. *Dyson Ltd v. Qualtex (UK) Ltd* [2006] RPC 31, [2006] EWCA Civ 166 (2006) 29(6) I.P.D. 29044.
43. [1995] R.T.R. 68, [1994] RPC. 545.
44. *Dyson Ltd v. Qualtex (UK) Ltd* [2006] RPC 31, [2006] EWCA Civ 166 (2006) 29(6) I.P.D. 29044.
45. *Bayerische Motoren Werke v. Round and Metal Ltd* [2013] Bus LR D30, [2012] EWHC 2099 (Pat), [2013] FSR 18.

Engineering v. Klucznik,[46] *Farmers Build v. Carrier Bulk Materials*,[47] *Dyson v. Qualtex*[48]).

In *Ocular Sciences v. Aspect Vision Care*[49] the Patent Court summarized 'commonplace' as conveying '*trite, trivial, common-or-garden or hackneyed*'. A methodology for determination of 'commonplace' designs based on counting similarities and the differences with the other articles in the field was set out by the Court of Appeal in *Farmers Build v. Carrier Bulk Materials*.[50]

[2] Novelty

UK Registered Designs and Registered and Unregistered Community Designs (but not UK unregistered designs), must be novel and have individual character over the state of the art. However, it is important to emphasize that the UK IPO, like WIPO and OHIM, does not perform novelty searches or examine novelty during the registration process (see DPN 1/06: Ending novelty examination)[51], and for Unregistered Community Designs there is no such process. These requirements, like other grounds of validity, can therefore only be raised after registration.

RDA section 1B(2), based on DD Article 4 (and equivalent to Article 5 CDR) defines the novelty requirement. *A design shall be considered to be new if no identical design has been made available to the public before the date of filing of the application for registration, or, if priority is claimed, the date of priority. Designs shall be deemed to be identical if their features differ only in immaterial details.* For Unregistered Community Designs, the date of first disclosure of the design in suit which would 'come to the attention of the circles specialized in the sector concerned' takes the place of the priority date.

The design has to differ from any preceding *design*. This goes further than meaning any design of the same product (see *Green Lane v. PMS*,[52] *Gimex v. Chill Bag*[53]). Thus (as in the previous UK case law) a design of, for example, a crankshaft could be anticipated by that of an electrical fuse.[54]

The comparison must be conducted on a one-for-one basis; it is not legitimate to combine references to make a novelty (or individual character) attack (*Karen Millen v. Dunnes)*[55].

46. *C&H Engineering v. Klucznik & Sons Limited* [1992] FSR 421.
47. [1999] RPC 461 p. 482.
48. *Dyson Ltd v. Qualtex (UK) Ltd* [2006] RPC 31, [2006] EWCA Civ 166 (2006) 29(6) I.P.D. 29044f.
49. [1997] RPC 289.
50. [1999] RPC 461.
51. Designs Practice Notice, DPN 1/06: Ending novelty examination, 20 Nov. 2007, Replaces DPN 01/03 and DPN 04/03.
52. *Green Lane Products Ltd v. PMS International Group Ltd & Ors* [2008] FSR 28, [2008] ECDR 15, [2008] EWCA Civ 358, [2008] ECC 28, [2008] Bus LR 1468.
53. *Gimex International Groupe Import Export v. The Chill Bag Company Ltd & Ors* [2012] EWPCC 31.
54. *Stenor v. Whitesides (Clitheroe) Ltd.* (1948), 65 RPC 1f.
55. Case C-345/13 *Karen Millen Fashions Ltd v. Dunnes Stores and Dunnes Stores (Limerick) Ltd* Opinion at ECLI:EU:C:2014:206, judgment at ECLI:EU:C:2014:2013, [2014] WLR (D) 273, [2014] ECDR 17.

A design is also required to have 'individual character'. This is not an originality-type requirement, but is based on a novelty-type comparison with the state of the art. RDA section 1B(3) & (4), based on DD Article 5 (and equivalent to Article 6 CDR) defines the 'individual character' requirement:

(1) A design shall be considered to have individual character if the overall impression it produces on the informed user differs from the overall impression produced on such a user by any design which has been made available to the public before the date of filing of the application for registration, or, if priority is claimed, the date of priority.
(2) In assessing individual character, the degree of freedom of the designer in developing the design shall be taken into consideration.

Again, the comparison is on a one-to-one basis. The 'informed user' is a user of the product in which the design is intended to be incorporated, not a designer, technical expert, manufacturer or seller; but a particularly observant user who has knowledge of the design state of the art, and of the design features normally included in the designs existing in the sector concerned (*Pepsico,*[56] applied in *Samsung v. Apple*[57]). In some cases, it may be a person who is involved in buying or selecting a product as well as in operating it (*Louver-Lite Ltd v. Harris)*[58].

In assessing the overall impression, it is necessary to take into account the '*nature of the product to which the design is applied, and the industrial sector to which it belongs*'. Thus, a design of apparently similar appearance may nonetheless create a significantly different overall impression if the differences have important implications for the nature of the product and the way it operates (*Gimex v. Chill Bag*)[59].

The subtest of 'design freedom' (RDA section 1B(4)) means that, where there is little scope for difference, then relatively little difference from the prior design will be needed to establish 'individual character', and conversely where design freedom is great, larger differences are required.

Features of a component part which are not visible in normal use cannot assist in establishing novelty or individual character (RDA section 1B(8) implementing DD Article 3(2), equivalent to Article 4(2) CDR). If no features of the design are visible, then it is invalid as lacking novelty.

The state of the art is defined by RDA section 1B(5), based on DD Article 6 (and equivalent to Article 7 CDR) to include everything '*published following registration or otherwise, or exhibited, used in trade or otherwise disclosed*'.

However, there are three explicit savings (for which the burden of proof probably falls on the proprietor of the design in suit) under RDA section 1B(6):

---

56. *PepsiCo Inc v. Grupo Promer Mon-Graphic SA* (C-281/10 P) ECLI:EU:C:2011:679, [2012] FSR 5, EU:C:2011:679, [2011] EUECJ C-281/10.
57. *Samsung Electronics (UK) Ltd v. Apple Inc* [2012] EWCA Civ 1339, [2013] FSR 9.
58. *Louver-Lite Ltd v. Harris Parts Ltd (t/a Harris Engineering)* [2012] EWPCC 53, [2012] All ER (D) 85 (Dec).
59. *Gimex International Groupe Import Export v. The Chill Bag Company Ltd & Ors* [2012] EWPCC 31.

(1) Disclosures '*under explicit or implicit conditions of confidentiality*' are excluded from the state of the art.
(2) Disclosures which '*could not reasonably have become known in the normal course of business to the circles specialised in the sector concerned, operating within the Community before the date of filing of the application for registration, or, if priority is claimed, the date of priority*' are excluded. Disclosures outside the EU may come to the attention of circles inside (Case C-479/12 *Gautzsch Großhandel*)[60].
(3) RDA section 1B(6)(c)-(e)), based on DD Article 6(2) and (3) (and equivalent to Article 7(2) and (3) CDR) introduces a twelve-month grace period prior to the priority or filing date, excluding disclosures '*by the designer, his successor in title, or a third person as a result of information provided or action taken by the designer or his successor in title*' or as a result of an abuse in relation to the designer or successor in title.

[a] Prior Use

As noted above, designs which are non-confidentially prior '*used in trade*' anywhere in the world are citable, provided none of the above exceptions apply. Secret prior use is no bar to validity.

[b] Prior Publication

As noted above, designs which are prior published anywhere in the world are citable, provided none of the above exceptions apply.

[c] Prior Application or Registration

A prior design registration which is published or made available to the public is treated like any other publication. Case law to date suggests that it will be seen as likely to come to the attention of the 'circles specialised in the sector concerned' since it was filed by a member of such circles, and is generally addressed to them.

A prior design registration which is not yet published or made available to the public at the date of filing or priority of the design in suit does not form part of the state of the art for novelty or individual character, but instead is treated as a 'prior right' which, if it would be infringed by the design in suit, can be used as a challenge by the proprietor of the prior design. It follows that where the prior design is in the same hands as the design in suit, it cannot be relied on as prior art by anyone else.

60. Case C-479/12, *H. Gautzsch Großhandel GmbH & Co. KG v. Münchener Boulevard Möbel Joseph Duna GmbH* [2014] EUECJ C-479/12, [2014] WLR(D) 66, [2014] RPC 28, [2014] Bus LR 391, EU:C:2014:75, [2014] ECDR 14, ECLI:EU:C:2014:75.

[C] Unobviousness (if Appropriate)

There is not, nor has there ever been, an 'obviousness' doctrine for UK designs. Prior references cannot be combined (*Karen Millen v. Dunnes*)[61], and the test person is the inexpert 'informed user' rather than a person of ordinary skill in the art.

§20.04 DURATION OF PROTECTION

UK domestic designs, Registered Community Designs, and International designs designating the EU or the UK all have a lifetime of up to twenty-five years from the application (not priority) date, by virtue of the effects of RDA section 8, DD Article 10 and Article 12 CDR. They are initially registered for a term of five years, extendable in further five year increments.

Unregistered Community Designs last for three years after the first disclosure in the EU which would have come to the attention of the sector concerned, operating within the EU (CDR Articles 11 & 110a).

UK Unregistered Design Right has a lifetime of the shorter of ten years from the end of the year of first sale or fifteen years from first recordal of the design in a design document or manufacture (CDPA section 216). During the last five years of its life, licences are available as of right.

Copyright extends for the life of the author plus seventy years. At the time of writing, this was curtailed to twenty-five years where a design was 'industrially applied' (section 52 CDPA), but the curtailment is prospectively repealed and will cease to have other than transitional effect on past acts of infringement in the near future.

§20.05 INFRINGEMENT

[A] Scope of Infringement Action

[1] Registered Designs and UCD

The acts which constitute a civil infringement are harmonized in Europe and are therefore the same for UK Registered Designs, Community Designs both Registered and Unregistered and International designs designating Europe or the UK. In addition to these, the UK provides criminal offences (see below), and there may also be ancillary civil wrongs or bases of liability, such as joint tortfeasorship, and crimes such as conspiracy (these are not dealt with here).

In general, any unauthorized 'use' of a design is an infringement, but an inclusive list of infringing acts in relation to products are given as examples in RDA section 7(2),

61. *Karen Millen Fashions Ltd v. Dunnes Stores and Dunnes Stores (Limerick) Ltd* (Case C-345/13) Opinion at ECLI:EU:C:2014:206, judgment at ECLI:EU:C:2014:2013, [2014] WLR (D) 273, [2014] ECDR 17.

based on Directive Article 12 (and equivalent to Article 19 CDR), namely *'the making, offering, putting on the market, importing, exporting or using of a product in which the design is incorporated or to which it is applied, or stocking such a product for those purposes'*. For registered designs, there is no additional requirement for copying, knowledge of the design or other state of mind.

The rights apparently extend to any product incorporating the design (see *Landor & Hawa v. Azure Designs;*[62] *Green Lane Products v. PMS*[63]). In particular, the indication of product filed is specifically stated not to affect the scope of protection (Article 36(6) CDR for RCDs, RDR Rule 5(5) for UK Registered Designs), and likewise the classification (Article 36(6) CDR). Thus, a registration of, say, a car is infringed by making toy cars in the same shape.

[2] UK UDR and Copyright

The acts infringing UK copyright and UDR are differently defined. They are usually divided into 'primary' and 'secondary' infringements.

'Primary' infringement of UDR is making articles to the design, or making a design document recording the design for the purpose of enabling such articles to be made (CDPA section 226). For copyright, the corresponding acts are copying, and issuing copies to the public (CDPA sections 16–18). No further knowledge or mental state is required.

'Secondary' infringement of UDR is importation into the UK, possession for commercial purposes, or selling, letting for hire, or offer or exposure for sale or hire, an infringing product in the course of a business (CDPA section 227), and likewise for copyright (CDPA sections 22–23). There is only an infringement where the defendant knows or has reason to believe, the product concerned is an infringing article. It is therefore normal to mark products (as discussed below), and to write to secondary infringers to 'fix them with knowledge' of the infringement, before taking action against them.

[B] Test for Infringement/Criteria to Constitute Infringement

For UK Registered Designs and Community Designs, Registered and Unregistered, the test for infringement (defining the 'scope of protection') is the same as that for 'individual character' (defining validity over the prior art), which is discussed above in the section on novelty; the discussion below reflects this.

Accordingly, the scope of a design extends to *'any design which does not produce on the informed user a different overall impression'*: RDA section 7(1) based on Directive Article 9(1) for UK registered designs, and equivalent Article 10(1) CDR for Community Designs).

62. *Landor & Hawa International Ltd v. Azure Designs Ltd* [2006] ECDR 31, [2006] EWCA Civ 1285, [2007] FSR 9.
63. *Green Lane Products Ltd v. PMS International Group Ltd & Ors* [2008] FSR 28, [2008] ECDR 15, [2008] EWCA Civ 358, [2008] ECC 28, [2008] Bus LR 1468.

The 'informed user' is a user of the product in which the design is intended to be incorporated, not a designer, technical expert, manufacturer or seller; but a particularly observant user who has knowledge of the design state of the art, and of the design features normally included in the designs existing in the sector concerned (*Pepsico,*[64] applied in *Samsung v. Apple*[65]). In some cases, it may be a person who is involved in buying or selecting a product as well as in operating it. (*Louver-Lite Ltd v. Harris*)[66].

As a practical matter, the 'informed user' is more knowledgeable and attentive than the 'average consumer' of trademark law, and therefore more likely to notice smaller difference. As a result, the scope of a design is narrower than that of an equivalent trade mark (*Procter & Gamble v. Reckitt Benckiser)*[67].

In assessing the overall impression, it is necessary to take into account the '*nature of the product to which the design is applied, and the industrial sector to which it belongs.*' Thus, a design of apparently similar appearance may nonetheless create a significantly different overall impression if the differences have important implications for the nature of the product and the way it operates (*Gimex v. Chill Bag*)[68].

The subtest of 'design freedom' (RDA section 7(2) based on Directive Article 9(2) for UK registered designs, and equivalent Article 10(2) CDR for Community Designs) means that, where there is little scope for difference, then relatively little difference from the design in suit will be needed to establish freedom from infringement, and conversely where design freedom is great, larger differences are required.

Those parts of the designs visible during normal use are given most weight.[69] It is an open question whether non-visible elements should be given any weight at all.[70]

[1] UK UDR and Copyright

For UK UDR, articles infringe if they are made exactly or substantially to the design (CDPA section 226(2)). A full examination of copyright infringement is beyond the scope of this chapter, but the difference between infringement of copyright and of UDR is given in *Woolley v. A&A.*[71]

64. *PepsiCo Inc v. Grupo Promer Mon-Graphic SA* (C-281/10 P) ECLI:EU:C:2011:679, [2012] FSR 5, EU:C:2011:679, [2011] EUECJ C-281/10.
65. *Samsung Electronics (UK) Ltd v. Apple Inc* [2012] EWCA Civ 1339, [2013] FSR 9.
66. *Louver-Lite Ltd v. Harris Parts Ltd (t/a Harris Engineering)* [2012] EWPCC 53, [2012] All ER (D) 85 (Dec).
67. *Procter & Gamble Company v. Reckitt Benckiser (UK) Ltd,* [2008] FSR 8, [2008] ECDR 3, [2007] EWCA Civ 936, [2008] Bus LR 801.
68. *Gimex International Groupe Import Export v. The Chill Bag Company Ltd & Ors* [2012] EWPCC 31.
69. *Shenzhen Taiden v. OHIM - Bosch Security Systems* (Communications Equipment) [2010] EUECJ T-153/08, EU:T:2010:248, ECLI:EU:T:2010:248.
70. *Biscuits Poult v. OHIM - Banketbakkerij Merba* (Biscuit), [2014] EUECJ T-494/12, EU:T:2014:757, ECLI:EU:T:2014:757.
71. *L Woolley Jewellers v. A&A Jewellery* [2002] EWCA Civ 1119, [2003] FSR 15, (2002) 25(11) IPD 25078.

[2] Additional Issues Specific to Unregistered Rights

Infringement of unregistered rights (copyright, UDR, UCD) involves two additional issues: the claimant/proprietor's freedom to select the parts of the design which form the basis of the action, and his need to prove copying.

[3] Selection of Parts of Product in Suit

Unregistered designs protect the design of '*the whole or [a] part of*' a product (Article CDR for UCD, CDPA section 213(2) for UK UDR, Article 3(a) CDR for UCD). Copyright is infringed by reproducing '*the work as a whole or any substantial part of it*' (CDPA section 16(3)(a)).

In either case, the proprietor/claimant in an infringement action is required to specify what the design in suit is, and is therefore able to select or 'trim' the part or parts of his own design for the action, and thereby the corresponding parts of the infringing product for comparison (*Ocular Sciences v. Aspect,*[72] *Fulton v. Totes Isotoner*[73]).

Likewise, when litigating UCD, Article 85(2) CDR requires the proprietor to indicate '*what constitutes the individual character of his Community design*' and by doing so he '*specifies the object of the protection which he is claiming*' *and* '*defines the scope of the comparison with the contested design and enables the defendant, if he so wishes, to prepare a suitably specific counterclaim*'.[74]

Adding (and, within limits, subtracting) features is therefore usually ineffective in avoiding infringement of unregistered rights.

[4] Copying

To infringe an unregistered design or copyright, a product must result from copying (Article 19(2) CDR for UCD, CDPA sections 16–17 for copyright, CDPA section 226(2) for UK UDR). Copying may be direct or indirect (CDPA section 16(3)(b) for copyright, CDPA section 226(4) for UK UDR), conscious or subconscious.

The burden of proof of copying lies with the proprietor. Disclosure of documents and cross-examination of witnesses are available to assist in establishing copying. Initially, close similarity in appearance coupled with access to the design in question are usually enough to throw the burden of proof onto the defendant (*Gautzsch*

72. *Ocular Sciences Ltd v. Aspect Vision Care Ltd (No. 2)* [1997] RPC 289, [1996] EWHC (Pat), (1997) 20(3) IPD 20022.
73. *A Fulton Company Ltd. v. Totes Isotoner (UK) Ltd* [2003] EWCA Civ 1514, [2004] RPC 16, (2004) 27(2) IPD 27010.
74. Opinion Of Advocate-General Wathelet in Case C-345/13 *Karen Millen Fashions Ltd v. Dunnes Stores, Dunnes Stores (Limerick) Ltd*, ECLI:EU:C:2014:206, judgment at ECLI:EU:C:2014:2013, [2014] WLR (D) 273, [2014] ECDR 17.

Großhandel)[75]. The closer the similarity, the stronger the inference of copying and the higher the burden of proof required for rebuttal (*John Kaldor v. Lee Ann Fashions*)[76]. 'Similar Fact' evidence of past copying is also admissible (*Mattel v. Woolbro*)[77].

[5] Criminal Offences Involving Copying a Registered Design

There are UK criminal offences of infringement of a registered design (RDA section 35ZA), either UK or Community, based closely on those applicable to copyright. By way of contrast with civil infringement of a registered design, the offence can only result from intentional copying, closely enough to show no 'immaterial differences' (the novelty standard). There are a variety of defences including belief in invalidity and in non-infringement.

[C] Standing to Sue

For a UK design, either the proprietor or an exclusive licensee can sue (RDA sections 15C and 24F). Likewise for an RCD, actions can be brought by the proprietor or by an exclusive licensee without the proprietor's consent, or other licensee with the proprietor's consent (CDR Article 32(3)), though it is necessary in each case for the relevant title or licence to be registered.

The rights of joint proprietors (and, presumably, licensees) are a matter of national, not EU, law and are not set forth in statute. It is likely but not certain that the consent of all is required in order to sue, but given that consent, one co-owner can sue without joining the others.[78] However, having started litigation, settlement (like any other form of dealing) appears to require the agreement of all co-owners.[79]

[D] Remedies for Infringement

For UK registered and unregistered designs, the remedies available are those generally available for infringement of intellectual property.[80] For Community designs (registered and unregistered), the remedies available under national law apply as they do to national registered designs.[81] These are harmonized in the EU by Directive

75. Case C-479/12, *H. Gautzsch Großhandel GmbH & Co. KG v. Münchener Boulevard Möbel Joseph Duna GmbH* [2014] EUECJ C-479/12, [2014] WLR(D) 66, [2014] RPC 28, [2014] Bus LR 391, EU:C:2014:75, [2014] ECDR 14, ECLI:EU:C:2014:75.
76. *John Kaldor Fabricmaker v. Lee Ann Fashions*, [2014] EWHC 3779 (IPEC).
77. *Mattel Inc v. Woolbro (Distributors) Ltd & Ors* [2003] EWHC 2414 (Ch), [2004] FSR 217.
78. *Ray v. Classic FM* [1998] ECC 488, [1998] EWHC Patents 333, [1998] FSR 622, (1998) 21(5) IPD 21047, [1999] ITCLR 256 at 35.
79. CDPA s. 258(2) for UDR, s. 173(2) for copyright, see for other rights Report Q194 The Impact of Co-Ownership of Intellectual Property Rights on their Exploitation United Kingdom, Bennet et al. 2007 (on AIPPI website).
80. RDA s. 24B(2) and CDPA S229(2) respectively.
81. CDR Article 89(1)(a)(iv).

2004/48/EC[82] ('the Enforcement Directive'), which was implemented in the UK (to the extent that the UK was not already compliant) by The Intellectual Property (Enforcement, etc.) Regulations 2006:[83]

- As design infringements are torts, the basic remedy at common law is an award of *damages*. These are essentially to compensate the loss suffered by the claimant, but the Enforcement Directive requires some additional factors to be taken into account. Except in low value cases or in the event of agreement between the parties, the enquiry as to the quantum of damages is a separate process taking place after the trial on substance.
- Where the infringement is flagrant, *additional damages* may be awarded in case of UK unregistered design right or copyright infringement (CDPA sections 229(3) and 97(2) respectively). As to 'innocence' as a defence to damages, see the section on marking below. If the case is brought in the Intellectual Property Enterprise Court (IPEC), there is a cap on recoverable damages (at present GBP 500,000). No limit applies in other competent UK courts.
- All UK courts competent for design cases can order an *injunction* in their inherent equitable jurisdiction Injunctions are also specifically provided for by section 24B(3) RDA in relation to registered designs and, for Community Designs, Article 89(1)(a)(i) CDR. Though it is discretionary, a final negative injunction almost always follows a finding of infringement of an intellectual property right.
- *Interim or interlocutory* injunctions are also available in principle, but must be justified by the balance of convenience.
- Other equitable remedies which are generally available to the courts of England and Wales in their inherent equitable jurisdiction include an *account of defendant's profits* as an alternative to damages.
- Seizure (for Community Designs, CDR Article 89(1)(ii)) or delivery up of infringing products is available (RDA section 24C; CDPA sections 230, 99 and100), as is an order for their disposal (RDA section 24D, CDPA sections 231 and 114).
- Finally, an award of *costs* is usually made to the winning party, reflecting some proportion of their actual expenses including attorney fees (although this is capped in the IPEC, currently at GBP 50,000).
- Regulation (EU) No. 608/2013[84] (the 'Customs Enforcement Regulation') provides for customs seizure of goods imported from outside the EU which infringe designs.

82. Directive of the European Parliament and of the Council of 29 Apr. 2004 on the enforcement of intellectual property rights (2004/48/EC) (OJ L157, 30/4/2004 p. 45, a corrigendum was published in OJ L195, 2/6/2004 p. 16).
83. SI 2006 No. 1028.
84. Regulation (EU) No. 608/2013 of the European Parliament and of the Council of 12 Jun. 2013 concerning customs enforcement of intellectual property rights and repealing Council Regulation (EC) No. 1383/2003, OJ L 181/15, 29/6/2013; implemented by Commission Implementing Regulation (EU) No. 1352/2013 of 4 Dec. 2013 establishing the forms provided for in Regulation

§20.06 DEFENCES TO INFRINGEMENT

[A] Validity

In the UK, an invalid right cannot be infringed. Validity is therefore generally a defence. In the past, difficulties arose when infringement actions were brought before inferior courts, as these did not have jurisdiction to cancel invalid designs and were not considered to have jurisdiction to entertain an invalidity defence either.

Currently, the UK courts with jurisdiction to hear an infringement action can also entertain an invalidity defence and a counterclaim to invalidate a registered design, whether it is a UK national design or a Registered Community Design, the result of which will appear on the relevant Register of Designs.

In relation to Community Designs, registered or unregistered, there is a statutory presumption of validity (Article 85 CDR) which is rebuttable only by bringing a defence or counterclaim to invalidate.

However, since there is little substantive examination (none in relation to prior art), the presumption is a formal one, requiring only the existence of a defence or counterclaim rather than signifying any particular evidential burden.

For Registered Community Designs, the defence can only be raised by way of a counterclaim to invalidate, except by the proprietor of an earlier unpublished design registration in relation to that ground (Articles 85(1) and 25(1)(d) CDR). For UK registered and unregistered designs, and Unregistered Community Designs, a defence without counterclaim is possible.

[1] Not Novel/Original

RDA sections 20 and 11ZA(1)(b) together with section 1B allow a UK registered design to be invalidated if it lacks novelty or individual character and likewise Articles 25(1)(b) together with 4(1) and 5-7 CDR for Community Designs.

A UK unregistered design or copyright work can be challenged on the basis of lack of originality (for an example, see *Interlego v. Tyco)*[85].

[2] Functional/Not Ornamental

As noted above, a design may be invalid where all features are dictated by function, and it can be cancelled on this ground: RDA section 1C(1) (based on Directive Article 7(1)) for UK Registered Designs, and equivalent Article 8(1) CDR for Community Designs.

The same applies if all features are 'must-fit' (RDA section 1C(2) based on Directive Article 7(2) for UK Registered Designs, and equivalent to Article 8(2) CDR for

(EU) No. 608/2013 of the European Parliament and of the Council concerning customs enforcement of intellectual property rights, OJ L 341/10, 18/12/2013.

85. (1989) AC 217, [1988] RPC 343, [1988] UKPC 3.

Community Designs), unless the product is a 'modular product' allowing assembly of multiple mutually interchangeable products such as Lego™ bricks (RDA section 1C(3) based on Directive Article 7(3), and equivalent Article 8(3) CDR for Community Designs).

[3] *Other Possible Criteria, Such as Non-obviousness*

Directive Article 11 (and equivalent Article 25 CDR for Community Designs) sets out an exhaustive list[86] of the grounds on which a design can be invalidated. Those that have been transposed into UK law and apply to UK Registered Designs are set out in RDA section 11ZA:

[a] *Not a Statutory Design*

RDA sections 11ZA(1)(a) and 1(2), based on Directive Articles 11(1)(a) and 1(a) for UK Registered Designs and equivalent Articles 25(1)(1) and 3(a) for Community Designs, allows invalidation of registrations for things that do not meet the definition of 'design'. Living organisms are probably excluded on this basis.[87]

[b] *Conflict with Prior Unpublished Design*

This ground (RDA section 11ZA(1A), based on Directive Article 11(1)(d) for UK Registered Designs, and equivalent Article 25(1)(d) CDR) for Community Designs) is usually only relevant where the earlier design (UK, Community or Hague) has not been published before the filing or priority date of the later, because otherwise the later would lack novelty or individual character.[88] It can be used only by the proprietor of the earlier design. Most designs are published quickly after they are filed, but some have publication deferred for up to thirty months. The 'conflict' is decided by determining whether the later design would infringe the earlier (*PepsiCo)*[89].

[c] *Conflict with Prior Distinctive Sign*

If a design uses a trademark protected in the UK (e.g., a national, International or Community Trade Mark), the proprietor of the latter can invalidate the design (RDA section 11ZA(3), based on Directive Article 11(2)(a), and equivalent Article 25(1)(e)

86. Article 84(1), *PepsiCo Inc v. Grupo Promer Mon-Graphic SA* (C-281/10 P) ECLI:EU:C:2011:679, [2012] FSR 5, EU:C:2011:679, [2011] EUECJ C-281/10 and at first instance (Case T-9/07 [2010] ECDR 7), 'Metal Rappers'
87. Appeal Board Decision of 18/2/2013, R 595/2012-3 Fruit & Vegetables, OHIM Alicante Newsletter January 2014.
88. Though OHIM's Boards of Appeal have held that it is also available even where the design is prior published: R 1703/2010-3.
89. *PepsiCo Inc v. OHIM - Grupo Promer Mon-Graphic SA* Metal Rappers, ECLI:EU:C:2011:679, [2012] FSR 5, EU:C:2011:679, [2011] EUECJ C-281/10.

CDR for Community Designs). The comparison is carried out in the same manner as for trademark infringement (General Court Cases T-148/08, T-608/11 *Beifa v. OHIM, Schwan-Stabilo* Instrument for Writing,[90] T-55/12 *Su-Chen v OHIM — AM Denmark* Cleaning devices[91]) so the design may be invalidated whether it is identical or merely similar to the earlier sign.

[d] Conflict with Prior Copyright Work

Similarly, infringement of a copyright work is a ground for attack by the owner of the work (RDA section 11ZA(4), based on Directive Article 11(2)(b) and equivalent Article 25(1)(f) CDR for Community Designs). An unsuccessful example is *Kastenholz v. OHIM - Quatchme;*[92] attacks succeeded in *Soler v. OHIM - Coquet.*[93] As copyright may last long after the author's death in some cases, even long-disappeared designs may still be protected by copyright.

[e] Immorality and Public Policy

RDA sections 11ZA(1)(b) and section 1D, based on Directive Articles 11(1)(b) and 8 for UK Registered Designs (and equivalent Articles 25(1)(b) and 9 CDR for Community Designs) rule out protection for designs which are *'contrary to public policy or to accepted principles of morality'*. The leading case under the previous UK law was *Masterman's Design*[94] concerning a doll with stylized male genitalia.

[f] Proprietor Not Entitled

The true proprietor can attack a design on this basis: RDA section 11ZA(2), based on Directive Article 11(1)(c) for UK Registered Designs (and equivalent Article 25(1)(c) CDR for Community Designs).

[g] Conflict with Prior National & International Symbols

A design cannot reuse certain symbols, specified in Schedule A1 to the RDA (see RDA section 11ZA(1)(c), based on Directive Article 11(2)(c) for UK Registered Designs, and equivalent to Article 25(1)(g) CDR for Community Designs). For example, national flags, Royal arms, private coats of arms, and the Olympic symbol are protected. The relevant public authority can raise this ground.

---

90. *Beifa v. OHIM, Schwan-Stabilo* [2010] EUECJ T-148/08, [2010] ETMR 42 GC, *Beifa v. OHIM, Schwan-Stabilo* Case T-608/11 [2013] ETMR 49 [2013] ECDR 17.
91. Case T-55/12 *Su-Chen v. OHIM — AM Denmark Cleaning devices* ECLI:EU:T:2013:219
92. *Kastenholz v. OHIM - Quatchme* watch dials, [2013] EUECJ T-68/11, EU:T:2013:298, ECLI:EU:T:2013:298.
93. *Soler Hispania (formerly Viejo Valle) v. OHIM - Etablissements Coquet* Cup and Saucer with stripes, [2013] EUECJ T-566/11, ECLI:EU:T:2013:549, EU:T:2013:549.
94. [1991] RPC 89 (RDAT).

[4] Faults in Application

There appears to be no explicit basis allowing the Court to strike down or refuse to enforce a duly registered UK design on formal grounds, but equally it is not explicitly prohibited from doing so. In relation to Community Registered Designs, however, the grant is the act of a European institution over which the UK Courts have only limited jurisdiction, so the grounds set out in Article 25 CDR (which do not include faults in an application) are an exhaustive list.

[B] Marking

Marking is not mandatory, or required to succeed in an infringement action. However, marking may assist in recovering damages.

For UK registered designs, RDA section 24B and for UK UDR, CDPA section 233 provide that damages cannot be awarded against an infringer who '*proves that at the date of the infringement he was not aware, and had no reasonable ground for supposing*', that '*the design was registered*' or that '*design right subsisted*'. Marking a product as protected will assist in removing an innocence defence (if, in the case of a registered design, it includes the registration number).

No innocence defence exists for Community designs (*Choo v. Towerstone*).[95] However, in the assessment of damages for any intellectual property infringement, it is relevant whether '*the defendant knew, or had reasonable grounds to know, that he engaged in infringing activity*' (Regulation 3, The Intellectual Property (Enforcement, etc.) Regulations 2006[96]). Thus, again, marking may assist in recovery.

False marking is, however, a criminal offence.

[C] Non-infringement

The following acts are not infringements.

[1] Crown Use

Use of a UK registered design (RDA section 12, Schedule 1) for the services of the Crown, by (or authorized by) the Government, is not an infringement. Likewise UK unregistered designs (CDPA sections 240–244) and, for defence purposes, Community Designs (Article 23 CDR). Such use may involve payment by the Government to the proprietor or exclusive licensee.

[2] Repair

In addition to whatever general right or licence to repair may exist at common law, the 'Europeanised' design laws contain a 'repair' clause which exempts from infringement any acts for the restoration of a 'complex product' to restore its original appearance.

95. *J Choo (Jersey) Ltd. v. Towerstone Ltd. & Ors* [2008] EWHC 346 (Ch), [2008] E.C.C. 20.
96. SI 2006 No. 102.

For Community registered and unregistered designs, Article 110 CDR provides that:

> protection as a Community design shall not exist for a design which constitutes a component part of a complex product used within the meaning of Article 19(1) for the purpose of the repair of that complex product so as to restore its original appearance.

It has been held to apply only to components which are 'dependent' in appearance, hence not, for example, to alloy wheels.[97] The scope of applicability, though not the legal effect, is therefore similar to the UK UDR 'must match' provisions which are discussed above.

For UK registered designs, a similar defence exists (RDA section 7A(5)):

> The right in a registered design of a component part which may be used for the purpose of the repair of a complex product so as to restore its original appearance is not infringed by the use for that purpose of any design protected by the registration.

It is unclear whether this too is limited to 'must match' spare parts.

For UK UDR, no such defence exists - indeed, the right was intended to offer some limited protection to spare parts - but as noted above, the exclusion of protection for 'must-match' features of components removes some of the need for it.

[3] Other Exclusions from Infringement

There are some standard exclusions from infringement in RDA section 7A(2) based on Directive Article 13 (for UK Registered Designs), with equivalent Article 20 CDR (for Community Designs) and CDPA section 244A & 244B for (UK UDR), namely:

- Private, Non-commercial Use.
- Experimental Use.
- Use for teaching or making citations.
- Use on ships & aircraft temporarily in the UK or for repair of the same.

[D] Other Defences

[1] Prior Use

Good faith use, or preparations for such use, prior to the filing or priority date, are a defence to infringement of a UK registered design (RDA section 7B) or a Registered Community Design (Article 22 CDR). This is a personal right, only transferrable with the business concerned.

97. *Bayerische Motoren Werke v. Round and Metal Ltd* [2013] Bus LR D30, [2012] EWHC 2099 (Pat).

[2] *Exhaustion of Rights*

Where goods are put on the market in the EU/EEA by or with the consent of the proprietor, further acts in relation to those goods are not infringements (Directive Article 15 and equivalent Regulation Article 21). This does not apply where the goods were first put on the market outside the EU (*Sony v. Pacific*)[98].

The same EEA-wide exhaustion of rights is provided for UK UDR by section 228(5) CDPA and for copyright by sections 18(2) and 27(5) CDPA.

[3] *Laches, Acquiescence, Delay, Estoppel*

These equitable defences in England and Wales apply, in principle, to designs as to any other civil litigation. However, in property cases, they rarely assist in the absence of some positive conduct from the proprietor. At most, they may affect the availability of equitable remedies. For a recent summary of the law, see *Fisher v. Brooker,*[99] the '*Whiter Shade of Pale*' case.

§20.07 PROCEDURES FOR FILING APPLICATION FOR REGISTRATION

The following concerns the procedures for filing domestic designs in the UK at the Intellectual Property Office (the 'IPO'). See elsewhere in this volume for details of filing procedures for RCDs and international applications having force in the UK. At the time of writing, the IPO intends to introduce new rules and procedures, and readers are therefore advised to check for amended legislation. Currently, the rules are set out in secondary legislation:

- The Registered Designs Rules 2006 (the 'RDR').[100]
- The Registered Designs Act 1949 and Patents Act 1977 (Electronic Communications) Order 2006.[101]
- The Trade Marks and Registered Designs (Amendment) Rules 2013.[102]
- The Registered Designs (Amendment) Rules 2014.[103]

98. *KK Sony Computer Entertainment and another v. Pacific Game Technology (Holding) Ltd* [2006] EWHC 2509 (Ch), (2006) 29(10) IPD 29076.
99. *Fisher v. Brooker & Ors* [2009] FSR 25, [2009] ECDR 17, [2009] 1 WLR 1764, [2009] Bus LR 1334, [2009] 4 All ER 789, [2009] UKHL 41, [2009] WLR 1764, [2010] EMLR 2. See also *Habib Bank Limited v. Habib Bank AG Zurich* [1982] RPC 1, [1981] 1 WLR 1265, [1981] 2 All ER 650 (CA).
100. SI 2006 No. 1975.
101. SI 2006 No. 1229.
102. SI 2013 No. 444.
103. SI 2014 No. 2405.

A Manual of Practice is expected in future; at present practice points are set out in Design Practice Notices, as follows:

- DPN 1/09: New procedure relating to convention country priority claims in design applications.[104]
- DPN 6/06: Colour in design applications.[105]
- DPN 4/06: Facilitating the restoration of lapsed design registrations.[106]
- DPN 3/06: Ending the withdrawal of some registered designs from public inspection.[107]
- DPN 2/06: Allowing multiple applications.[108]
- DPN 1/06: Ending novelty examination.[109]
- DPN 2/04: US priority documents in support of UK design applications.[110]
- DPN 1/04: Filing requirements for design applications.[111]
- DPN 5/03: Designs dictated by their technical function.[112]
- DPN 3/03: Divide-outs.[113]
- DPN 2/03: Applying to register part of a product.[114]

The basic registration process is brief and simple. On filing, a filing receipt is issued and an examination as to form follows. There is no examination for novelty, or most other matters of substance, but the Registry do examine for, and can refuse.

- Under RDA section 3A(3) if the applicant does not appear to be entitled to apply to register under RDA section 3(2) or (3) or 14; or

Under RDA section 3A(4) on the following substantive grounds:

- Design does not meet the statutory definition under RDA section 1(2).
- Design consists only of features dictated by technical function under RDA section 1C(1) (see DPN 5/03: Designs dictated by their technical function).[115]

104. Designs Practice Notice, DPN 1/09: New procedure relating to convention country priority claims in design applications, 28 Jul. 09.
105. Designs Practice Notice, DPN 6/06: Colour in design applications, 4 Jan. 2008, Amends DPN 6/03.
106. Designs Practice Notice, DPN 4/06: Facilitating the restoration of lapsed design registrations, 20 Nov. 2007.
107. Designs Practice Notice, DPN 3/06: Ending the withdrawal of some registered designs from public inspection, 20 Nov. 2007.
108. Designs Practice Notice, DPN 2/06: Allowing multiple applications, 20 Nov. 2007.
109. Designs Practice Notice, DPN 1/06: Ending novelty examination, 20 Nov. 2007, Replaces DPN 01/03 and DPN 04/03.
110. Designs Practice Notice, DPN 2/04: US priority documents in support of UK design applications, 3 Sep. 2003.
111. Designs Practice Notice, DPN 1/04: Filing requirements for design applications, 3 Sep. 2006.
112. Designs Practice Notice, DPN 5/03: Designs dictated by their technical function, 18 Aug. 2003.
113. Designs Practice Notice, DPN 3/03: Divide-outs, 21 Mar. 2003.
114. Designs Practice Notice, DPN 2/03: Applying to register part of a product, 7 Jan. 2003.
115. Designs Practice Notice, DPN 5/03: Designs dictated by their technical function, 18 Aug. 2003.

- Design consists only of features which must be exactly reproduced to allow interconnection under RDA section 1C(1).
- Design is contrary to morality or public policy under RDA section 1D (based on Directive Articles 11(1)(b) and 8 (and equivalent to Articles 25(1)(b) 9 CDR)), e.g., *Masterman's Design.*[116]
- Design involves a protected national or international symbol listed in RDA Schedule A1.

The Registry will also object if the necessary fees are not paid, or if the various formal requirements are not met.

If there are no objections, the application will be registered typically within around a month. If objections are raised, a time limit is set for response, typically two months. The application must be in order for allowance within an acceptance period of twelve months (RDA section 3(5), RDR Rule 10(1)) or it will be deemed abandoned. On registration, the application is published on the IPO website, and a registration certificate is issued.

Amendment of the design is possible, but where the identity of the design depicted changes, the design may be redated to the date of the amendment, losing the previous filing date (and perhaps also priority date).

If the application is found to contain more than one design, the additional design(s) can be cancelled. If so desired, they can then be filed in a divided out application, retaining the parent filing date (and any priority), which will then be separately examined, see RDA section 3B(3) and 3C(3)(b), and DPN 5/03.

There are two variants to the process. First, a single application can contain multiple designs, at a reduced marginal fee per design. The multiple designs do not need to be linked in any way whatsoever, except in having a common proprietor. On filing, each is allocated a different application number and, other than checking that the fees paid cover all, each is separately examined, registered or refused, and published, see DPN 2/06.

Second, the applicant can request that publication of the design (or, in a multiple application, any design) be deferred for up to a maximum period, currently twelve months from filing. Having done so, the applicant must pay the fee for publication in good time before the maximum period, following which the design will be registered and published as above.

An applicant has the right to a hearing before refusal (RDA section 29). A refusal can be appealed (RDA section 3D), either to the High Court or to an Appointed Person (RDA section 27A). Appeals to the latter are governed by The Appointed Person (Designs) Rules 2015,[117] and provide a cheaper route with no further appeal (though judicial review is possible in principle).

At the time of writing, significant changes of procedure are contemplated. The reader is advised to check more recent secondary legislation for these.

116. [1991] RPC 89 (RDAT).
117. SI 2015 No. 169.

[A] Priority

Priority can be claimed from a foreign design application, within a period of six months from the filing of that application (RDA section 14). A valid priority claim is treated as the filing date for the purposes of determining validity, but not for the purpose of calculating the duration or renewal dates of the design.

The countries from which priority can be claimed ('Convention countries') vary from time to time and are listed in an Order under RDA section 13[118] - at present they include all Paris Convention and WTO countries, and some others with which the UK has bilateral relationships. The UK is not itself a convention country so domestic priority claims are not possible. However, regional and international applications which apply to a convention country are treated as applications in a convention country, so applications for RCDs, or International applications, do give rise to a priority right.

Although the point has not recently been tested in the UK, it is likely that priority cannot be claimed from a patent application (in other words a 'utility patent' application, but not including a 'design patent' in the US or China).[119] The position for Utility Models has likewise not been recently tested, but a priority claim should be allowable, in view of the explicit mention of the possibility in the Paris Convention.[120] If this is the case, it is possible that a priority claim from an international patent application under the PCT (which by default includes a designation of utility model protection in various countries) might likewise be valid.

The adequacy of a priority document in the UK is judged on the basis of the law applicable where it is filed.[121] However, the normal presumption is that the representations should be identical or almost so. Where this is the case, a self-certified declaration can be filed instead of filing a certified copy of the priority application.[122] As the primary focus is on the drawings, filing a translation of the priority application is rarely necessary.

[B] Drawings

The IPO practice is to accept a wide range of different types of representations, or under some circumstances, a sample.[123] Representations may be line drawings, photographs or computer-generated rendered images. There is no maximum limit on the number which can be filed. Nor is there a minimum, but '*There should be a sufficient number*

118. Currently The Designs (Convention Countries) Order 2007 SI 2007 No. 277, as amended by The Designs (Convention Countries) (Amendment) Order 2009 SI 2009 No. 2747 and The Designs (Convention Countries) (Amendment) Order 2013 SI 2013 No. 539.
119. See, by analogy, *Agfa Gevaert's Application* [1982] RPC 441, [1983] IIC 411.
120. Paris Convention, Article 4E(1).
121. *Deyhle's Design Applications*, [1982] RPC 526 (RDAT).
122. See UK IPO Designs Practice Notice DPN 1/09, 28 Jul. 2009, New procedure relating to convention country priority claims in design applications.
123. See UK IPO Designs Practice Notice DPN 1/04: Filing requirements for design applications (and Annex), 3 Sep. 2006.

of different views of the design so as to leave no doubt as to the precise design features for which protection is sought'. This does not require all views of the *article* concerned, as the design may not be visible in some.[124]

[C] Design of Part of Another Article/Environment of Design

It is possible to file for the design of a part of an article even where that part is integral not separable.[125] There is also no per se exclusion on showing environment around the design, although it is not encouraged.

Where the representations do show parts which are not the design claimed, '*[t]he parts in the representation which are not intended to form a part of the design may be indicated by such means as dotted lines, blue wash or as circled areas in red ink*'.[126] However, another visual convention and/or a written disclaimer can be used (see below).[127]

[D] Description of Design (Where Required)

It is conventional to provide a label next to each picture, e.g., 'Front Elevation' etc., so a listing of drawings is not required. Where the design is of only part of the depicted article, it is necessary to indicate the part in writing,[128] and preferable also to do so graphically (see below). There is no requirement for a description of any kind, but the form allows:

- A description,[129] which does not limit the scope of protection.
- A disclaimer, which does.[130]

There is little case law or other guidance on the effect of these under the current law. Disclaimers or descriptions can be used to indicate that the design is of the 'shape and configuration' or the 'pattern and ornament' of the product shown, and to deal with the status of colour,[131] tonal contrast and text which may be present in the drawings.

Where the design is of a repeating surface pattern, it is advisable to specifically indicate this in the application.

124. *Ford Motor Co Ltd's Design Application* [1971] FSR 314, [1972] RPC 320 (RDAT).
125. See UK IPO Designs Practice Notice DPN 2/03: applying to register part of a product, 7 Jan. 2003.
126. *Ibid.*, see also UK IPO Designs Practice Notice DPN 1/04: Filing requirements for design applications (and Annex), 3 Sep. 2006.
127. *Ibid.*
128. RDR Rule 6(b).
129. RDR Rule 4(5).
130. RDR Rule 6(a).
131. See DPN 6/06.

§20.08 EXPUNGING/CANCELLING OR VARYING REGISTRATION

Registered Community Designs, and International designs designating Europe, can only be invalidated centrally by applying to OHIM in Alicante,[132] or by raising a counterclaim in an infringement action.[133]

Parties may apply to invalidate UK registered designs (and, in future, international designs designating the UK) at the IPO,[134] or to rectify the register before the competent UK Court[135] which, within England and Wales, is the Patent Court or the IPEC.

It is also possible to counterclaim for rectification (of a UK design) or invalidation (of a Registered Community Design)[136] in an infringement action:

- *The proprietor* may apply to cancel or amend his UK registered design (RDA section 11).
- *An entitlement challenger* may do likewise after a judgment finding them to be the owner.
- *The owner of, or other relevant person in relation to, an earlier right* (e.g., a copyright, trademark, unpublished design application or national or international symbol) may apply to cancel a UK registered design based on 'relative grounds' (i.e., that use of the design would infringe that earlier right).
- *Any person* may apply to cancel a UK registered design on 'absolute grounds' (e.g., lack of novelty etc.).

A full list of available grounds of attack is set out above in the invalidity section.

Invalidation procedure before the IPO and rectification procedure before the Court are significantly different. At the IPO, parties file statements of case and evidence sequentially. If requested by any party, there is a Hearing, and the IPO has wide procedural powers where necessary, but many cases are determined solely on paper.

The IPO then issues a written decision, which can be challenged by appeal (RDA section 11ZF) to either the Court or an Appointed Person, as indicated above in section G. If one party appeals, the other can file a Respondent Notice resisting the appeal on grounds other than those given by the IPO at first instance. New evidence is rarely admitted on appeal. The IPO may appear or be represented on appeal, but rarely chooses to do so.

Amendment after an IPO invalidation decision is possible, provided the identity of the design remains unchanged. (RDA section 11ZD).

In Court, a trial hearing on the merits is almost always held. Expert witnesses may be called where issues of functionality are in dispute, and there may also be witnesses of fact. There is then an appeal to the Court of Appeal and, ultimately, to the Supreme

132. CDR Articles 24, 52.
133. CDR Articles 81, 85.
134. RDA s. 11ZB.
135. RDA s. 20.
136. Articles 81(d), 84, 85(1) CDR.

Court if an important point of law is to be decided. As in relation to infringement, an award of legal costs to the winning party is usual.

§20.09 OTHER

[A] Interface between Industrial Design and Trademarks

The Trade Marks Act 1994 ('TMA'),[137] implementing EU Directive No. 89/104/EEC[138] (the 'Trade Mark Directive', 'TMD') extended the scope of trade mark protection to include three-dimensional shapes (TMA 1994, section 1(1) corresponding to Article 2 TMD) and so there is an important overlap with design law.

However, TMA section 3(2) (corresponding to TMD Article 3(e) and CTMR Article 7(1)(e)) excludes protection for signs consisting exclusively of:

(i) the shape which results from the nature of the goods themselves;
(ii) the shape of goods which is necessary to obtain a technical result;[139] and
(iii) the shape which gives substantial value to the goods.[140]

These absolute grounds for refusal cannot easily be overcome – in particular, they cannot be overcome by evidence of acquired secondary meaning through use.

[B] Relationship between Industrial Design and Copyright/Design Copyright

Recital 8 of the Design Directive/ Recital 32 CDR establishes the 'principle of cumulation' of protection under design and copyright law. It is implemented in Article 17 DD/ Article 96(2) CDR, allowing Member States to determine the extent to which, and the conditions under which, such a protection is conferred, including the level of originality required.

The UK operates a system of 'partial cumulation' under the CDPA 1988. Surface decoration designs are protectable as artistic works irrespective of artistic quality,[141] if they meet the low 'not copied' originality threshold.[142]

However, section 51 CDPA drastically curtails the copyright protection for designs of the shape or configuration of articles, by providing a wide-ranging defence

137. 1994 c. 26.

138. First Council Directive of 21 Dec. 1988 to approximate the laws of the Member States relating to trademarks, OJ L40/1, 11 Dec. 1989, consolidated as EU Directive 2008/95/EC, OJ L299/25.

139. See *Lego v. OHIM*, C-48/09P, [2010] ECR I-8403, [2010] ETMR 63 (ECJ), ECLI:EU:C:2010:516; *Philips v. Remington* [2003] 2 WLR 294, ECLI:EU:C:2002:377, EU:C:2002:377, [2002] ECR I-5475, [2002] EUECJ C-299/99, [2002] 2 CMLR (52) 1329, [2003] RPC 2), [2002] ETMR 955.

140. *Hauck v. Stokke* C-205/13, [2014] WLR(D) 399, EU:C:2014:2233, ECLI:EU:C:2014:2233, [2014] Bus LR 1284, [2014] EUECJ C-205/13, *Benetton Group SpA v. G-Star International BV* (Case C-371/06), [2007] All ER (D) 118 (Sep), [2007] ECR I-7709; *Bang & Olufsen v. OHIM* (Representation of a loudspeaker), T-508/08, ECR, EU:T:2011:575.

141. Section 4 CDPA.

142. The leading case is *Designers Guild v. Russell Williams* [2000] 1 WLR 2416, [2000] UKHL 58, [2000] WLR 2416, [2001] 1 All ER 700, [2001] ECDR 10, [2001] FSR 11, [2001] FSR 113.

against infringing copyright in any document or model recording such a 'shape and configuration' design where the infringement is to make a product or to copy a product.

The only copyrights which survive this defence are those in documents or models which record the design of an 'artistic work'. And the only relevant statutory categories of artistic works defined by section 4 CDPA (i.e., those which could have a 'shape and configuration', rather than being 'surface decoration') are:

(1) Sculptures, irrespective of artistic quality.
(2) Engravings, irrespective of artistic quality.
(3) Works of artistic craftsmanship.

As to sculptures, the leading case of recent times[143] (concerning Storm Trooper armour and helmets featuring in the film Star Wars™) indicates that a necessary ingredient is an intention to create a work of art to be appreciated visually for its own sake, and that articles with a utilitarian purpose are therefore unlikely to qualify. A multifactorial test for subsistence was employed, for the details of which the reader should review the case.

The same may now be true of engravings, notwithstanding earlier cases protecting utilitarian articles with engraved features.[144]

The leading case[145] concerning works of artistic craftsmanship (relating to a mass-produced sofa) would seem to require both an intention to create a work of art, and the involvement of a craftsman.

Recently, in Case C-168/09 *Flos*[146] the EU Court of Justice suggested a broad interpretation of the principle of cumulation according to which: (i) all registered designs should be entitled to copyright, (ii) for the full term of seventy years after the author's death. The latter finding caused the UK to revoke the existing twenty-two-year term limit for industrially exploited works (section 52 CDPA)[147]. The revocation will come into effect in the near future. At present there appear to be no plans to alter section 51, so few three-dimensional designs will benefit from the repeal of section 52.

[C] Relationship between Industrial Design and Unfair Competition

The UK has no general tort of 'unfair competition'. The one form available is known as 'passing off'. This tort was developed by case law out of the tort of deceit, and has no basis in statute.

---

143. *Lucasfilm v. Ainsworth* [2011] 4 All ER 187, [2011] FSR 41, [2011] UKSC 39, see also the Court of Appeal judgment at [2009] EWCA Civ 1328.
144. *Hi-Tech Autoparts Ltd. v. Towergate Two Ltd.* (No. 1), [2002] FSR 15; (No. 2), [2002] FSR 16, for example.
145. *George Hensher v. Restawhile Upholstery (Lancs) Ltd*, [1975] RPC. 31 (HL), [1976] AC 64, see also the first instance decision in *Lucasfilm v. Ainsworth* [2008] EWHC 1878 (Ch).
146. Case C-168/09 *Flos SpA v. Semeraro Casa e Famiglia SpA* ECLI:EU:C:2011:29, 2011 I-00181, RPC (2011) 128 (5): 303–328.
147. As applied by the Copyright (Industrial Processes and Excluded Articles) (No. 2) Order SI 1989 No. 1070.

In the leading design-related passing off case of modern times, concerning a lemon juice container the shape, size and colour of a lemon, in finding reluctantly for the claimant the House of Lords restated the 'classical trinity' of elements required for success:[148]

> 1. *goodwill or reputation* attached to the goods ... in the mind of the purchasing public by association with the identifying 'get-up' ... under which his particular goods ... are offered to the public, such that the get-up is recognised by the public as distinctive specifically of the plaintiff's goods ...
> 2. a *misrepresentation* by the defendant to the public (whether or not intentional) leading or likely to lead the public to believe that goods ... offered by him are the goods ... of the plaintiff. ... and
> 3. *damage* by reason of the erroneous belief.

Where shape of products is concerned, claimants need to prove '... *that the shape ... is the "crucial point of reference"*' for their buyers - which is generally difficult, particularly so with functional products (in that case, a cushion for medical use).[149]

The problem was explained in a case concerning a popular, and somewhat anthropomorphic, vacuum cleaner:[150]

> It is recognised that it is more difficult to acquire a sufficient reputation and goodwill in the shape or get-up of a product. Whilst the principal function of a brand name is to denote origin, the shape and get up of a product are not normally chosen for such a purpose. A member of the public seeing a product which looks identical to another (a red cricket ball is an example) does not necessarily, or even normally, conclude that they come from the same source.

In that case, survey evidence was successful in showing that the product shape and get-up had acquired goodwill. Designs which are well-known and popular in the UK, and non-functional, may therefore on occasion be protected by passing off, if the defendant has taken insufficient steps to avoid deception, but in general, unless some specific intellectual property right is in play, copying is regarded in the UK as fair rather than unfair competition.

---

148. *Reckitt & Colman Ltd v. Borden Inc* [1990] 1 All E.R. 873, [1990] RPC 341 (the 'Jif Lemon' case), Lord Oliver at p. 880.
149. *Hodgkinson & Corby v. Wards Mobility* [1994] 1 WLR 1564, [1995] FSR 169 (the 'Roho cushion' case), Jacob J (as he then was).
150. *Numatic International Ltd v. Qualtex UK Ltd* [2010] RPC 25, [2010] EWHC 1237 (Ch) (the 'Henry Hoover' case) Floyd J (as he then was) at para. 39.

CHAPTER 21

Industrial Design Rights: United States

George D. Raynal & Perry J. Saidman

§21.01 INTRODUCTION

This chapter focuses on U.S. design patents, which are part of the broader U.S. patent system governed by Title 35 of the U.S. Code. A description of some of the important aspects of design patent practice are provided, as well as their interface between Trademarks and Unfair Competition and Copyright.

[A] Highlights of New Developments in This Jurisdiction since Last Edition

Since the last edition, the realization of economic value of design and design intellectual property has grown significantly, and accordingly the U.S. design patent system and design patent litigation have received more attention. Beginning with the Egyptian Goddess decision in 2008, in which it reset the infringement test for design patents, the U.S. Court of Appeals for the Federal Circuit (Federal Circuit) has addressed a number of key issues and this chapter provides updates regarding anticipation, obviousness, functionality, injunctive relief and infringement based on Federal Circuit jurisprudence. Also since the last edition, the U.S. has changed from a first-to-invent to a first-to-file patent system, adopted several post-grant proceedings to address questions regarding validity following issuance, and reformed patent marking.

§21.02 CONVENTIONS AND LEGISLATION

[A] Conventions

Effective January 24, 1978, the U.S. is a contracting party to the Patent Cooperation Treaty.

Effective May 13, 2015, the U.S. is a contracting party to the Geneva Act of the Hague Agreement concerning the International Registration of Industrial Designs, July 2, 1999.

Effective March 1, 1989, the U.S. is a contracting party to the Berne Convention for the Protection of Literary and Artistic Works of September 9, 1886, as revised at Paris on July 24, 1971, and amended on October 2, 1979.

Effective November 2, 2003, the U.S. is a contracting party to the Madrid Protocol for International Registration of Trademarks.

[B] U.S. Design Patent Law

The provisions of the statutory U.S. patent law, Title 35 of the United States Code ("35 U.S.C."), relating to patents for utilitarian inventions also apply to U.S. patents for ornamental designs, except as otherwise provided.[1] For statutory exceptions specific to design patents, *see* 35 U.S.C. §§ 171, 172, 173 and 289.

Likewise, the U.S. Patent and Trademark Office ("U.S. PTO") Rules, Title 37 Code of Federal Regulations ("37 C.F.R."), relating to applications for utilitarian inventions also apply to applications for U.S. patents for ornamental designs, except as otherwise provided.[2] For rule exceptions specific to design patents, *see* 37 C.F.R. §§152–155.

Similarly, the U.S. PTO Manual of Patent Examining Procedure, Ninth Edition, March 2014, applies to design patents and Chapter 1500 is specific to design patents.[3] The U.S. PTO also publishes a helpful guide to filing a design patent application.[4]

The Leahy-Smith America Invents Act (AIA) was signed into law on September 16, 2011, and section 3 of the AIA amends the patent laws pertaining to the conditions of patentability to convert the U.S. patent system from a first-to-invent system to a first-to-file system. The changes in section 3 went into effect on March 16, 2013.

[C] U.S. Copyright and Trademark Law

Copyright law is set out in title 17 of the U.S. Code and Trademark law is set out in title 15.

1. 35 U.S.C. s. 171.
2. 37 C.F.R. s. 1.151.
3. The M.P.E.P. is available at http://www.uspto.gov/web/offices/pac/mpep/.
4. A Guide to Filing A Design Patent Application, available at http://www.uspto.gov/web/offices/com/iip/pdf/brochure_05.pdf.

§21.03 SUBJECT MATTER

[A] Definition/What Constitutes a Protectable Design?

Whereas statutory subject matter for utility patents is defined by 35 U.S.C. §101, for designs, a substitute provision is found in the first paragraph of 35 U.S.C. §171. According to this paragraph, to be patentable in the U.S., a design must be new, original, ornamental and for an article of manufacture.

The requirements of 35 U.S.C. §171 have been judicially interpreted as requiring that a design meet:

(1) the novelty requirements of 35 U.S.C. §102;
(2) the non-obviousness requirements of 35 U.S.C. §103, but not the utility requirement of 35 U.S.C. §101; and
(3) the requirement for a written description of the manner and process of making and using under the first paragraph of 35 U.S.C. §112, which can be satisfied by what the drawings show.

[1] Shape, Configuration

Unlike the provision in 35 U.S.C. §101 that a utilitarian invention must be "useful" in order to be patentable, the comparable provision in 35 U.S.C. §171 governing patentability of designs instead contains a requirement for ornamentality.

Ornamentality can reside in the configuration (shape) of an article or in the surface ornamentation (decoration) of the article or in a combination of both.[5]

[2] Ornamental or Functional

It is commonly understood that the article of manufacture which is the subject of a design may have a utilitarian purpose and use. Design patent protection is focused on the overall appearance of a claimed design. A design patent does not protect how a design works or general design concepts. In this context, the adjectives ornamental and functional are often used as opposites, however, the law clearly acknowledges that a design may be ornamental such that is patentable and also functional or useful at the same time.

If the appearance of an article is dictated (completely required) by its function, its design is not "ornamental" and it cannot be the subject of a valid U.S. design patent.[6] The standard is not whether any particular feature is functional, but whether the overall appearance of the design as a whole, is dictated by the utilitarian purpose of the

5. M.P.E.P. 1504.01; *Gorham v. White*, 81 U.S. (14 Wall.) 511, 20 L. Ed. 731 (1871); In re Schnell, 46 F.2d 203, 8 USPQ 19 (CCPA 1931); Ex parte Donaldson, 26 USPQ2d 1250 (Bd. Pat. App. & Int. 1992).
6. *In re Carletti*, 328 F.2d 1020, 140 U.S.P.Q. (ENA) 653 (C.C.P.A. 1964).

article.[7] The existence of alternate designs performing substantially the same function has been said to be an important factor in the conclusion that a design is primarily ornamental.[8] A design can be ornamental despite the fact that an article bearing the design may be expected to remain hidden during some significant period of its existence.[9] If there is a question as to ornamentality because the design may be hidden at some state in the interval after manufacture of the article has been completed and before it has been consumed, the applicant must be given the opportunity to provide evidence (in the form of literature and declarations) that at some point during the lifetime of the article, during any period of time between when the manufacture of the article is completed and the design becomes lost due to wear, consumption or other destruction, its appearance is a "matter of concern."

[B] Originality/Novelty

[1] Test for Originality Generally

Although Courts have not often addressed the requirement for "originality" it generally excludes protection for a design derived from any source or person other than the named inventor (designer), including, e.g., a design which simulates an existing, well-known or naturally occurring thing. The Federal Circuit has stated that the purpose of incorporating the originality requirement is unclear.

A U.S. PTO Examiner may reject a claimed design which simulates a well-known or naturally occurring object or person under 35 U.S.C. §171 as non-statutory subject matter in that the claimed design lacks originality by identifying the allegedly "simulated" thing and, if possible, providing evidence of the appearance of the thing so a comparison may be made to the claimed design.[10]

[2] Novelty

Until, March 16, 2013, the U.S. operated under a first-to-invent patent system and applicants enjoyed a one-year grace period from public disclosure to file.

Now under the AIA, patents, including design patents, are evaluated on a "first-inventor-to file" basis, under which any pre-filing disclosure, including prior public use, prior publication and prior registration, constitutes prior art.

Although there are exceptions for disclosures made one year or less before the effective filing date of a claimed design, the exceptions should not be relied on as a grace period because the exceptions are limited and allow for intervening Applicants.

7. *Avia Group International Inc. v. L.A. Gear California Inc.*, 853 F.2d 1557, 7 U.S.P.Q.2d (ENA) 1548 (Fed. Cir. 1988). *L.A. Gear, Inc. v. Thom McAn Shoe Co.*, 988 F.2d 1117 (Fed. Cir. 1993).
8. *Avia Group International, Inc. v. L.A. Gear California, Inc.*, 853 F.2d 1557, 1563, 7 U.S.P.Q.2d (ENA) 1548 (Fed. Cir. 1988).
9. *In re Webb*, 916 F.2d 1553, 16 U.S.P.Q.2d (ENA) 1433 (Fed. Cir. 1990).
10. M.P.E.P. 1504.01(d).

Moreover, the exceptions, including the scope of derivation, have not yet been addressed by courts and expounded in case law.

Section 102 sets forth the novelty requirement. In summary, an applicant shall be entitled to a patent unless the claimed design was patented, described in a printed publication or in public use, on sale, or otherwise available to the public before the effective filing date of the claimed invention. There are limited exceptions for disclosure made one year or less before the effective filing date, e.g., if the subject matter was made or disclosed by the inventor or the subject matter of the disclosure was obtained directly or indirectly from the inventor.[11]

---

11. *(35 U.S.C. s. 102:*

a) *Novelty; Prior Art* — A person shall be entitled to a patent unless —
   (1) the claimed invention was patented, described in a printed publication, or in public use, on sale, or otherwise available to the public before the effective filing date of the claimed invention; or
   (2) the claimed invention was described in a patent issued under section *151*, or in an application for patent published or deemed published under section *122 (b)*, in which the patent or application, as the case may be, names another inventor and was effectively filed before the effective filing date of the claimed invention.

(b) Exceptions —
   (1) *Disclosures made 1 year or less before the effective filing date of the claimed invention* — A disclosure made 1 year or less before the effective filing date of a claimed invention shall not be prior art to the claimed invention under subsection (a)(1) if —
      (A) the disclosure was made by the inventor or joint inventor or by another who obtained the subject matter disclosed directly or indirectly from the inventor or a joint inventor; or
      (B) the subject matter disclosed had, before such disclosure, been publicly disclosed by the inventor or a joint inventor or another who obtained the subject matter disclosed directly or indirectly from the inventor or a joint inventor.
   (2) *Disclosures appearing in applications and patents* — A disclosure shall not be prior art to a claimed invention under subsection (a)(2) if —
      (A) the subject matter disclosed was obtained directly or indirectly from the inventor or a joint inventor;
      (B) the subject matter disclosed had, before such subject matter was effectively filed under subsection (a)(2), been publicly disclosed by the inventor or a joint inventor or another who obtained the subject matter disclosed directly or indirectly from the inventor or a joint inventor; or
      (C) the subject matter disclosed and the claimed invention, not later than the effective filing date of the claimed invention, were owned by the same person or subject to an obligation of assignment to the same person.

(c) *Common Ownership Under Joint Research Agreements*— Subject matter disclosed and a claimed invention shall be deemed to have been owned by the same person or subject to an obligation of assignment to the same person in applying the provisions of subsection (b)(2)(C) if—
   (1) the subject matter disclosed was developed and the claimed invention was made by, or on behalf of, 1 or more parties to a joint research agreement that was in effect on or before the effective filing date of the claimed invention;
   (2) the claimed invention was made as a result of activities undertaken within the scope of the joint research agreement; and
   (3) the application for patent for the claimed invention discloses or is amended to disclose the names of the parties to the joint research agreement.

(d) *Patents and Published Applications Effective as Prior Art*— For purposes of determining whether a patent or application for patent is prior art to a claimed invention under subsection (a)(2), such patent or application shall be considered to have been effectively filed, with respect to any subject matter described in the patent or application —

In order to anticipate a patent claim under section 102, each element of the claim must be found in a single prior art reference.[12] In the case of design patents, although absolute identity to the claimed design is not necessary to constitute an invalidating reference, the reference "must be identical in all material respects."[13] A prior design may be anticipatory even if is shown on an article of different use or in a non-analogous art.[14]

The Federal Circuit has held that the ordinary observer test of "substantially the same" for infringement applies to design patent anticipation, however, this logic has been challenged because the infringement test encompasses a scope that contradicts established law in which the application of the anticipation standard has been significantly more stringent.[15]

[C] Unobviousness (if Appropriate)

Under section 103, if the differences between the subject matter sought to be patented and the prior art are such that the subject matter as a whole would have been obvious at the time the design was made, to a person having ordinary skill in the art to which the subject matter of the design pertains, then that subject matter is not patentable.

Obviousness analysis for all U.S. patents involves three steps: (1) determining the scope and content of the prior art; (2) identifying the differences between the prior art and the claims at issue; and (3) ascertaining the level of ordinary skill in the art.[16]

"Obviousness" is determined from the perspective of a designer of ordinary skill who designs the articles of the type involved.[17]

To reject a design based as obvious, a U.S. PTO Examiner must first identify a single, primary prior art reference which has design characteristics which are basically the same as the claimed design.[18] Without a primary reference, it is improper to reject a claimed design on grounds of obviousness.[19] Once a proper primary (*Rosen*) reference is found, secondary references may be used to modify the primary reference to create a design that has the same overall visual appearance as the claimed design, provided that the primary and secondary references are so related that the appearance

---

(1) if paragraph (2) does not apply, as of the actual filing date of the patent or the application for patent; or

(2) if the patent or application for patent is entitled to claim a right of priority under section *119*, *365 (a)*, or *365 (b)*, or to claim the benefit of an earlier filing date under section *120*, *121*, or *365 (c)*, based upon 1 or more prior filed applications for patent, as of the filing date of the earliest such application that describes the subject matter.

12. *Crown Operations Int'l, Ltd. v. Solutia Inc.*, 289 F.3d 1367 (Fed. Cir. 2002).
13. *Hupp v. Siroflex of America Inc.*, 122 F.3d 1456, 43 USPQ2d 1887 (Fed. Cir. 1997).
14. *In re Glavas*, 230 F.2d 447, 109 U.S.P.Q. (BNA) 50 (C.C.P.A. 1956).
15. *See* Design Patents Sunk in International Seaway, Perry J. Saidman, 83 PTCJ 278 (Dec. 23, 2011).
16. *Graham v. John Deere Co.*, 383 U.S. 1, 17–18 (1966).
17. *Apple*, 678 F.3d at 1329 citing *Titan Tire Corp. v. Case New Holland, Inc.*, 566 F.3d 1372, 1375 (Fed. Cir. 2009).
18. *In re Rosen*, 673 F.2d 388 (C.C.P.A. 1982).
19. *Durling v. Spectrum Furniture Co.*, 101 F.3d 100, 103-04 (Fed. Cir. 1996).

of certain ornamental features in one would suggest the application of those features to the other.[20]

Secondary considerations of non-obviousness may also be considered, such as commercial success, long-felt but unmet need, failure of others, and professional approval or skepticism.

§21.04 DURATION OF PROTECTION

The statutory term for a design patent which issues from an application filed before May 13, 2015 is fourteen years from the issue date.

The statutory term for a design patent which issues from an application filed after May 13, 2015 is fifteen years from the issue date.

The term of the patent which issues from a continuing application (i.e., a divisional application, a continuation application or a continued prosecution application (CPA)) is measured from the issue date for that application, not its preceding U.S. application(s).

§21.05 INFRINGEMENT

[A] Scope of Infringement Action

A U.S. design patent is a federal government grant which allows the owner to exclude others from making, using, offering to sell or selling in the United States, or importing into the U.S., an article of manufacture bearing the patented design.

A design patent may be enforced through a civil action filed in a U.S. District Court, with appeals made to the U.S. Court of Appeals for the Federal Circuit, and then the Supreme Court of the United States.

The body of case law regarding the litigation of utility patents is procedurally and substantively well developed in the U.S. civil court system, and may generally be applied to design patent litigation. In some instances this works fine, in others, design patents are unique. The body of case law which specifically addresses design patent litigation is limited.

[B] Test for Infringement/Criteria to Constitute Infringement

Typically, the figures of an asserted design patent are directly compared to photographs of the accused article of manufacture, however it may be acceptable to compare an article of manufacture that clearly embodies (is identical to) the patented design to the accused article of manufacture.

Section 289 which speaks to profits as a measure of damages for design patent infringement, discussed below, frames design patent infringement analysis as whether

20. In re Glavas, 220 F.2d 447, 450 (C.C.P.A. 1956).

the defendant has applied the patented design, or a "colorable imitation thereof" to an article of manufacture.

The U.S. Supreme Court set for the basic test for design patent infringement in 1871 in *Gorham v. White*:

> We hold, therefore, that if, in the eye of the ordinary observer, giving such attention as a purchaser usually gives, two designs are substantially the same, if the resemblance is such as to deceive such an observer, inducing him to purchase one supposing it to be another, the first one patented is infringed by the other.[21]

Note that it is not necessary to introduce evidence of actual confusion or deception; design patent infringement analysis involves a simple comparison of the designs.[22] In *Apple v. Samsung*, the Federal Circuit addressed actual deception and confirmed that infringement does not require that any purchasers actually were deceived or confused.[23]

In *Egyptian Goddess v. Swisa*, the U.S. Court of Appeals for the Federal Circuit addressed several issues regarding design patent infringement analysis including claim construction and the role of prior art in infringement analysis.

Design patents are subject to the *Markman* requirement that a patent's claim be construed as a matter of law prior to determining infringement.[24] Before *Egyptian Goddess*, the parties would submit competing verbal descriptions of the claimed design and the district court would adopt a claim construction. This process lent itself to gamesmanship by the parties and often resulted in an unduly narrow claim construction. In *Egyptian Goddess*, the Federal Circuit put this practice to rest, stating:

> Given the recognized difficulties entailed in trying to describe a design in words, the preferable course ordinarily will be for a district court not to attempt to construe a design patent by providing a detailed verbal description of the claimed design.[25]

Although the court did not strictly prohibit design patent claim verbalization, it clearly discouraged the practice, and since 2008, most district court elect to construe the claim by simply referring to what the illustrations show, perhaps with some brief guidance or description.[26]

Also in *Egyptian Goddess*, the Federal Circuit addressed the role of prior art in infringement analysis. Beginning in 1985, the test for design patent infringement added

---

21. 81 U.S. 511 (1871).
22. *Braun Inc. v. Dynamics Corp. of America*, 975 F.2d 815, 24 U.S.P.Q.2d (BNA)1121 (Fed. Cir. 1992).
23. *Apple Inc. v. Samsung Electronics Co., Ltd.*, 786 F.3d 983, 999-1000 (Fed. Cir. 2015).
24. *See Elmer v. ICC Fabricating*, Inc., 67 F.3d 1571, 1577 (Fed. Cir. 1995).
25. *Egyptian Goddess* at 679.
26. The Federal Circuit did suggest certain things a court might do as part of *Markman* claim construction that might be helpful to the trier of fact, including: (1) pointing our various features of the claimed design as they relate to the accused design and the prior art; (2) describing the role that conventions play in design patent drafting (e.g., broken lines and surface shading); (3) describing the effect of the prosecution history of the patented design; and (4) distinguishing between those features of the claimed design that are ornamental and those that are purely functional. *Egyptian Goddess* at 680.

a "point of novelty" test following the "ordinary observer" test of *Gorham*.[27] Under this second prong, to infringe, an accused design had to copy the "point of novelty" that distinguished the claimed design from the prior art. In *Egyptian Goddess*, the Federal Circuit held that the "point of novelty" test should no longer be used and confirmed that the "ordinary observer" test should be the sole test for determining whether a design patent has been infringed.[28] Throughout the opinion, the Federal Circuit offered guidance for lower courts to consider regarding the role of prior art in design patent infringement which is distilled into the following guidelines:

(1) When the claimed design is close to the prior art designs, small differences between the accused design and the claimed design are likely to be important to the eye of the hypothetical ordinary observer.
(2) If the accused design has copied a particular feature of the claimed design that departs conspicuously from the prior art, the accused design is naturally more likely to be regarded as deceptively similar to the claimed design, and thus infringing.
(3) If the claimed design consists of a combination of old features that creates an appearance deceptively similar to the accused design, even to an observer familiar with similar prior art designs, a finding of infringement would be justified.
(4) In some instances, the claimed design and the accused design will be sufficiently distinct that it will be clear without more that the patentee has not met its burden of proving the two designs would appear "substantially the same" to the ordinary observer.
(5) [W]hen the claimed and accused designs are not plainly dissimilar, resolution of the question ... will benefit from a comparison of the claimed and accused designs with the prior art.
(6) Where there are many examples of similar prior art designs ... differences between the claimed and accused designs that might not be noticeable in the abstract can become significant to the hypothetical ordinary observer who is conversant with the prior art.
(7) [E]xamining the novel features of the claimed design can be an important component of the comparison of the claimed design with the accused design and the prior art. But the comparison of the designs, including the examination of any novel features, must be conducted as part of the ordinary observer test, not as part of a separate test focusing on particular points of novelty that are designated only in the course of litigation.

In *Apple v. Samsung*, the Federal Circuit confirmed that the fact finder must consider the prior art admitted at trial in determining infringement.[29]

27. *See Litton Systems, Inc. v. Whirlpool Corp.*, 728 F.2d 1423, 1444 (Fed. Cir. 1985).
28. *Egyptian Goddess* at 678.
29. *Apple*, 786 F.3d at 1000.

[C] Standing to Sue

The following parties have standing to file a design patent infringement action filed in a U.S. District Court:

The patentee, who may be the inventor(s) or the assignee of the inventor(s);
The patentee's successor in title; and
An exclusive licensee of the patent, whose license grants "all substantial rights" in the asserted patent.

[D] Remedies for Infringement

Title 35 Chapter 29 addresses the remedies available for patent infringement, including injunctions under section 283, damages under section 284, and attorney's fees under section 285, and an additional remedy for design patent infringement is provided in section 289.

[1] Injunctions under Section 283

To prevail on a motion for an injunction, including a preliminary injunction, the party seeking the injunction must show all of the following: (1) a likelihood of success on the merits; (2) a likelihood of suffering irreparable harm without injunction; (3) the balance of equities favors the party seeking the injunction; and (4) the public interest would be served by the injunction.[30]

In 2012, the Federal Circuit held that a patentee seeking injunctive relief must establish a "causal nexus" relationship between the alleged infringement and the alleged harm in order to satisfy the irreparable harm requirement.[31] For an injunction against design patent infringement, the party seeking injunction must make a clear showing that the infringement of its design was responsible for an actual loss or harm. Design dilution, otherwise known as erosion of design distinctiveness, is a viable theory for irreparable harm, under which the plaintiff must provide concrete evidence of design dilution as it relates to the patentee's reputation for innovation.

[2] Damages under Section 284

Upon finding for the claimant the court shall award the claimant damages adequate to compensate for the infringement, but in no event less than a reasonable royalty for the use made of the invention by the infringer, together with interest and costs as fixed by the court. When the damages are not found by a jury, the court shall assess them. In either event the court may increase the damages up to three times the amount found or assessed. Increased damages under this paragraph shall not apply to provisional rights

30. *Apple Inc. v. Samsung Elecs. Co.*, 678 F.3d 1314, 1319 (Fed. Cir. 2012).
31. *Ibid.*

under section 154(d). The court may receive expert testimony as an aid to the determination of damages or of what royalty would be reasonable under the circumstances.

[3] Attorneys Fees under Section 285

The court in exceptional cases may award reasonable attorney fees to the prevailing party.

[4] Design Patent Infringement under Section 289

Section 289 explicitly authorizes the award of total profit from the article of manufacture bearing the patented design. The award of total profit is not subject to causation or apportionment, for example, when the accused product comprises more than the subject article of manufacture.[32]

Section 289 states in full:

> Whoever during the term of a patent for a design, without license of the owner,
> (1) applies the patented design, or any colorable imitation thereof, to any article of manufacture for the purpose of sale, or
> (2) sells or exposes for sale any article of manufacture to which such design or colorable imitation has been applied shall be liable to the owner to the extent of his total profit, but not less than $250, recoverable in any United States district court having jurisdiction of the parties.
> Nothing in this section shall prevent, lessen, or impeach any other remedy which an owner of an infringed patent has under the provisions of this title, but he shall not twice recover the profit made from the infringement.

§21.06 DEFENSES TO INFRINGEMENT

[A] Validity

All patents, including design patents, issue with a presumption of validity.[33] To overcome this presumption of validity, the party challenging a patent must prove facts supporting a determination of invalidity by clear and convincing evidence.[34]

[1] Not Novel /Original

A party challenging the novelty of a claimed design must provide clear and convincing evidence that a single prior art reference shows a design that is identical in all material respects.

32. *Apple*, 786 F.3d at 1001-02.F.3d – (Fed. Cir. 2015).
33. 35 U.S.C. s. 282 (2000).
34. *Rosco, Inc. v. Mirror Lite Co.*, 304 F.3d 1373, 1377-78 (Fed. Cir. 2002).

A party challenging the originality of a claimed design must provide clear and convincing evidence that the claimed design is not original.

[2] Functional/Not Ornamental

A design patent can be declared invalid if there is clear and convincing evidence that the claimed design is dictated by the utilitarian purpose of the article instead.[35]

The determination of whether the patented design is dictated by the function of the article of manufacture rests on an analysis of its overall appearance.[36]

When performing this assessment, "the claimed design should be viewed in its entirety, for the ultimate question of not the functional or decorative aspect of each separate feature, but the overall appearance of the article ... ."[37]

A leading Federal Circuit case on functionality is *Best Lock v. Ilco Unican*, in which a three-judge panel determined that a design patent for a key blade blank was invalid because the blank design was dictated solely by the key blade's function of having to fit within its corresponding keyway.[38] In other words, the patent was invalid because it did not satisfy the statute's ornamentality requirement, i.e., every feature of the design was dictated by functional considerations, i.e., there were no alternate designs that would perform the function of fitting in the corresponding lock's keyway.[39]

When there are several ways to achieve the function of an article of manufacture, the design of the article is more likely to serve a primarily ornamental purpose.[40]

During enforcement, it is not uncommon for non-infringement to be found based in part on asserting that at least aspects of the claimed design are functional and should not be considered in comparison against the accused design. In some instances this has occurred when the claimed and accused designs do not share substantially the same overall appearance but share a design concept, and instead of making a determination based on appearance, the Court diverges into the subject of functionality.[41] In *Apple v. Samsung*, the Federal Circuit confirmed there is no rule to eliminate entire elements from the scope of a patented design.[42]

---

35. *High Point Design, LLC v. Buyers Direct, Inc.*, 730 F.3d 1301, 1314 (Fed. Cir. 2013); L.A. Gear, 988 F.2d at 1123.
36. *Berry Sterling Corp. v. Prescor Plastics, Inc.*, 122 F.3d 1452, 1456 (Fed. Cir. 1997).
37. L.A. Gear, 988 F.2d at 1123.
38. 94 F.3d 1563, 40 U.S.P.Q.2d 1048 (Fed. Cir. 1996).
39. *Id.* at 1565–1566.
40. *Berry Sterling Corp. v. Pescor Plastics, Inc.*, 122 F.3d 1452 (Fed. Cir. 1997), *Rosco, Inc. v. Mirror Lite Co.*, 304 F.3d 1373 (Fed. Cir. 2002).
41. *See Richardson v. Stanley Works, Inc.*, 597 F.3d 1288 (Fed. Cir. 2010); *Lee v. Dayton-Hudson Corp.*, 838 F.2d 1186, 1188 (Fed. Cir. 1998).
42. *Apple*, 786 F.3d at 998-99.

[3] Other Possible Criteria, Such as Non-obviousness

Obviousness is a question of law based on the following underlying factual inquiries: (1) the scope and content of the prior art; (2) the level of ordinary skill in the art; (3) the differences between the claimed invention and the prior art; and (4) objective evidence of non-obviousness.

When assessing the potential obviousness of a design patent, two steps are employed: first, there must exist be a single reference, something in existence, the design characteristics of which are basically the same as the claimed design; second, once such a primary reference is found, other references may be used to modify it to create a design that has the same overall visual appearance as the claimed design.[43] Secondary references used to modify a primary reference must be visually so related to the primary reference that the appearance of certain ornamental features in one would suggest the application of those features to the other.[44]

Under the first step, a court must both "(1) discern the correct visual impression created by the patented design as a whole; and (2) determine whether there is a single reference that creates 'basically the same' visual impression."[45]

A court should add sufficient detail to its verbal description to evoke a visual image consonant with that design, and should compare the designs side-by-side to provide its reasoning as to whether there is a single reference that creates "basically the same" visual impression.[46]

The ultimate inquiry in an obviousness analysis is "whether the claimed design would have been obvious to a designer of ordinary skill who designs the articles of the type involved."[47]

[4] Faults in Application

During the pendency of a U.S. design patent application, faults or errors in inventor and applicant information may corrected by filing an appropriate response to a notice identifying the fault or by petition, e.g., to correct inventorship. Upon examination, a U.S. PTO Examiner may object to the specification (figure descriptions) or drawings, and require the applicant to make amendments to address the examiner's concerns. More serious concerns, particularly regarding the drawings, may cause the Examiner to reject the claimed design under 35 U.S.C. §112. Such concerns might be inconsistency between figures or inadequately disclosed contours. Drawing issues are discussed further below.

43. High Point Design, 730 F.3d at 1311; citing *Durling v. Spectrum Furniture Co.*, 101 F.3d 100, 103 (Fed. Cir. 1996); *Apple*, 678 F.3d at 1329.
44. In re Rosen, 673 F.2d. 388, 39 (CCPA 1982); In re Borden, 90 F.3d 1570 (Fed. Cir. 1996).
45. *Durling*, 101 F.3d at 103.
46. *High Point Design* at 1314.
47. Durling, 101 F.3d at 103; High Point Design at 1313.

[B] Marking

Title 35 Chapter 29 addresses marking under section 287 and false marking under section 292.

[1] Marking under Section 287

Marking is not essential to maintaining or enforcing patent rights, however, a party enforcing patent rights can only collect damages for infringement if notice of the patents at issue is provided. Notice may be actual notice to the accused party, e.g., by demand letter, or constructive by traditional or virtual patent marking. With constructive notice by patent marking, damages can be recovered for infringement since the date the patent issued for a period up to six years prior to when the patent infringement action was filed in court.

Traditional patent marking consists of marking a patented article of manufacture with the word "patent" or the abbreviation "pat." and the patent number. If marking the patented article is not practical, marking the packaging, labels or tags may suffice.

The 2011 Leahy-Smith America Invents Act ("AIA") permits virtual patent marking, which constitutes constructive notice when a patented article of manufacture is marked with the word "patent" or the abbreviation "pat." and an internet address for a website at which the patented article of manufacture is associated with one or more patents. The website should be regularly updated as patents issue or expire.

[2] False Marking under Section 292

The AIA amended this section to address rampant false marking litigation.

False marking remains prohibited:

> Whoever, without the consent of the patentee, marks upon, or affixes to, or uses in advertising in connection with anything made, used, offered for sale, or sold by such person within the United States, or imported by the person into the United States, the name or any imitation of the name of the patentee, the patent number, or the words "patent," "patentee," or the like, with the intent of counterfeiting or imitating the mark of the patentee, or of deceiving the public and inducing them to believe that the thing was made, offered for sale, sold, or imported into the United States by or with the consent of the patentee; or Whoever marks upon, or affixes to, or uses in advertising in connection with any unpatented article, the word "patent" or any word or number importing that the same is patented, for the purpose of deceiving the public; or Whoever marks upon, or affixes to, or uses in advertising in connection with any article, the words "patent applied for," "patent pending," or any word importing that an application for patent has been made, when no application for patent has been made, or if made, is not pending, for the purpose of deceiving the public — Shall be fined not more than $500 for every such offense.

Since the AIA, only the United States may sue for the penalty authorized by this subsection.

A private plaintiff may allege false marking in a civil action, but must demonstrate a competitive injury.

Finally, marking an article of manufacture with an expired patent no longer constitutes false marking.

[C] Non-infringement

In the U.S., patent infringement is a strict liability cause of action, with no exceptions for fair use or innocent infringement. A party which has been accused of infringing a U.S. design patent may affirmatively seek a declaratory judgment of non-infringement from an appropriate U.S. District Court. A party which has been accused of infringing a U.S. design patent in a formal civil action may answer the complaint by asserting non-infringement according to the criteria discussed above, and by requesting that the District Court grant summary judgment of non-infringement. In litigation since *Egyptian Goddess*, it has become common for a party denying infringement to allege that as a threshold matter, the accused design or product and the asserted patent are plainly dissimilar or sufficiently distinct in overall appearance such that a decision of non-infringement may be reached without consideration of the prior art. art.

[1] Inequitable Conduct

A court may determine that a patent which is otherwise valid, is nevertheless invalid and unenforceable due to inequitable conduct on behalf of the patentee. A party seeking to establish inequitable conduct must prove that the applicant breached its duty of candor and good faith to the U.S. PTO by failing to submit or misrepresenting information that is "material" to patentability, with the intent to deceive the U.S. PTO. Information is material to patentability if the U.S. PTO would not have issued the patent had it been aware of the information or misrepresentation.[48]

§21.07 PROCEDURES FOR FILING APPLICATION FOR REGISTRATION

A U.S. design patent application must include a specification, which generally comprises illustrations of the claimed design and some brief description thereof, a claim in standard format ("I / We claim the ornamental design for a [name of the article of manufacture] as shown and described"), and a declaration from the inventor(s). A U.S. design patent application may optionally include a transmittal form, which identifies the application parts being submitted, and an Application Data Sheet (ADS) form, which provides information about the inventor(s)/designer(s), and the applicant/assignee, as applicable, as well as the representative of the inventor inventor(s)/designer(s) or applicant/assignee. The identity of the inventor(s)/designer(s) and the applicant/assignee, the ADS, the declaration of the inventor(s)/designer(s) and the

48. *Therasense, Inc. v. Becton, Dickinson & Co.*, 649 F.3d 1276, 1288–1289 (Fed. Cir. 2011).

required fees, as well as other supporting documents (e.g., a power of attorney) may be supplied after the initial filing. The specification of a U.S. design patent application may also include a separate Appendix document comprising additional disclosure in the form of illustrations and descriptions which are not claimed.

Inventors, applicants and their representatives have a duty to disclosure information material to patentability to the U.S. PTO under 37 C.F.R. 1.56. To fulfill that duty, an Information Disclosure Statement may be filed either with the initial application or during prosecution, to disclose known references which are material to the patentability of the claimed design. Such references might include the closest known prior art and co-pending design and utility patent applications of the same inventors and/or applicants, as well as references cited by a foreign patent office in a counterpart application. There is no duty to submit information which is not material to the patentability of any existing claim. A reference is material to the patentability of the claimed design when it is not cumulative to information already of record or being made of record in the application, and: (1) it establishes, by itself or in combination with other information, a prima facie case of unpatentability of a claim; or (2) it refutes, or is inconsistent with, a position the applicant takes in: (i) opposing an argument of unpatentability relied on by the Office, or (ii) asserting an argument of patentability.

Multiple Embodiment Applications

It is permissible to file a U.S. design patent application with multiple embodiments (which may be the same design shown with varying scope or designs that are alternatives or variants), however, a U.S. design patent will issue with only one claim. During prosecution an examiner must determine whether different embodiments or designs claimed in the same application are "basically the same" (i.e., obvious in view of each other) in which case they can remain in the same application, or patentably distinct (i.e., nonobvious in view of each other) in which case the examiner will issue an office action requiring the applicant to elect one embodiment for prosecution and cancel the others.[49] The applicant is entitled to file divisional applications on the non-elected, cancelled embodiments.

The decision whether to file multiple embodiments in one design patent application versus separate applications for each embodiment has taken on increased importance in view of recent U.S. case law.

If the multiple embodiments have minor differences, the decision warrants careful attention. An accused product may not be exactly like any of the embodiments, differing in minor ways. A determination that the multiple embodiments appear basically the same as each other provides good basis for an argument that the infringer's product is also substantially the same, and thus infringing. If the multiple embodiments with minor differences are deemed patentably distinct, i.e., not basically the same as each other, then during later enforcement an accused infringer could argue that the differences between the multiple embodiments are equivalent to or greater than the differences between the asserted patent and the accused design, and therefore

49. 35 U.S.C. s. 121. *See* M.P.E.P. 1504.05.

the accused design does not infringe. This has occurred in an actual litigation context, and the district court concluded no infringement.[50] Moreover, in a similar situation, the Federal Circuit has recently confirmed that prosecution history estoppel applies to design patents and may bar a claim for infringement when the applicant surrenders an embodiment in response to a restriction requirement and the accused design is within the scope of surrender.[51]

[A] Priority

In order to be entitled to the benefit of the earlier filing date in a non-U.S. Paris-convention member country or an international design registration application at WIPO, the U.S. design patent application must be filed within six months of the non-U.S. filing.[52]

A claim for priority based on a prior non-U.S. application generally involves identifying the prior application and claiming priority rights from it in the applicant's declaration for the application, identifying the priority claim in an Application Data Sheet form filed with the U.S. PTO, incorporating the priority document by into the specification by reference, and filing a certified copy of the prior application as filed, by no later than the date the issue fee of an allowed application is paid.

[B] Drawings

A design patent claims a design "as shown and described." The description of a design is generally brief, and most often simply describes the orientation of the figures and explains the use of broken lines, boundary lines, enlargements, cross sections, surface shading and the like, and does not generally describe the appearance of the design itself or its features.

The drawings themselves represent the claimed design and are therefore a critical part of the application. Therefore, all efforts should be made before filing in the U.S. to ensure that the drawings clearly illustrate the claimed design to meet the "written description" and "enablement" requirements of section 112. An Examiner might reject the claimed design where the Examiner finds it difficult to discern the shape and contour of the design from the illustrations. If the illustrations are alleged to be indefinite because the shape and contour is not adequately shown, it may be possible to overcome the rejection by amending the drawings to clearly illustrate or disclaim the allegedly indefinite portions. If the illustrations are alleged to be inconsistent, it may be possible to overcome the rejection by amending the drawings for consistency, or to disclaim the allegedly inconsistent portions. When amending the drawings after filing, care must be taken to ensure that no new matter is introduced.

50. *See Great Neck Saw Mfg. v. Star Asia*, U.S.A., LLC, 727 F.Supp.2d 1038, 1056 (W.D. Wash. 2010).
51. *Pacific Coast Marine Windshields Ltd. v. Malibu Boats, LLC*, 739 F.3d 694 (Fed. Cir. 2014).
52. 35 U.S.C. s. 119, as modified for design applications by 35 U.S.C. s. 172.

Line drawings, photographs and digital images (renderings) are all acceptable means to illustrate a claimed design. Color is permitted when color is an integral part of the claimed design, which must be stated in a petition to accept color illustrations, along with a fee.

During enforcement, the patent drawings will often be compared to photographs of a physical embodiment of the actual accused design, therefore it is important that the patent drawings convey the look of a physical product. For this reason, CAD or engineering drawings which have detail such as contour lines at edges and radii of curvature are generally not preferred. Line drawings suitable for design patents may be prepared from CAD or engineering drawings or 3D models, removing unnecessary detail and conveying curvature to make the design appear life-like.

Although a U.S. design patent application may include multiple embodiments of a design illustrated in different formats, a design patent may issue with only one illustration format.

The drawings should show the overall appearance of the claimed design from all sides. If a side generally is not regarded as part of the claimed design, e.g., the bottom side of an automobile, it may be excluded, and a statement may be added to the specification to explain that the excluded side forms no part of the claimed design. There is no steadfast rule about what views must be shown, and it may be unduly limiting to illustrate views that are unimportant or do not add to an understanding of the claimed design.

Many U.S. PTO examiners expect to see the standard six orthogonal views, as well as one or more perspective views which may help show contour and depth and inter-relate the six orthogonal views. A requirement for missing views may be challenged with arguments that the claimed design is fully disclosed in the provided views and any part of the design not shown forms no part of the claimed design.

Surface shading is not required, however, the drawings should reveal the shape and contour of the claimed design, and should distinguish apertures from solid surfaces and depressions and protrusions, and surface shading is useful in this regard. Surface shading may also be used to claim color contrast, transparency or translucency and material or finish (e.g., plastic, metallic or glossy). Many U.S. PTO examiners expect to see surface shading and may require it as a matter of course, however a requirement for surface shading may be challenged with arguments that the contour of the claimed design is clear from the totality of the drawings.

Cross-sectional views are permissible, but not necessarily required. One or more cross-sectional views may help reveal the shape and contour of the claimed design, particularly for designs with complex curvature or features of which the depth ore relative location is not apparent from the standard views. However, cross-sectional views may be unduly limiting if used when not necessary or not carefully prepared to avoid claiming more than necessary or desired. Even if an aspect of the claimed design is not readily understood in any single view, it might nevertheless be taught by the other views taken together such that the design satisfies the requirements of section 112.

As noted above, the specification of a U.S. design patent application may include a separate Appendix document comprising additional disclosure that is not claimed.

This disclosure may include additional illustrations of the claimed design, e.g., photographs or digital images, which may provide a basis for amending the claimed design to overcome a rejection under section 112 or a prior art rejection, as well as disclosure of additional embodiments as well as alternative designs. The use of an Appendix may be advisable where there is uncertainty about the clarity or comprehensiveness of the application drawings. An Appendix may also provide disclosure that may be pursued in continuing applications, illustrating various features of the design or related designs to be claimed later.

[C] Design Part of Another Article/Environment of Design

An applicant should be wary of unduly limiting the claimed design by including minor details or features as part of the claim. As a U.S. design patent claim simply references the design as shown in the drawings (or as shown and described), it is important that the drawings do not show any feature in solid lines unless such feature is part of the design desired to be protected. An issued patent may not disclaim details or features by description in the specification if those features are shown in solid lines in the drawings, however upon filing an application, an applicant may use descriptive language in the specification to enable amendment of the drawings during prosecution.

All details or features that do not constitute the essence of the overall design may be illustrated but disclaimed by showing them in broken lines.[53] During prosecution, broken lines may be converted to solid lines and solid lines may be converted to broken lines, so long as it can be successfully argued that the resulting design finds support in the original disclosure. Broken lines may also be used to represent unclaimed environment to provide context. If the illustrations of the claimed design are photographs or digital images, it is permissible to show unclaimed structure, features or environment in broken lines, or by significantly fading the unclaimed portions relative to the claimed design and providing a statement that the faded portions form no part of the claimed design. Occasionally, it is helpful to employ a boundary line (e.g., a dot-dash line) around faded portions for clarity.

[D] Description of Design (Where Required)

There must be only one claim in a design patent application and it must be in standard format: "I / We claim the ornamental design for a [name of the article of manufacture] as shown and described." Any such description of the claimed design is typically brief, comprising basic figure descriptions (e.g., "FIG. 1 is a front perspective view showing our claimed design; FIG. 2 is a front view thereof; FIG. 3 is a right side view thereof."), a description of the use of broken lines, if any, (e.g., "The broken lines illustrate structure, features or environment that forms no part of the claimed design") and perhaps a brief description to clarify surface shading (e.g., "The oblique surface shading represents transparency." or "The drawings include surface shading which

53. *In re Zahn*, 617 F.2d 261, 204 U.S.P.Q. (BNA) 988 (C.C.P.A. 1980).

represents contour, not surface ornamentation or decoration.") or other aspects of the illustrations. In other words, it is really the illustrations that are described, not the claimed design itself. If the illustrations of the claimed design are photographs or digital renderings, it may be advisable to provide additional description of the illustrations, e.g., that shadows, background and reflection represent contour, not surface ornamentation or decoration, and that background, labels, logos etc. form no part of the claimed design.

Beyond these simple statements, description of the claimed design may not be necessary and may not be advisable, as it is generally understood that the drawings themselves are the best representation of the claimed design and superfluous description could be unduly limiting.

Although common in many foreign countries, the use of a characteristic feature clause is rare and not generally recommended. However, a characteristic feature clause may sometimes be helpful to express an important feature of the claimed design which may not be easily illustrated.

As noted above, the specification of a U.S. design patent application may include a separate Appendix document comprising additional disclosure that is not claimed. This disclosure may include more expansive description, including description of the disclosed design and description of alternative designs.

§21.08 EXPUNGING/CANCELLING OR VARYING REGISTRATION

[A] Certificate of Correction

After a design patent has issued, faults or errors may be corrected by Certificate of Correction pursuant to 35 U.S.C. 254 for mistakes made by the U.S. PTO and 35 U.S.C. 255 for mistakes made by the applicant.

[B] Reissue

Under 35 U.S.C. 251, a patentee may file an application for the U.S. PTO to reissue a design patent by alleging that the patent is invalid by reason of a defective specification or drawing, or by reason of the patentee claiming more or less than he had a right to claim in the patent. No new matter shall be introduced into the application for reissue. A reissue application which attempts to broaden the claim must be filed within two years of the issue date of the original patent. The term of a reissue patent is the remaining unexpired part of the term of the original patent.

[C] Post-Grant Review

A U.S. design patent with an effective filing date after March 15, 2013 may be the subject of a petition for post-grant review based on any statutory ground. A post-grant review is a trial-like proceeding before the Patent Trial and Appeal Board (PTAB). In a post-grant review, the subject patent is not presumed valid. The claims of a patent are given their broadest reasonable interpretation. To prevail, a petitioner must unpatentability by a preponderance of the evidence.

[D] Inter Partes Review

Also at the PTAB, *any* U.S. design patent may be subject to an Inter Partes Review (IPR) for patentability under sections 102 (novelty) or 103 (obviousness), and only on the basis of prior art consisting of patents or printed publications. An IPR may be instituted upon a showing that there is a reasonable likelihood that the petitioner would prevail with respect to at least one claim challenged. If the proceeding is instituted and not dismissed, a final determination by the Board will be issued within one year (extendable for good cause by six months).

[E] Supplemental Examination

Under the Supplemental Examination provisions of the AIA, the owner of a U.S. design patent may request that the U.S. PTO consider, reconsider, or correct information believed to be relevant to the patent. A request for supplemental examination may be filed by a patent owner at any time during the term of the patent. A request for supplemental examination may be based on any information (up to twelve items per request) that the patent owner believes to be relevant to the patent and is not limited to patents and printed publications. Within three months of the request, the U.S. PTO Director shall conduct the supplemental examination and shall conclude such examination by issuing a certificate indicating whether the information presented in the request raises a substantial new question of patentability, and if it does, the U.S. PTO will require re-examination of the patent. Generally, a patent may not be held unenforceable based on conduct relating to information considered by the U.S. PTO during a supplemental examination proceeding.

§21.09 OTHER

[A] Interface between Industrial Design and Trademarks

Protection of trademarks and trade dress arises from the law of unfair competition. Unfair competition law forbids firms operating in the marketplace from committing unfair acts against competitors which would be likely to deceive or confuse the public.

A trademark or service mark is a word, design, or symbol, including trade dress, used by a manufacturer or merchant to identify its goods or services and to distinguish them from those of others. A trademark protects the owner of the mark and the public from unfair competition, i.e., unfairly passing off goods or services as those of another. Trademark rights may be obtained in the U.S. by using the mark in commerce, and those rights may be confirmed and enhanced by registering the mark with the U.S. PTO for a ten-year renewable term. Trademark rights last as long as the mark is used in commerce.

A trademark having a three dimensional design is a form of trade dress, namely, a product configuration mark. Product configuration trade dress may be protectable as a trademark or under broader principles of unfair competition law, provided the trade dress has developed a secondary meaning such that the significance of the trade dress to ordinary consumers is to indicate the source of the product.[54] In order to show that product configuration trade dress has developed a secondary meaning, it may be useful to show success in sales, significant advertising expenditures, consumer recognition of the product, and a lengthy and exclusive period of use of the trade dress in the marketplace.

Trade dress may be protected only if it is nonfunctional. As explained by the Federal Circuit:

> Functionality reflects a tension between two fundamental principles of trademark law. [*See In re Morton-Norwich Prods., Inc.*, 671 F.2d 1332 at 1336-37 (CCPA 1982).] On the one hand, 'there exists a fundamental right to compete through imitation of a competitor's product, which right can only be temporarily denied by the patent or copyright laws' [at 1336]. Because trademark protection is potentially perpetual in duration, protection of a functional trademark would defeat that fundamental right. On the other hand, the functionality doctrine also reflects the individual's right to protect symbols which identify the source of particular goods [at 1337].[55]

Thus, in order to avoid undermining utility patent law, functional trade dress may not be registered. Additionally, a design which is ornamental without a source-indicating effect cannot be protected as trade dress.

[B] Relationship between Industrial Design and Copyright/Design Copyright

Copyright law seeks to promote literary and artistic creativity by protecting original works of authorship from unauthorized reproduction, distribution, importation, preparation of unauthorized derivatives and (as to certain classes of works) public performance and public display. Limited moral rights are also protected in certain works. Because protectable works include the non-utilitarian design features of useful articles, copyright law can be used to protect product designs in certain cases.

A copyright is obtained by creating a work and fixing it in a tangible medium of expression. Registration of the copyright confirms and enhances those rights. Copyright protection lasts for the author's lifetime plus seventy years. However, if the work is anonymous or created within the scope of the author's employment, the copyright lasts the shorter of ninety-five years from the date of publication or 120 years from the date of creation.

---

54. *Wal-Mart Stores, Inc. v. Samara Brothers, Inc.*, 529 U.S. 205, 120 S. Ct. 1339, 54 U.S.P.Q.2d (BNA) 1065 (2000).
55. *Brunswick Corp. v. British Seagull Ltd.*, 35 F.3d 1527 at 1530, 32 U.S.P.Q.2d (BNA) 1120 (Fed. Cir. 1994) *cf. Kohler Co. v. Moen Inc.*, 12 F.3d 632, 29 U.S.P.Q.2d (BNA) 1241 (7th Cir. 1993), note also dissent by Judge Cudahy.

A design for a useful article is protectable by U.S. copyright law only to the extent that the design is physically or at least conceptually separable from its utilitarian aspects. A design is not separable from its utilitarian aspects if the form and function of the article are intertwined.[56] Most products of industrial design have as a goal the blending of form and function. As a result, only original, ornamental aspects of a design that are clearly separable from its functional aspects will be protected under copyright law, which effectively limits the use of copyright for industrial designs to a small group of products which are highly non-functional (e.g., toys, jewelry, fabric designs).

[C] Relationship between Industrial Design and Unfair Competition

A product design may be protected under section 43(a) of the Lanham Trademark Act which codifies the federal law of unfair competition. This section of the statute does not require the design to be registered. The design may also be registered in the Trademark Office as product configuration trade dress. Either way, in order for protection to obtain the design must be proven to be (a) distinctive; and (b) non-functional.

The U.S. Supreme Court has held in the 2001 *Wal-Mart* case that to prove distinctiveness for a product design, the owner must show that it has achieved "secondary meaning" in the eyes of consumers, namely that consumers recognize the source of the product merely by looking at it. This may require survey evidence.

The Court in the 1995 *Qualitex* case defined functionality for trade dress as follows: "A product feature is functional if it is essential to the use or purpose of the article or if it affects the cost or quality of the article." This is a particularly difficult burden for trade dress owners, much more difficult than proving non-functionality of a design patent. The rationale of the Supreme Court is that trademark law allows for a perpetual monopoly, and thus great care must be taken before granting trademark/trade dress rights to product designs. A design patent, on the other hand, has a term limited to fifteen years, and thus non-functionality is much easier to establish, i.e., the generally accepted test is whether the design is dictated solely by functional considerations. The best form of proof of non-functionality in a design patent case is the existence of alternate designs that perform substantially the same function, which is proof positive that the design patent owner is not monopolizing the function of his particular design.

An alternate trade dress functionality test is used in some courts, and in the U.S. Trademark Office. This alternate test is from the seminal *Morton-Norwich* case. It looks at the following four factors (no one factor is determinative):

(1) Does the product feature provide a utilitarian advantage?
(2) Are there alternative designs?
(3) Does the advertising tout utilitarian advantages?
(4) Are there more economical methods of manufacture?

56. *Brandir International, Inc. v. Cascade Pacific Lumber Co.*, 834 F.2d 1142, 5 U.S.P.Q.2d (BNA) 1089 (2nd Cir. 1987).

Suffice it to say that generally trade dress protection is much more difficult to establish than design patent protection, except perhaps for the most famous of product designs, e.g., the Absolut vodka bottle, the Apple iPhone, where trade dress protection is ideal.

CHAPTER 22
Industrial Design Rights: Venezuela

Luis F. Elias

§22.01 INTRODUCTION

[A] Highlights of New Developments in This Jurisdiction since Last Edition

When the first edition of this book was published, the Cartagena Agreement, also known as Andean Pact, and lately as Comunidad Andina de Naciones (CAN) was in force in Venezuela, and the Common Regime on Industrial Property was applied through Decision 344. This regime provided a unitary view in which Industrial Design included the concept of industrial model and industrial design as one, among other features that differ from the actual, applicable law, that is the Venezuelan Industrial Property Law (1955).

However, in 2006, Venezuela denounced the Agreement, and after five years, in November 19, 2011 the country retired completely from the Comunidad Andina de Naciones (CAN), returning to the regime provided by the applicable law, contained in the Industrial Property Law (IPL), which splits the industrial design into industrial design in the strict sense of the phrase, and industrial model.

§22.02 CONVENTIONS AND LEGISLATION

The following conventions and legislation apply in Venezuela:

(1) Paris Convention for the Protection of Industrial Property. 1995.
(2) Trade-Related Aspects of Intellectual Property Rights (TRIPS) contained in section 1 C of the Marrakech Agreement for the Creation of the World Trade Organization. 1994.

(3) Bilateral treaties:

(a) Trademark Convention between the United States of Venezuela and the Republic of France. 1879:

> Wishing the Government of the United States of Venezuela and the Government of the French Republic ensure full and effective manufacturing of national protection of both States, the undersigned, being duly authorised for that purpose, have agreed upon the following provisions:
> The citizens of each of the two contracting nations shall enjoy in the territories and possessions of the other, the same rights as nationals in all that relates to brands or trade, must comply with the formalities prescribed for this purpose by the respective laws of both countries.
> The provisions of the preceding two paragraphs shall also apply to industrial designs of all kinds.

(b) Agreement between the Kingdom of Spain and the Republic of Venezuela on the reciprocal promotion and protection of investments. 1997, which provides that:

> Intending to create favorable conditions for investments made by investors of either Contracting Party in the territory of the other Party...
> [...]
>
> Article I
> DEFINITIONS
> For the purposes of this Agreement,
> 2. The term "investments" means any kind of assets invested by investors of one Contracting Party in the territory of the other Contracting Party and in particular, although not exclusively, the following assets:
> (d) Any rights connected with intellectual property, including expressly patents of invention and trademarks, as well as manufacturing licenses, technical knowledge, and goodwill
> [...]
>
> Article III
> PROTECTION
> 1. Each Contracting Party shall provide full protection and security in accordance with international law to investments made in its territory by investors of the other Contracting Party and shall not obstruct by arbitrary or discriminatory means the management, maintenance, development, use, enjoyment, extension, sale or, where appropriate, liquidation of such investments.
> [...]
>
> Article IV
> TREATMENT
> 1. Each Contracting Party shall guarantee in its territory fair and equitable treatment, in accordance with international law, of investments made by investors of the other Contracting Party.

The applicable law is the Venezuelan Industrial Property Law, published in October 14, 1955, which regulate separately the industrial design and the industrial model.

§22.03 SUBJECT MATTER

[A] Definition/What Constitutes a Protectable Design?

Article 22 IPL, says that by industrial design is meant any disposition or union of lines, of colors and of lines and colors, destined to give any industrial object a special appearance. At the same time, this concept is separated from the notion of Industrial Model, which stipulates that by industrial model is meant any plastic form combined or not with colors, and any industrial, commercial or domestic object or utensil which can serve as pattern for the production or manufacture of others and which differ from others of like nature through its form or different configuration. Containers are comprised amongst the articles which can be protected as industrial models.

According to Article 14 IPL, any new model or design for industrial use may be subject to a patent that may be granted for five or ten years as per applicant request (Article 9 IPL).

The patentability of an industrial design depends on its ornamentation, whilst the protectable matter in the industrial model is constituted by its shape and configuration.

In addition, both industrial design and model must present characteristics of novelty and originality that confer them their own physiognomy.

Furthermore, industrial design and model have some similarities, which are:

(a) Characteristics of novelty and originality (Article 22 paragraph 4 IPL).
(b) They can be registered, only if they are for products that are to be manufactured in Venezuela. (Article 26 IPL).
(c) The protection afforded just refers to the exterior aspect and is not extended to the product itself, nor to its utility (Article 24 IPL).
(d) The Industrial Design or the Industrial Model do not comprise artistic works protected by the Copyright Law (Article 22 paragraph 5).
(e) The privilege is forfeited if within the next two years following the date of grant of patent, the product has not been manufactured in Venezuela, or if at any time its owner imports it from abroad (Article 26 paragraph 2).
(f) Shall be issued for five or ten years depending on the applicant's choice (Article 9 IPL).
(g) Cannot be renewed (Article 10 IPL).
(h) Can be assigned (Article 89 IPL) or licensed.

Exclusions from registrability include:

(1) Industrial designs and models do not comprise artistic works protected by the Copyright Law, nor the products of clothing of whatever nature.

(2) The following cannot be registered as industrial designs:
   (a) Words, phrases or signs that suggest immoral ideas, or serve to distinguish immoral objects or merchandise of prohibited production or trade and those that are used in illicit business or on harmful articles.
   (b) The National Flag, the nation's Coat of Arms or any other emblem of Republic, of its States, its Municipalities, and in general of any public Venezuelan entity.
   (c) Signs, emblems and badges of Red Cross International, and of any other entity of the same nature.
   (d) The flag, Coat of Arms or another insignia of foreign nations, except when their commercial use is duly authorised by certificate issued by corresponding office of interested nation.
   (e) Geometrical figures which do not show novelty.
   (f) Caricatures, portraits, drawings or expressions that tend to ridicule ideas, persons or objects worthy of respect and consideration.

All the compositions represented by a drawing with union of lines, colors or both, manufactured in the country, can be registered as Industrial Design. The Industrial Design must contain an element of novelty, without which the application cannot be accepted.

[1] Shape, Configuration

The protectable object with regard to the industrial model is its shape and configuration, as explained in its legal definition. *See above.*

[2] Ornamental or Functional

The protection for an industrial design refers only to its exterior or formal aspect and not to the product nor to its utility. Therefore its registrable character is ornamental. *See above.*

[B] Originality/Novelty

Originality and novelty are both required for the grant of an application (Article 22 fourth paragraph IPL). If a design or a model is neither original not new, or does not have its own physiognomy, as required by law, then it cannot be registered.

[1] Test for Originality Generally

It will suffice for the examiner to know that the physiognomy of a design or model does not derive from any other design or model.

The first motive for the requirement of novelty is the need for knowing the object of protection. If the design or model fails to show any novelty, then it loses grounds for registration. Lack of novelty can create a risk of confusion and affect the registration of another patent.

[2] Novelty

[a] Prior Use

Prior use by applicant of a model or design is not an obstacle for their patentability. (Article 25 IPL).

[b] Prior Publication

Prior publication does not destroy novelty (*see also* prior use, above).

[c] Prior Application or Registration

Whoever that has obtained a registration of an industrial design or model in a foreign country, will have a preference to obtain it in Venezuela within the next twelve months following the date of the foreign registration (Article 11 IPL).

[C] Unobviousness (if Appropriate)

Not applicable

§22.04 DURATION OF PROTECTION

The duration of a patent for an industrial design or for an industrial model, shall be for five or ten years, as per the request of applicant. (Article 9 IPL)

§22.05 INFRINGEMENT

[A] Scope of Infringement Action

Those who in order to prejudice the rights of the legitimate titleholder or beneficiary (licensee) of an industrial design or industrial model, manufacture, execute, transmits or use with industrial or lucrative aims, either of the industrial design or model, or any other that might be confused with these, shall be sentenced to prison from one to twelve months (Article 99 IPL).

Likewise, those who, without having obtained a patent for an industrial design or model, invoke it as if they reap benefit therefrom, and those who use an industrial design or model without having the corresponding letter-patent, and indicate that the design or model are patented shall be punished with a fine from VEB 50–1,000. (Article 102 IPL).

On a complementary fashion, the Venezuelan Criminal Code provides that anyone who forfeits or alters industrial designs or models, and whoever makes use of them thus forfeited or altered, although the falsity be the work of a third party, will be punished with prison from one to twelve months. The judicial authority may order that the sentence thereof be published on a journal indicated by it, at the expense of defendant. (Article 337).

As a consequence of all the foregoing pecuniary sanctions and penalties of prison concerning industrial designs or models, all sentences condemning any offense against industrial property, will order the destruction of the instruments that have served or would be prepared for the perpetration of the offense, as well as the objects derived therefrom. (Article 105 IPL).

[B] Test for Infringement/Criteria to Constitute Infringement

There are no special rules that may orient a legal expert in determining the resemblance between designs or models. Notwithstanding, the Venezuelan Civil Procedure Code provides that "*the judge may base his decision in facts of common knowledge that are found in common experience or maxims of experience.*" (Article 12).

[C] Standing to Sue

Either, the owner, or a possessor of precarious tenure (licensee) of an industrial design or model, can take legal actions against any infringement. (Article 98 IPL).

[D] Remedies for Infringement

Besides, any civil action initiated for damages and losses, the following remedies are provided by IPL:

(1) Those who make any attempts against the legitimate owner or the possessor of a patent by manufacturing, executing, transmitting, or using with industrial or lucrative aims deceitful and fraudulent copies of patent object without explicit or implicit consent shall be sentenced to prison from one to twelve months. (Article 98).
(2) The same penalty provided in Article 98 shall be applicable to anyone who in order to injure the rights or interests of the legitimate possessor use, manufacture or execute registered marks, *models or designs,* which are confused with these.
(3) In any verdict of guilty for an offense against industrial property, it shall be ordered the destruction of the instruments that may have served or were prepared for the commission of the offense, as well as the objects derived therefrom.

§22.06 DEFENSES TO INFRINGEMENT

[A] Validity

The Industrial Property Registrar, is the authority duly qualified for assessing the validity of an industrial design or an industrial model, applying the Industrial Property Law.

[1] Not Novel/Original

It is the absence of novelty (not originality) that disqualifies an industrial model or design for patentability (Article 62 IPL), see Article 14 - 7° IPL: any new model or design for industrial use must be ratified that both originality and novelty are required for the grant of protection for an industrial model or design as provided by Article 22, fourth paragraph IPL. Also see above in §22.03[B].

[2] Functional/Not Ornamental

The protection of both the industrial model and design is centered on its exterior aspects. It is not extended to the manufactured object, nor to its utility or functionality. (Article 24 IPL).

[3] Other Possible Criteria, Such as Non-obviousness

Not applicable

[4] Faults in Application

If applicant does not fulfill the requisites required for its admission (see below), then application will be returned to applicant, so that any fault or omission is cured within the next thirty workdays. Extensions for up to three months may be requested. (Article 61 IPL).

[B] Marking

Not applicable

[C] Non-infringement

Not applicable

[D] Other Defenses

4.1. During the publication period in the local press, and during sixty work days once the excerpt of application is published in the Official Bulletin, any person may object the application and oppose the granting of the patent before the patent office (with the exception of 4.1. 1c):
   (1) On account of considering that the object of the patent: (a) is not a new model or design for industrial use; (b) has already been the subject of registration as a trademark or is included in ordinals 1, 2, 3, 4, 7, 8, 11 and 12 of Article 33 referring to marks; and (c) or on account of the opposer considering himself the creator of the design or with a better right than applicant. For example, opposer may invoke as a better right his/her being heir to a previous creator of the design, or his/her being first applicant, or first user in time. Oppositions based on better rights are to be decided by Courts of First Instance in Civil matters. In the meantime, the administrative proceedings shall be suspended until the Court produces a decision, and the interested party renews the prosecution of this matter.

4.2. The annulment of a model or design registration that might have been granted in detriment of a third party right may be demanded before the competent courts if the interested party should not have opposed it as indicated in 4.1.

This action can only be attempted until the end of two years counted from the date of the certificate (Letters-Patent).

§22.07 PROCEDURES FOR FILING APPLICATION FOR REGISTRATION

Any person aspiring to obtain a patent for an industrial design or an industrial model must fulfill the following requirements:

(1) Article 59 IPL. First numeral.
(2) Submit the corresponding application to the Industrial Property Registry, either personally or through an industrial property agent, wherein the applicant will set forth:
   (a) Name, domicile and nationality of creator.
   (b) The agent's identification when acting by proxy.
   (c) That applicant is really the creator of the industrial model or design.
   (d) The originality of the model or design.
(3) Article 59 IPL: Second numeral.
(4) Enclosures:
   (a) A description in Spanish of the industrial object to be protected by the patent.

(b) Drawings and samples of the object.
(c) A cliché of the design or model not exceeding of 8 x 10 cm. in size.
(d) A certified copy legalised or stamped with the Apostille, and officially translated into Spanish of the letters-patent or registration issued in the country of origin, in case it be an application for an industrial design or model already patented or registered in another country.
(e) Power-of-attorney notarised, and subsequently legalised or stamped with the Apostille.
(f) Evidence that the foreign patent or registration for an industrial model or design is in force, and of the time left for its lapsing in the country of origin, in case of a patent application for an industrial model or design already patented or registered in another country. In our professional opinion, a certification issued by the local authority of the country in question, subsequently legalised or stamped with the Apostille would suffice.

If application should have been filed according to law, the Registrar shall order its publication within the next two months counted from the official date of publication in the Industrial Property Bulletin IPB, at the expense of the interested party in one of the newspapers of daily circulation in the country's capital, Caracas, for three times during thirty days, with intervals of ten days between one publication and the other, and thereafter in the IPB once the three previous publications are received. (Article 60 IPL). Failure to do so will bring about the lapsing of the procedure due to applicant's inactivity.

As from the date of publication of the application in the IPB, the opposition period of sixty workdays shall commence to run.

In case of opposition, applicant once notified in the IPB will have fifteen workdays to appear before the Registrar to acquaint itself thereof. At the expiration of said term an additional period of fifteen work days will commence to run in order that applicant present those arguments that it deems convenient to its rights. (Article 63 IPL).

Upon the expiration of the term to which Article 63 IPL refers without there having been any opposition, or having rejected it, Registrar shall proceed to resolve upon the issuance of the patent if it be in agreement with the law, will effect its registration, and shall issue the corresponding certificate, or letters-patent. (Article 65 IPL).

[A] Priority

About the priority rights, Venezuela applies the six months priority right for the deposit of industrial design or model applications, in accordance with the Paris Convention.

[B] Drawings

See above.

[C] Design Part of Another Article/Environment of Design

Not applicable

[D] Description of Design (Where Required)

Article 59 IPL, second numeral, literal A demands that a clear and complete description of the industrial design and model to be filed attached to the application.

§22.08 EXPUNGING/CANCELLING OR VARYING REGISTRATION

(1) The registration/patent is expunged when:
   (a) It is annulled through decision of the competent courts declaring it to have been issued in detriment to a better right.
   (b) When it be annulled in conformity with Article 21 of IPL, that says: "At any time the Minister of Commerce, based on a previous report of the Industrial Property Registrar, may annul per detailed resolution the registration of industrial model or designs obtained in violation of IPL. The affected party may appeal of this decision before the Supreme Court within the period of three months counting from the date of its publication in the Official Gazette."

The registration/patent may be cancelled:

(a) The privilege to exploit the patent will lapse if within the period of two years further to the date of issuance of patent the products (design/model) are not manufactured in Venezuela, or if at any time the patent holder imports it from abroad. (Article 26 IPL. Second paragraph).

In all the above cases the industrial design or model will pass into the public domain.

§22.09 OTHER

[A] Interface between Industrial Design and Trademarks

Both, industrial designs and models are related to trademarks in as much as that they can be affected by the same exclusions from registrability mentioned in Article 33 IPL (which contains the prohibition of adoption and registration of a trademark). Likewise, an industrial design or model can be registered as a three-dimensional mark, as long as applicant does not try to protect them simultaneously by means of a patent.

The legal definition of a trademark has close similarities with the legal definition of industrial models and designs: "Under the denomination of trademark is included any sign, figure, drawing (plan), word or combination of words, legend and another token showing novelty, used by a natural or juridical person for distinguishing the articles which he produces, those he trades in or his own Enterprise" Article 27 IPL.

The registration of the industrial design or model as trademark, for those bi-dimensional and tri-dimensional forms whose descriptions also fit within the legal definition of a trademark, sometimes is the best choice, because of the advantage that represents the indefinite renewal of its registration. However, in the words of Miguel Arteaga-Bracho, one should always keep in mind that applicant will be exchanging "The protection of the design itself, for the protection of its function on the goods and services that use it as a distinctive element."[1]

[B] Relationship between Industrial Design and Copyright/Design Copyright

Article 1 of the Venezuelan Copyright Law says: "The provisions of this law protect the rights of authors on all works of genius of creative nature, be they literary, scientific or artistic, whatever their kind, form, of expression, merit or destiny." Likewise, Article 2 states that among the works of genius mentioned before will be included: "The works of applied art that are not mere Industrial Models or Industrial Designs." This definition excludes industrial designs and industrial models considered works of applied art from copyright protection, for the scope of protection of copyright laws is originality as a form of creative expression, not its industrial character.

Therefore, for the registration as Work of Applied Art, the industrial design or model will have artistic qualities.

The benefit of the registration as Works of Genius is based on its duration, because it will last the lifetime of the author and sixty years after his death (Article 25 Copyright Law).

[C] Relationship between Industrial Design and Unfair Competition

Industrial Property Law:

> Anyone who deceitfully avails himself of the advantages of an industrial or commercial reputation acquired through the effort of another who holds his property under the protection of the Industrial Property Law, shall be punished with imprisonment from one to twelve months. (Article 101 IPL).

Law for the Promotion and Protection of Free Competition (1992):

1. M. Arteaga-Bracho. "Protección Legal del Diseño Industrial en Venezuela" (1995). (Propiedad Intelectual. Ediciones EPI-ULA. Año 1. N° 1.Mérida.Venezuela) pp. 72–73.

Section Three
About Unfair Competition

- Article 17. (This law prohibits) the development of commercial policies designed to eliminate competitors through unfair competition and, in particular, the following are prohibited:

  1 Misleading or false advertising to prevent or limit free competition;
  2 The promotion of products and services based on false statements concerning the disadvantages or risks of any product or service from competitors, and
  3 Commercial bribery, violation of trade secrets and simulation products.

- Article 49: those who engage in prohibited conduct and practices identified in the First, Second and Third Sections of Chapter II of Title II of this Act, shall be punished by the Superintendency of Protection and Promotion of Free Competition with fines of up to ten percent (10%) of the value of the sales of the offender, this quantity can be increased to twenty percent (20%). In case of recurrence, the penalty shall be increased to forty percent (40%). The calculation of the amount of sales to which this article refers, shall correspond to the fiscal year previous to the date of decision that fixes the fine.
- Article 50: the amount of the penalty referred to in the preceding article, shall be fixed according to the seriousness of the offense, for which it will be considered:

  1 The form and scope of the restriction of competition;
  2 The size of the relevant market;
  3 The market share of the corresponding subject;
  4 The effect of the restriction of competition on other actual or potential competitors, on other parts of the economic process and on consumers and users;
  5 The duration of the restriction of competition; and
  6 Recidivism in carrying prohibited conducts.

- Article 33: except for violations of the provisions of Section III of Chapter II of Title II of this Act, which lapse after six (6) months.

  In order to compute it time shall begin to run from the date of the offense; and for continued or permanent infringements, starting the day this continued or permanent infringement has ceased.

- Article 44: the penalties provided in this Title shall apply without prejudice of those established in other laws (as the ones established in IPL).

Index

B

C

N